FORENSIC PROCEDURES FOR BOUNDARY AND TITLE INVESTIGATION

T0335423

Other titles by Donald A. Wilson

Deed Descriptions I Have Known But Could Have Done Without
Easements and Reversions
Interpreting Land Records
Brown's Boundary Control and Legal Principles (with Walter G. Robillard)
Evidence and Procedures for Boundary Location (with Walter G. Robillard)

FORENSIC PROCEDURES FOR BOUNDARY AND TITLE INVESTIGATION

Donald A. Wilson, LLS, PLS, RPF
LAND BOUNDARY CONSULTANT

John Wiley & Sons, Inc.

For general information about our other products and services, please contact our Customer Care
Department within the United States at (800) 762-2974, outside the United States at (317) 572-3993 or
fax (317) 572-4002.

Wiley publishes in a variety of print and electronic formats and by print-on-demand. Some material
included with standard print versions of this book may not be included in e-books or in print-on-demand.
If this book refers to media such as a CD or DVD that is not included in the version you purchased,
you may download this material at http://booksupport.wiley.com. For more information about Wiley
products, visit www.wiley.com

Library of Congress Cataloging-in-Publication Data:

Wilson, Donald A., 1941–
  Forensic procedures for boundary and title investigation / Donald A. Wilson.
    p.   cm.
  Includes index.
  ISBN 978-0-470-11369-1 (cloth : alk. paper)
 1. Land titles.   2. Boundaries (Estates)   3. Forensic sciences.   4. Forensic surveying.   I. Title.
II. Title: Forensic procedures for boundary and title investigation.
  HD1181.W55   2008
  363.25—dc22
                                                                              2007023229

10 9 8 7 6 5 4 3 2

CONTENTS

FOREWORD

Today, as a result of the multitude of television programs and the advancements of science, the term "forensics" has been applied to many fields of endeavor-dentistry, accounting, art, psychiatry and many more, some never before considered to be classed as such. However, there are some fields which have been considered forensic, according to its definition, all along even though the word may never have been mentioned or considered. Since the term as properly used includes anything scientific which is of interest to the legal system, most techniques and procedures relative to criminalistics are included, but evidence of the location of land boundaries and the investigation of title documents are also a part of the broad field, by their very nature. Since land location, the stability of property lines and the sanctity of title documents are of utmost interest to the legal system in general, and the court system in particular, land surveying and title investigation have been very much a part of forensic science since their very beginning, and the people who work in these fields follow accepted scientific procedures on a regular basis.

Don Wilson has explored these disciplines from a forensic science point of view, making application to problems in both areas which have been solved through the application of scientific procedures. Relying on numerous examples, whether personal, regularly encountered, or reported by the court system, he has identified a number of problem areas that occur in regular practice and offers approaches and solutions throughout the text.

This is a new title to the field of forensic studies. Although there are numerous texts of a general nature, and others dealing with various aspects of criminal investigation, this is the first in the civil arena dealing with boundary evidence and other title issues or concerns. Many standard operating procedures for the protection of scenes, evidence collection and photography, and reporting are very similar in all the disciplines, and Mr. Wilson has applied appropriate ones to questions and problems confronting the land surveyor, the title examiner and the real estate lawyer.

The legal standard of following the footsteps of previous and early surveyors is detailed in several chapters, including one of its own. The importance of boundary line stability is stressed, and a discussion of numerous court decisions on the subject are included, emphasizing the importance of the role of the retracement surveyor.

A chapter on profiling, knowing as much as possible about the education, tools and techniques of a previous surveyor, is sometimes the best clue to deciphering where he went and finding what he left behind.

Search techniques, notekeeping, making diagrams, photography and scientific analyses are detailed leading to the identification of evidence and its application to questions of boundary lines, property corners, road locations, changes in water bodies and the interpretation of words, numbers and other characters found in documents. Interviewing witnesses and former owners or residents is part of the evidence collection process and knowing how to weight people's memory, and working with their descriptions and scenarios is a serious art in itself. Both of these areas occupy their own chapters.

Evidence does not lie, but it must be found and evaluated to be useful to the investigator or the scientist. This text devotes considerable space to uncovering facts, and learning to interpret trace evidence such as wood fragments, bits of fence wire and subtle changes in the landscape. Handwriting and the use of ancient descriptive terms often provide pitfalls for the investigator. Methods for working with these are presented so that hopefully no document will be cast aside as worthless.

Measurements and mathematical adjustments are mostly a matter of choice for the surveyor in order to deal with or eliminate error through the use of accepted routines. Determining what a previous individual did with their numbers will sometimes spell the difference between success and failure in reproducing original lines on the surface of the earth.

Extensive appendix material addresses tree, wood and stump identification, fences, and retracement terminology. Of particular interest are those procedures and interpretations which lend insight to the past.

Sprinkled throughout the text are quotes from a variety of sources, technical, scientific, and wisdom from such fictional investigators as Sherlock Holmes and the characters on modern television shows dealing with crime investigation. Examples taken from several authorities in the general field of forensic science lend credibility and enhance the discussion of the science and its application to the topics at hand.

This extensive treatment is expected to become a standard reference work for professionals in many fields related to land investigation. It is likely to also become an accepted text for training investigators in the evidence recovery and interpretations processes leading to successful property location and ownership.

by Dr. Henry C. Lee

PREFACE

SOLVING THE SEEMINGLY UNSOLVABLE

> We are continually faced with a series of great opportunities
> brilliantly disguised as insoluble problems.
>
> —*John W. Gardner,*
> U.S. administrator (1912–2002)

A surveyor, testifying at a hearing, was once asked the question "How long have you been preparing for this case?" His reply without hesitation was "All my life." There couldn't have been a better answer.

Like the surveyor, this work was undoubtedly born earlier in my life, back in 1962 when, as a undergraduate in college, my recreational reading, which I had little time for but was tremendously intrigued by, consisted of Sherlock Holmes stories by Sir Arthur Conan Doyle. I believed then, and believe even more strongly now, that that part of my education helped me learn how to think and to be inquisitive. I also studied some principles of logic as part of one of my English courses, and I found that equally intriguing. Sometime later, well into my consulting practice, which consisted more of detective work and trying to solve what others apparently couldn't, didn't have time for, or had given up on for one reason or another, I was once asked how I learned to do what I do since there was no book or course to rely on. My immediate answer was "I read Sherlock Holmes stories." Realizing that the answer brought looks of doubt, I followed with "It's true, they teach you how to think and to figure things out." I thought about that afterward and realized that it certainly is true. Being successful at "figuring things out" demands a grasp of principles of logic and detective science.

It may come as somewhat of a surprise to the reader to learn that Dr. Edmond Locard, one of the great pioneers in forensic science, was a great admirer of Sir Arthur Conan Doyle. In fact, Conan Doyle once visited with Dr. Locard at his laboratory in Lyons, France, during a return trip from Australia to England.

Dr. Locard maintained that Sherlock Holmes stories were of prime importance in the development of the science of criminal detection. He is credited with what is today known as Locard's Exchange Principle: *that every contact leaves a trace*. Dr. Locard once stated his opinion of the Holmes stories: "I hold that a police expert, or an examining magistrate, would not find it a waste of time to read Doyle's novels. . . If, in the police laboratory at Lyons, we are interested in any unusual way in this problem of dust, it is because of having absorbed ideas found in Gross and Conan Doyle."[1]

Dr. Locard took what Conan Doyle wrote and what Sherlock Holmes had to say very seriously. In one of the novels Holmes mentioned that he had published a monograph, *Upon the Distinction of the Ashes of Various Tobaccos*. In this treatise, Holmes claimed that the examination of tobacco ash at a crime scene could furnish clues that would help solve a mystery, a claim that many readers still regard as preposterous. However, Locard took the idea and published his own scientific paper on the significance of tobacco ash found at the scene of a crime.

Today there are many learning aids: courses, college programs in forensics, television programs and movies—some true and scientific, others "Hollywood" fables—and countless books on many aspects of forensic science. But, until now, there have been none dealing specifically with land, its ownership, its records, and its boundaries. This book is an attempt to fill that void, and hopefully it will become a reliable tool for those title investigators and retracement surveyors who really "want to figure things out."

Many lessons can be learned from the current television—approaches to investigations for evidence, evidence in the form of documents, corner markers and information leading to other approaches, solutions, and answers. Perhaps one of the most valuable lessons that can be learned is the fact that others, the successful investigators, have their rules. And they stick to them. Napolean Hill (1883–1970), Writer, Teacher, and Author of *Think and Grow Rich*, said about success: "Before success comes in any person's life, one is sure to meet with much temporary defeat, and, perhaps, some failure. When defeat overtakes a person, the easiest and most logical thing to do is to quit. That is exactly what the majority of people do." In fact, Gil Grissom said just that in a recent *CSI: Crime Scene Investigations* episode: "I'm wrong all the time. That's how I eventually get to right." (*Justice is Served*)

Many of problems are attributable to:

- Things don't come out the way we want.
- Things don't come out the way we think they should be.
- We find things we don't expect.
- We fail to find things we want or think we need.

Indeed, problems *are* solvable, but you must have patience and perseverance. There is no place for haste or allowing outside influences like time and money to dictate if you are to be successful in solving problems of this nature. If you cannot

---

[1] Hans Gross (1847–1915), an Austrian magistrate and professor of criminology, was the author of *System der Kriminalistik*, the first comprehensive work on forensic science, published in 1891. It was translated into English in 1907, under the title of *Criminal Investigation*.

do the job the way it should be done, perhaps you should think twice about it before bungling the problem and making it worse for the next person on the scene. And make no mistake about it, someday there will be another person on the scene, trying to follow our footsteps or unravel what we have done.

If one learns the basics, almost anything can be solved. Every set of circumstances, every property, every retracement is unique, and therefore each one is different from every other one encountered, and different from the examples included here, although some of the facts or characteristics may be the same, similar, or appear to be the same or similar.

Learn to think like an investigator, be open-minded, and don't try to put square pegs in round holes. Don't spend a lot of valuable time trying to make numbers fit; they are only means to an end.

Perhaps the greatest reward and satisfaction in the investigative world is to solve the seemingly unsolvable—that which people believe cannot be solved. Sherlock Holmes believed in that, and speaking to his associate Dr. Watson in "The Adventure of the Blue Carbuncle" Holmes said, "Chance has put in our way a most singular and whimsical problem, and its solution is its own reward." Perhaps he spoke for many of us when he remarked in "The Adventure of the Bruce-Pardington Plans," "I play the game for the game's own sake."

Newfields, New Hampshire
May 2007

ACKNOWLEDGMENTS

Without the aid and advice of others, this work would not be what it is. The author would like to express appreciation to mentors, fellow professionals, and leaders in other fields:

My good friend and colleague, George Butts, who shared some of his fence photos.

The folks at the Geocaching Web site, www.geocaching.com, for being so generous as to provide information as well as allow me to copy selected material.

Henry Grissino-Mayer, one of the great leaders in dendrochronolgy, allowed me the use of text material from his Web site, the Ultimate Tree-Ring Web Pages, http://web.utk.edu/~grissino, along with providing a wonderful photograph of the extraction of a tree core.

Dr. Henry C. Lee unhesitatingly granted permission to quote key phrases from one of his several books, *Cracking Cases*. As busy as he is, he agreed to write the foreword for this book, for which I am grateful. Dr. Lee has an uncanny insight into the solution of seemingly unsolvable cases. Dr. Lee also convinced me that forensic surveying is a very real science, not just something that surveyors like to do.

The staff at The National Oceanic Atmospheric Administration (NOAA) and especially Susan McLean, for assistance in the use of the declination Web site.

My good friend William Rohde, Wisconsin retracement surveyor extraordinare, for supplying the photograph of the remains of a wooden post.

The staff at New Hampshire Supreme Court Law Library—Mary, Erin, and Brian—for their ready assistance in locating reference material and the appropriate credits.

John Wiley & Sons, Inc., and my editor, Jim Harper, for having patience with my shortcomings and for giving me the opportunity to express my thoughts and ideas on real property investigation.

And finally, thanks to Sir Arthur Conan Doyle, whose creations not only entertained me and provided a diversion from routine college studying, but gave me a perspective I found nowhere else and started me down a fascinating and challenging road.

CHAPTER 1

FORENSIC SCIENCE

> Problems may be solved in the study which have baffled all
> those who have sought a solution by the aid of their senses.
> To carry the art, however, to its highest pitch, it is necessary
> that the reasoner should be able to utilize all the facts which
> have come to his knowledge, and this in itself implies, as
> you readily see, a possession of all knowledge, which, even
> in these days of free education and encyclopaedias, is a
> somewhat rare accomplishment.
>
> —*Sherlock Holmes*, "The Five Orange Pips"

> Problems cannot be solved by the same level of thinking
> that created them.
>
> —*Albert Einstein, genius*

"Forensic science," often shortened to "forensics," is the application of a broad spectrum of sciences to answer questions of interest to the legal system. It may be in relation to a crime or to a civil action.

Definition

"Forensics" is science of interest to the legal system. Surveyors and many title people deal with the system on a daily basis, as decisions and procedures must be in accordance with law. Science—instrumentation, radio waves, light waves, declination, soils, wood processes, astronomy, and many others—have their bases in the sciences. Title examiners and investigators may deal with forensics in finding missing heirs, analyzing questionable documents, and collecting extrinsic evidence to explain the meaning of items in documents.

The term "forensics" is derived from the Latin *forensic* (meaning "public"), from *forum,* the principal meeting place in ancient Roman cities. This is where legal disputes were settled at that time. The modern equivalent is court. According to some, the word "forensics" is misused, since the politically correct term is "forensic science."

Forensic science usually is concerned with finding out what happened in the usually recent past.

History of Forensic Science

Forensic science may have begun back with Archimedes (287–212 BC). Legend has it that he cried "Eureka" (I have found it) when, as a result of examining the displacement of water, he proved that a crown was not made of gold due to its density and buoyancy.

The earliest known use of fingerprints to establish identity was during the seventh century, where a debtor's fingerprints, according to an Arabic merchant, were affixed to a bill of sale given to the lender. The bill then was proof of the debt.

The first written account of using medicine and entomology to solve (separate) criminal cases is attributed to the book *Xi Yuan Ji Lu* (translated as "Collected Cases of Injustice Rectified") written in China in 1248 by Song Ci (1186–1249). One of the accounts had to do with a murder with a sickle. All the farmers were told to bring their sickles to one place, and the one with the blood on it attracted flies, whereupon the murderer confessed. The book also contained advice on distinguishing drowning (water in the lungs) and strangulation (broken neck cartilage).

In sixteenth-century Europe, medical practitioners began to gather information on cause and manner of death. Surgeons documented the results of violent deaths and diseases on the human body. Writings on the topics appeared in the late 1770s.

TYPES OF FORENSIC SCIENCE

Some of better-known branches of forensic science are described next.

Definitions

Forensic Accounting. With regard to the legal system, concern is about money in cases such as fraud, embezzlement, and other misuses or identification of funds.

Forensic Anthropology. This area of anthropology concerns itself with the identification of human remains, which may lead to identification and cause of death.

Forensic Archaeology. Involving the application of archaeological methods, this branch seeks to recover human remains and interpret their spatial relationships.

Forensic Art. The application of scientific procedures to the identification of art forms and particularly paintings in cases of potential fraud.

Forensic Ballistics. Deals with the investigation of firearms and ammunition.

Forensic Botany. The understanding of the use of botanical evidence as evidence in the judicial system. Included are species identification, tree ages, wood fragment analysis, and the application of wood products to boundary identification.

Forensic Economics. The study and interpretation of evidence of economic damage, including present-day calculations of lost revenue: earnings, benefits, business, profits, and replacement costs.

Forensic Engineering. An extremely wide area including several branches of engineering, such as bioengineering and biomedical, chemical, civil, electrical, mechanical, and metallurgical engineering. It has application in both civil and criminal investigations. While accident investigation may sometimes fall under this category, forensic surveying providing locations of certain objects may also apply.

Forensic Entomology. Primarily, the identification of insects, their stages, and their role in solving crimes through observations on dead bodies. However, certain destructive evidence with regard to wood evidence may be helpful regarding its age or deterioration

Forensic Evidence. Scientific evidence from a scene, generally applied to a crime scene.

Forensic Geology. Concerned with geological information and techniques that may be of interest to the legal system.

Forensic Odontology. Also known as forensic dentistry, this field has to do with anything related to the teeth.

Forensic Palynology. The science of pollen and spore evidence useful in legal cases. It also includes legal information derived from the analysis of a broad range of microscopic organisms found in both fresh and marine environments, and may involve soil, dirt, or dust.

Forensic Pathology. This area has to do with medicine and primarily deals with the investigation of causes of sudden and unexpected death.

Forensic Photography. Sometimes referred to as *forensic imaging* or *crime scene photography*. The art of producing an accurate reproduction of a scene or an accident for the benefit of a court or preservation of evidence. It is part of the process of evidence collection.

Forensic Psychiatry. The assessment and evaluation of people who are involved in legal matters, either civil or criminal.

Forensic Psychology. The application of psychological findings to the legal system. Handwriting analysis and the formation of written legal documents may fall under this category (e.g., the interpretation of wills and possibly other documents).

Forensic Soil Science. Also known as *forensic pedology*. Involves information relative to the properties of soils and related materials having application to the legal system.

Forensic Taphonomy. Concerning environmental aspects to postmortem as well as occasionally other studies. It has to do with the history of the body after death.

Forensic Toxicology. The analysis of drugs and poisons, often postmortem.

Questioned Documents. The science of analyzing the authenticity of origin of a particular document. Handwriting analysis, examination of paper, inks and other aspects are of concern.

For purposes of this treatise, I add these definitions:

Forensic Surveying. While surveying is often considered a branch of engineering, particularly *civil* engineering, due mostly to the measurements aspect, there is far more to the field than numbers and their application. Property line location, which may be important regarding jurisdiction in a criminal case or location of items in an accident investigation, has its foundation in real property law. Wherever location or pattern is a concern, a survey can depict the spatial relationship of objects.

Forensic Title Examination. Examining title documents to find missing heirs, locate unrecorded instruments, decipher handwriting, interpret words and phrases, and study family relationships to compile or complete chains of title within the requirements of a set of standards.

Knowledge of the procedures utilized in these forensic areas can provide insight to investigating other fields of endeavor, including title examination and boundary retracement. Some of the tools and techniques are directly applicable.

This book is about finding the traces that can lead to the discovery of the truth about past events—specifically, past events that are of interest to the law. It is related to other sciences, such as history and archaeology, the aim of which is to discover the course of events that took place long ago. The time scale may be much shorter in forensic science, but the thought processes involved are much the same. *Finding a wooden stake that turns out to be marking a property corner is a matter of law. If placed in the recent past, it is a question of law and of survey; if placed 100 or more years ago, and if nothing or only traces are left, it is of historical interest as well.* Forensic science and history merge at the edges, for where one ends, the other begins. It should be borne in mind that forensic science has an applicability far beyond the area of the law, for its techniques and thought processes are widely used to interpret facts—the traces available to us—in other historical subjects. These traces are often our only clues to the past, so we must make the most of them, knowing that they could easily mislead us. Our task is not only to find these traces, but to interpret them correctly and to attribute to them their true significance.

In the words of Dr. Zakaria Erzinçlioglu "The techniques of forensic science are the techniques of reconstructing the past, whether that past is of legal interest or not." Deed research, whether for title examination or boundary research, physical evidence marking a property or property rights or interests, and genealogical studies all fall into that category. Many of these studies are, or may become, of legal interest.

Learn to Be a Good Investigator and a Successful Retracement Surveyor—Read Everything You Can Get Regarding Sherlock Holmes

Reading about Sherlock Holmes has never been easier: Books on tape, stories online, videotapes of movies in addition to the many reprints of the stories by Sir Arthur Conan Doyle and his followers are entertaining and instructive.

Study Selected Forensics Texts and Criminal Investigative Texts for Proper Procedures

A host of titles on many aspects of forensic science and investigations can be found by perusing the Internet. Some lend insight into recent high-profile cases, while others are highly technical in nature but offer instruction in techniques and considerations that must be made when investigating a scene. Dr. Henry C. Lee has numerous titles to his credit, along with a television program.

Watch Television

Mentally filter out the "Hollywood" aspect and concentrate on the thought processes, the team effort, the analysis of evidence, and the rules, especially for the protection and presentation of evidence in court. Keep in mind that on television cases are solved in an hour, while in real life they may take weeks, months, even years, to solve depending on the circumstances.

Attend Seminars

Seminars and short courses are offered by professional societies, seminar companies, and colleges and universities. Some are general while others are very specific and very technical. With little research you can find specialized courses in almost every aspect of forensic science.

Participate in Online Programs

Online programs contain more offerings almost by the day. While not necessarily advertised as forensic courses, many of them contain techniques that can be extremely useful.

Enroll in an Academic Program

Many college and universities now have two-year and four-year programs in forensic science. In addition, private institutions also offer training programs.

Get Involved in Geocaching

People wishing to learn retracement skills or a retracement specialist wishing to hone his or her skills can get involved in one of the latest activities using GPS to locate hidden objects. It is not as simple as it might first sound, and some of the caches will tax the abilities of the very best and most experienced retracement individuals.

There are many available books on the subject along with several Web sites. To get started, visit *www.geocaching.com* and read the instructions.

An added feature of the geocaching sites is the reporting by searchers of nearby benchmarks and other control points. One needing information regarding nearby control, or if there is difficulty in finding an existing one, additional information may readily be available for assistance. Of the 736,425 markers in the database, nearly 100,000 have been recovered and reported on by geocachers.

At this writing, there are nearly 400,000 geocaches worldwide.

CHAPTER 2

THE NATURE OF EVIDENCE

> There is nothing like firsthand evidence.
>
> —*Sherlock Holmes, A Study in Scarlet*

Evidence is what investigators collect, either to lead them to a conclusion or to support a theory advanced. And evidence is what surveyors and title people collect—evidence of boundary location and evidence of title elements, ownership, rights, and interests, in land. Physical items, documents, and a whole host of other pieces of information tell a story, or at least part of a story.

Definitions

The legal definition of "evidence" is "any species of proof, or probative matter, legally presented at the trial of an issue, by the act of the parties and through the medium of witnesses, records, documents, concrete objects, etc., for the purpose of inducing belief in the minds of the court or jury as to their contention." *Black's Law Dictionary.*

The layperson's definition is "an outward sign; indication; also, that which furnishes any mode of proof." *Webster's Dictionary.*

"Evidence" may also be defined as "grounds for belief."

Evidence. Any species of proof, or probative matter, legally presented at the trial of an issue, by the act of the parties and through the medium of witnesses, records, documents, concrete objects, etc., for the purpose of inducing belief in the minds of the court or jury as to their contention. *Black's Law Dictionary.*

The term "evidence" is to be distinguished from its synonyms, "proof" and "testimony."

"Proof" is the logically sufficient reason for assenting to the truth of a proposition advanced. In its juridical sense, it is a term of wide import, and comprehends everything that may be adduced at a trial, within the legal rules, for the purpose of producing conviction in the mind of judge or jury, aside from mere argument; that is, everything that has a probative force intrinsically, and not merely as a deduction from, or combination of, original probative facts. But "evidence" is a narrower term, including only such kinds of proof as may be legally presented at a trial, by the act of the parties, and through the aid of such concrete facts as witnesses, records, or other documents. *Black's Law Dictionary.*

"Testimony," again, is a still more restrictive term. It properly means only such evidence as is delivered by a witness on the trial of a cause, either orally or in the form of affidavits or depositions. Testimony is one species of evidence. But the word "evidence" is a generic term that includes every species of it.

Testimony is the evidence given by witnesses. Evidence is whatever may be given to the jury as tending to prove a case. It includes the testimony of witnesses, documents, admissions of parties, and so on. *Black's Law Dictionary.*

FACT VERSUS EVIDENCE VS. EVIDENCE OF A FACT

> There is nothing more deceptive than an obvious fact.
>
> ——*Sherlock Holmes, "The Boscombe Valley Mystery"*

Definitions

Fact. A circumstance, event, or occurrence, as it actually takes or took place; a physical object of appearance, as it actually exists or existed. An actual and absolute reality, as distinguished from mere supposition or opinion; a truth, as distinguished from fiction or error.

The word is much used in phrases that contrast it with law. Law is a principle; fact is an event. Law is conceived; fact is actual. Law is a rule of duty; fact is that which has been according to or in contravention of the rule.

The terms "fact" and "truth" are often used in common parlance as synonymous, but they are widely different. A fact is a circumstance, act, event, or incident; a truth is the legal principle that declares or governs the facts and their operative effect.

Evidence. The state of being evident, plain, apparent, or notorious. Proof in a strictly accurate and technical sense is the result or effect of evidence, while evidence is the medium or means by which a fact is proved or disproved.

The "facts" and the "evidence" are quite different since facts can neither be added to nor taken from, while evidence may be added to, weakened, or even destroyed.

Case #1

Confusing Evidence with Facts

Cherry v. Slade's Administrator

3 Murph. 82 (N.C., 1819)

In the case of *Cherry v. Slade's Administrator,* this was an issue, Judge Henderson stated:

> I think a *venire facias de novo*[2] should be awarded, because the Jury, instead of finding the facts have only found the evidence. That the line C.D. is Ward's line or a line of a tract of land belonging to Ward, is a matter of *evidence.* That it is the line of Ward called for in Hislop's patent, is a question of *fact*, for the Jury to find from the evidence: and this fact may depend upon a variety of circumstances, all proper for the consideration of a Jury. This error has become too common, from confounding the evidence with the facts. A line, when once established to be the one called for, no matter by what evidence (if it be legal evidence,) whether it be artificial or natural, will certainly control course and distance, as the more certain description. A natural boundary, such as a water course, is designated from other water courses by its name, or by its situation, or by some other mark. One of those means of identifying the water course cannot control all the rest, if those other means are more strong and certain. A name, for instance, is the most common mean of designating it; and this in general is sufficiently certain: but it cannot control every other description; and where there are two descriptions in compatible with each other, that which is the most certain must prevail. Cases might be put, where it must be evident that the parties were mistaken in the name, and therefore the name must yield to some other description more consistent with the apparent intent of the parties. It is true, that in cases of water courses or other natural boundaries, and in some cases of artificial boundaries, which are of much notoriety, and have therefore obtained well known names, the other descriptions must be very strong: but if they be sufficiently so, the name must give away, and be accounted for from the misapprehension of mistake of the parties. This doctrine was fully illustrated in the famous suit relative to the Cattail branch, between Bullock's heirs and Littlejohn, which was more than once in this Court, and finally decided on the Circuit, to the entire satisfaction of the bench and the bar.

---

[2] A fresh or new appearance, which the court grants when there has been some impropriety or irregularity in returning the Jury, or where the verdict is so imperfect or ambiguous that no judgment can be given upon it, or where a judgment is reversed on error, and a new trial awarded. *Black's Law Dictionary*.

Types of Evidence

Evidence can be classified in a number of different ways depending on its application. While different people view evidence in different ways, it is important to define what it is in a legal sense and how it is used. Since we seldom have firsthand knowledge about boundary and survey information, we rely on evidence of past events. What some may believe is evidence may turn out not to be, and too often people equate, evidence with proof. The other consideration is that there may be information available in the form of documents and physical features that is evidence of something but not relevant to the issue at hand. And even if it is valuable evidence, it may not serve to prove anything.

Judicial Evidence. The means, sanctioned by law, of ascertaining in a judicial proceeding the truth respecting a question of fact.

Extrajudicial Evidence. That which is used to satisfy private persons as to facts requiring proof.[3] Surveyors and title people are involved in both the former when called to testify, and the latter in the course of regular work. Finding rights of way, locating lost parcels of land, and tracking unknown heirs are part of the investigator's duty.

Primary Evidence. The kind of evidence that, under every possible circumstance, affords the greatest certainty of the fact in question. For example, a written instrument is itself the best possible evidence of its existence and contents.

Secondary Evidence. That which is inferior to primary evidence. A copy of an instrument, or oral evidence of its contents is secondary evidence of the instrument and its contents.[4]

Investigators seldom see the original documents relating to land parcels. The usual method is to obtain copies from the courthouse or other repository, and many times that is satisfactory. However, copies are secondary evidence, and only as good as their source. Today's documents are generally direct copies (microfilm, photocopy, or digital images) of their source. However, that can be totally misleading to a researcher, since the source document may be based on some other source which is faulty. We are misled today in that direct copies are reliable, but must remember that if an original was handwritten and less than perfectly legible, the copy of it may be faulty, and therefore everything taken from it from that point forward is also faulty. *There is no substitute for getting as close to the original document or at least the original copy as possible.* Relying on recent copies is often exceedingly risky at best.

Direct Evidence. That means of proof which tends to show the existence of a fact in question, without the intervention of the proof of any other fact. It is distinguished from circumstantial evidence, which is often called "indirect evidence." *Direct evidence* is that which immediately points to the question at issue, while *indirect*

---

[3] Black's Law Dictionary.
[4] Ibid.

or circumstantial evidence is that which only tends to establish the issue by proof of various facts sustaining by their consistency the hypothesis claimed.[5]

On occasion, the only evidence available is circumstantial, which would be acceptable under the *best evidence rule*. While not the best evidence one could hope for, it may be the best anyone will ever obtain under the circumstances.

Intrinsic Evidence. That which is derived from a document without anything to explain it.

Extrinsic Evidence. External evidence, or that which is not contained in the body of an agreement, contract, and the like.

Extrinsic evidence may be very important to an investigator. Where there is ambiguity[6] in a document or in a description, extrinsic evidence may be sought to explain away the ambiguity. If the writing is clear, additional evidence is not necessary since there is no confusion and the words speak for themselves.

Circumstantial Evidence. Consists in reasoning from facts that are known or proved to establish such as are conjectured to exist. See **indirect evidence.**

In *The Boscombe Valley Mystery*, Sherlock Holmes stated that "circumstantial evidence is a very tricky thing. It may seem to point very straight to one thing, but if you shift your own point of view a little, you may find it pointing in an equally uncompromising manner to something entirely different."

Conclusive Evidence. That which is incontrovertible, either because the law does not permit it to be contradicted or because it is so strong and convincing as to overbear all proof to the contrary and establish the proposition in question beyond any reasonable doubt.[7]

Corroborative Evidence. Evidence that is strengthening or confirming. It is additional evidence of a different character adduced in support of the same fact or proposition.[8] Documents may serve to locate a parcel of land or a right of way on the surface of the earth. Markers and evidence of use or occupation may serve as further evidence as to the location. The physical evidence serves to support, or corroborate, the story told by the documents.

Documentary Evidence. That evidence supplied by writings and documents of every kind in the widest sense of the term; evidence derived from conventional symbols by which ideas are represented on material substances.[9] Documents come in many forms and are found in a wide variety of locations. While the courthouse is a convenient repository for searching title documents, numerous

---

[5] Ibid.

[6] Subject to more than one meaning.

[7] *Black's Law Dictionary.*

[8] Ibid.

[9] Ibid.

relevant documents are located at many other repositories and locations, public and private. Do not limit your search merely to public records at a public location, especially if document is not found at the public office, or any time further search elsewhere is warranted.

Expert Evidence. Testimony presented in relation to some scientific, technical, or professional matter by experts, that is, persons qualified to speak authoritatively by reason of their special training, skill, or familiarity with the subject.[10] Surveyors and title abstractors are investigative experts in searching and examining land records and title documents, such that they may drawn certain conclusions or render an opinion on some aspect.

Hearsay Evidence. Evidence not proceeding from the personal knowledge of the witness, but from the mere repetition of what he or she has heard others say. The very nature of the evidence shows its weakness, and it is admitted only in specified cases from necessity. There are, however, some 24 exceptions to the rule that hearsay evidence is inadmissible. Many of the exceptions apply to boundary and title matters.

Opinion Evidence. Evidence of what the witness thinks, believes, or infers in regard to facts in dispute, as distinguished from his or her personal knowledge of the facts themselves; not admissible except (under certain limitations) in the case of experts.[11] Experts are allowed, because of their background and expertise setting them apart from the ordinary layperson, to render opinions based on the evidence they have collected and from their knowledge and expertise. Laypeople are allowed to testify only from their senses, unless a special exception is made under certain circumstances.

Oral Evidence. The evidence give by word of mouth; the oral testimony of a witness.[12] Oral evidence may be that which is presented in a court of law, or may be outside of court, appearing in the form of affidavits and oral presentations sworn and unsworn to be the truth.

Original Evidence. An original document, writing, or other material object introduced in evidence as distinguished from a copy of it or from extraneous evidence of its content or purport.[13] Most times, with deeds and related title documents, originals are not available or even in existence, so copies from the courthouse or other appropriate respository, sometimes certified as authentic copies made from the record, will suffice.

Parol Evidence. Oral or verbal evidence; that which is given by word of mouth; the ordinary kind of evidence, given by witnesses in court.

Prima Facie Evidence. Evidence good and sufficient on its face; such evidence as, in the judgment of the law, is sufficient to establish a given fact, or the group or chain of facts constituting the party's claim or defense, and which, if not rebutted

---

[10]Ibid.

[11]Ibid.

[12]Ibid.

[13]Ibid.

or contradicted, will remain sufficient.[14] A document may be prima facie evidence as to the names and identities of the parties involved, but may not necessarily be prima facie evidence of its subject matter, that is, the land it purports to describe.

Real Evidence. Evidence furnished by things themselves, on view or inspection, as distinguished from a description of them by the mouth of a witness.[15] Things such as the physical appearance of a parcel of land, the direction taken by a right-of-way, markers at the corners of a property, fences, and other physical items are examples of real evidence.

Not all evidence is equal in value. Some provides specific and detailed information for the investigation, while other evidence provides only limited assistance. Based on specificity, characteristics of all physical evidence can be classified in two categories: class characteristics or individual characteristics.

Class characteristics are traits or characteristics of evidence that allow the item to be compared with a group, or "characteristics that are common to several objects."[16] They serve forensics best in eliminating possibilities. Depending on similarities or differences of the class characteristics, items often can be excluded as belonging to a group. If a surveyor stated that he set iron pipes at the corners of a property, and we find an iron pin, it is obviously subject to question. It is likely a replacement, and will not deserve the value that the iron pipes do. By comparing class characteristics of a questioned item to a known item, the investigator can establish that they are not of common origin. This is of tremendous value in comparing original evidence with secondary, or replacement, evidence.

Individual characteristics allow the investigator to compare the item with a specific object or person and include or exclude it as having originated from it. Tuthill and George define individual characteristics as "unique," resulting from natural variation, damage, or wear.[17] Individualism, the examination of individual characteristics, is a primary goal of forensics. While there are patterns depending on education, or who a surveyor may have apprenticed under, each person is an individual, with individual equipment, techniques, and style. Until the advent of sophisticated software compartmentalizing drafting methods, maps and plans in particular were identifiable because of an individual's drafting style. Frequently monumentation is identifiable as to who set it although whether it was put in the right place or is at the location originally set remains to be proved.

Trace Evidence. Consisting of a wide variety of materials, trace evidence may include bits of rust, soil discoloration, wood fragments, remnants of blazed trees, and rotted stumps that once were witness trees. By using proper instruments, tools, and techniques, the forensic investigator often can compare trace amounts to known samples.

[14]Ibid.

[15]Ibid.

[16]H. Tuthill and G. George, *Individualization: Principles and Procedures in Criminalistics* (Jacksonville, FL: Lightning Powder, 2002).

[17]Ibid.

Case # 2

Hemlock Stump

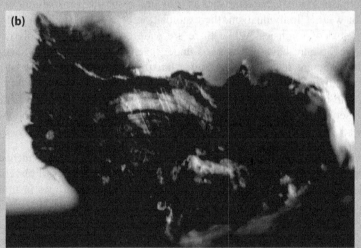

Figure 2.1 and 2.2   The remains of a hemlock stump (a) called for as a corner monument in a 1927 deed description. This wood was later verified by extracting a piece of solid root (b) and examining it microscopically.

Example: Wood post staining in soil

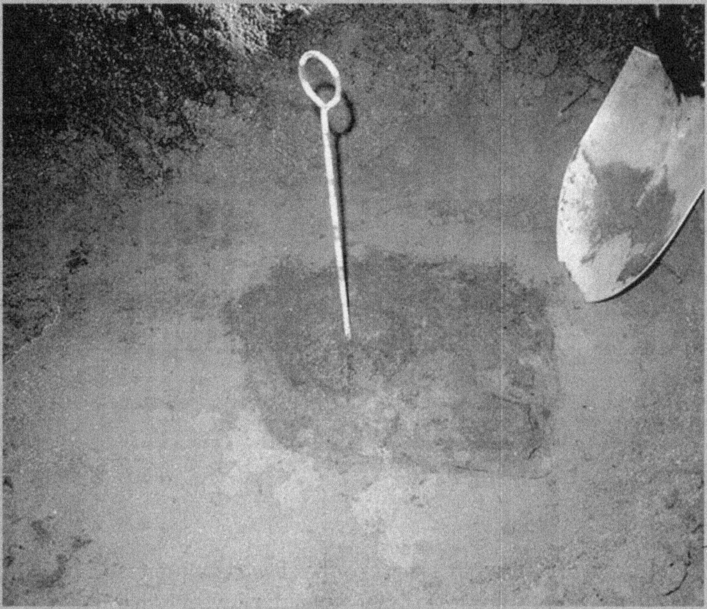

Figure 2.3   A wooden post set in the late 1800s has rotted away but has left behind a recognizable stain in the soil. When found and identifiable, this evidence usually starts out at the surface as a relatively large, round stain, gradually tapering to a point as one digs deeper. Some posts are square at the tip, resulting in the stain gradually changing from a round feature to a smaller, square feature. Photo courtesy of William C. Rohde.

Soil

Since the late 1800s, the analysis of soil has grown into a multidisciplinary field of forensic study. Soil by itself has a value in forensic studies, but more so to land investigators as a repository for nonsoil contaminants, such as rust residue and flakes, wood slivers, and fibers. Soil is considered trace evidence, along with those minute objects that may show up within it.

Physical evidence has a greater power and ability than testimonial evidence in defining what happened at any site. Testimonial evidence should not be ignored, but an understanding of physical evidence will provide an objective foundation for any subsequent theory. Physical evidence is what it is, and it is tangible.

Understanding this, testimonial evidence is not excluded but corroboration must be sought through some objective means. The means is called *scene analysis*, which is the end result of any effort directed at scene processing.

CHAPTER 3

THE SCENE

> If a herd of buffaloes had passed along, there could not be a
> greater mess.
>
> —*Sherlock Homes*, A Study in Scarlet

By and large, a scene of interest to a surveyor has been compromised, contaminated, and altered to some extent since it was first created. However, that does not stop the investigator from preventing further compromise. One of the biggest offenders is the soil testing equipment that invades a scene to insure that the site is suitable for its intended purpose. This should not be done until a surveyor at least has an opportunity to do a reconnaissance of the site to mark with flagging anything that should not, under any circumstances, be disturbed.

Evidence collection, scene protection and processing, paperwork, and photography are important procedures regardless of the type of investigation being conducted. Survey evidence does not make big news, but murder investigations generally hit the front page and the evening news, with some eventually becoming bestselling novels and television or movie hits.

Once evidence is lost, opportunities are lost, and lost opportunities are usually extremely difficult to overcome. Reconstructing a scene from scattered surveys, if they indeed exist, photographs, and recollections of witnesses is exceedingly difficult and often questionable. The basic problem is, once a scene has changed, you cannot change it back.

> There is no substitute for a well-preserved [crime] scene, absolutely none. And
> there are many ways for well-intentioned investigators to compromise the phys-
> ical integrity of a [crime] scene.
>
> —*Dr. Henry C. Lee*, Cracking Cases

Case #3

BARBED WIRE CUT FROM TREE

Figures 3.1 and 3.2 A valuable lesson was learned in this dispute. Photographs were taken on the first visit to the site (a). A few weeks later the site was revisited with the attorney, only to find that the evidence had been tampered with and attempts made to cut the fence wire from the trees (b). With litigation, evidence such as this should be examined again just prior to trial.

Case #4

THE MISSING STONE

Figure 3.3 Similar to the previous example, only a matter of days after the initial discovery of an important stone bound, the surveyor brought a second surveyor to the site to point out the evidence. The stone was missing, but the hole remained, and red paint on the leaves was still visible. A pen was placed in the photograph for size comparison. Fortunately, the stone had been located by survey before it was removed.

Case #5

A STONE MYSTERIOUSLY APPEARS

Figure 3.4 A stone recited in a 1837 land description was the subject of an intensive search by a number of persons, including surveyors, but without success. Several months later, a call was received from the attorney for the abutting landowner that the lost stone had at last been found. Examination of the stone revealed fresh leaves under it at the bottom of the hole.

Forensic Team

The team may consist of one person, or it may be involve a group of specialists. A team, per se, is likely to contain these members:

Investigator (reconnaisance)
Evidence team—measurers
Evaluator
Scientist
Legal team (where and when necessary)

SCENE OF THE CRIME VS. THE CRIME SCENE

The *scene of the crime* and the *crime scene* are not the same. The scene of the crime is the place where the crime actually took place, while the crime scene may be anything and everything that relates to scene of the crime. A crime scene is not only the actual location of the crime, it is also all the other areas that relate to it. For example, the total crime scene for a modern-day crime might include international considerations as well as dozens of physical locations and individuals, hundreds if not thousands of exhibits, and many witness statements.

Consider a homicide where the victim was slain in her home, the site cleaned up, and the body disposed of in a remote location. Then the murder weapon was disposed of at a different time at an entirely different location. The scene of the crime is the home; the crime scene is, at the very least, the locations of the body and the weapon, and may include much more, depending on what might be uncovered in the investigation.

Now consider a simple retracement survey. How many scenes are involved in addition to the physical site of the property? At the least, the abutting properties and their surveys, the various chains of title with their individual descriptions, testimony of former owners and other knowledgeable individuals, related properties sequentially or simultaneously created, conflicts of evidence, and highway records are all involved, along with perhaps many more. Potentially many other scenes in addition to the physical site are routinely part of a survey investigation.

Identifying the Scene

What is the scene? Is it the property itself, the frontage, where certain improvements are, one corner, one line, or some other part or related item? It may even be a land description purporting to identify the premises, or some part of it. There may be several scenes, depending on the situation.

Assessing the Scene

The process of assessing the scene involves identifying the scope of the scene(s), the extent of compromise, the team approach, methods of search, and protective measures.

Scene Protection

The investigation of the scene of the crime usually is protected by surrounding it with yellow crime scene tape, so it is not contaminated or compromised. Most property boundary scenes have already been compromised, some long ago, before the retracement investigator arrives. However, that does not preclude protection to prevent further destruction of the scene and its evidence. At the site, corner markers may have been moved, destroyed, deteriorated, and otherwise compromised. Even as far away as the county courthouse or other repository, relevant records may be altered, stolen, or destroyed.

Protecting or Preserving the Scene

An effort should be made to disturb the scene as little as possible in assessing the scene or its situation. Many times the arrival of additional personnel can cause problems. Only those people who are necessary should be present. At first, three people are generally the maximum allowable on a site: the owner or his or her agent, an investigator if called in, and the surveyor. After an assessment is made, additional experts and survey field crews can be allowed on the site. In a records office, generally one or two persons is the maximum necessary at any one time to accomplish the necessary documentary research.

One way to prevent, or at least minimize, destruction of evidence at a scene is to search it in stages.

Processing the Scene

The goal of scene processing is to collect as much evidence as possible in as pristine a condition as possible. Any on-site investigation must consider how any actions may affect the integrity of the scene. Soils analyses, timber harvesting and clearing, flooding, and related activities may seriously compromise a site and should not be considered until the site is released after all evidence has been identified, located, and processed.

There is no one-and-only "right" way to process a scene. There is a clear and specific purpose for why we process a scene: It is to collect as much evidence as possible in a functional and pristine a condition as is possible.

Scene Reconstruction

Crime scene reconstruction includes scientific scene analysis, interpretation of scene pattern evidence, and laboratory examination of physical evidence. It also involves systematic study of related information and the logical development of a theory. Property study and retracement is a close parallel, in that the site is analyzed from its appearance and the interpretation of existing evidence of location and occupation. Related information in the form of title documents and elements is a necessity, while peripheral information in the form of previous surveys, abutting surveys, highway and other utility information, and related information is often helpful and sometimes necessary. Generally a theory is developed as to how the title and boundaries were first established and how either may have changed since their origin. Conclusions are drawn based on whether the evidence supports the theory of what the property is or looks like.

In short, we are interested in the reconstruction of what the first surveyor at the scene did—the surveyor who first established the boundaries that created the beginning title, or a part of the title. Of interest are the surveyor's equipment, techniques, resulting maps, and any and all evidence he or she left behind, at the scene or otherwise.

Well-known criminologist Dr. Henry Lee states that reconstruction not only involves the scientific scene analysis, interpretation of scene pattern evidence, and laboratory examination of physical evidence, but also involves systematic study of related information and the logical formulation of a theory.[18]

Despite its importance to investigative and legal venues, crime reconstruction often is performed inappropriately by persons who are unknowledgeable and overconfident, such as those with little knowledge of, or training in, the peculiarities of physical evidence and the forensic sciences. Forensic analysis in general, and crime reconstruction in particular, is concerned with those conclusions that can be logically drawn from the evidence as well as with those that cannot.[19] The physical evidence

[18] Henry Lee, ed., *Crime Scene Investigation* (Taoyuan, Taiwan: Central Police University Press, 1994).

[19] John I. Thornton, "The General Assumptions and Rationale of Forensic Identification," in David L. Faigman, David H. Kaye, Michael J. Saks, and Joseph Sanders, eds., *Modern Scientific Evidence: The Law and Science of Expert Testimony*, vol. 2 (St. Paul, MN: West Publishing Co., 1997).

left behind at the crime scene plays a crucial role in reconstructing the events that took place surrounding the crime. The collection and documentation of physical evidence is the foundation of a reconstruction.[20]

Written Documentation at the Scene

Written notes may be useful later in constructing a report, which is required nowadays in most cases that are to be litigated. Notes and reports should be done in chronological order and include no opinions, no analysis, or no conclusions, only facts. The investigator should document what he or she sees, not what he or she thinks. In a survey investigation, this would be equivalent to the reconnaissance. Much will be recorded later in the form of field notes when items of evidence found in the reconnaissance are located mathematically. Photographs should be taken during the reconnaissance as necessary, to help tell the story and to perpetuate what the investigator saw during the initial visit.

Sketching the Scene

A sketch should always be made, or, in the case of a survey or survey-related problem, a survey and plan made. Photographs are two-dimensional representations of three-dimensional objects. Because of this, they can distort the spatial relationships of the objects, causing some to appear closer together while others appear farther apart than they actually are.

A sketch is usually made as if the investigator is looking straight down. The most common mapping methods used are:

Rectangular coordinates
Triangulation
Baseline coordinates
Polar coordinates
Triangulation or rectangular coordinates on a grid
Triangulation on a baseline
Total-station systems

Accepted and conventional mapping symbols should be used wherever possible, so that others, particularly nonscientists, judges, and juries, will understand the sketch or map.

[20]R. Saferstein, *Criminalistics: An Introduction to Forensic Science*, 6th ed. (Upper Saddle River, NJ: Prentice-Hall, 1998).

Photographing the Scene

Photographing a scene to preserve the conditions and characteristics at a point in time can be done in a variety of ways, from simple photographs taken with a cheap or a sophisticated camera, or digital photography, to videotaping. Photographs can be existing, or can be taken anew, or both. It is often a wise idea to take photographs on the first visit to a scene, even if it has changed since it was created.

Any photograph an investigator takes may find itself into court. Keep this in mind, and take photographs accordingly. If necessary, bring a professional photographer with you, but by all means get good photographs. They will be indispensable later for refreshing the memory, showing others what evidence was found, documenting the site, and court presentation if necessary.

Do not disturb the scene. Do not move anything, and take your photographs before any thing is moved. After the scene has been photographed, a second series of photos can be taken, documenting any changes.

Videotaping

Some scenes should be videotaped. Videotape can provide a perspective that cannot be perceived as easily in photographs and sketches.

Still Photography

Most photographs taken of sites are still photographs. Almost any type of camera will do the job, so long as the operator knows how to use it. Black-and-white photography will show things that color photographs will not, and vice versa. Select the type of film, filter, and conditions to accomplish the goals intended.

Procedure

Add measuring scales, background color, flash, and remove obstacles, but only after the scene has already been photographed once, and the pictures have been viewed to insure they are acceptable.

Get a complete series, not just one photograph. Take shots at different angles, and take a series of three photos: an overview, a midrange shot, and a close-up. Remember, a close-up of an iron pipe could be a photo of an iron pipe anywhere in the state, not necessarily at the scene in question.

The overview should cover the entire range to bring out the relationships be-tween the objects. Put something in the background that will be recognizable later: a

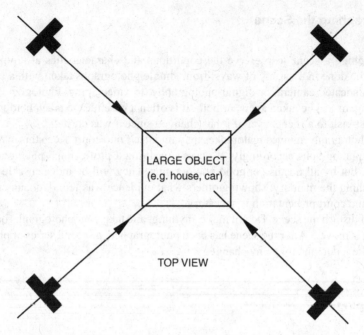

Figure 3.5   Four-corner approach to evidence photography.

building, a power pole, some identifiable object. The midrange shot should show the important object and its immediate surroundings. Each close-up should show a key detail clearly.

Camera Angle

Relationship of size and distance may be distorted by the wrong viewpoint. Examine the scene in the view finder to understand the scene as the camera will see it. Shoot most pictures at eye level. This is the height from which most people normally see things and that makes it easier to judge perspective. One practical method to ensure complete coverage and to provide complete perspective is to follow the four-corner approach.

Record All Data

Take careful notes or mark the print itself.

Sketch Map

You may wish to support your prints on a sketch map of the scene and indicate the camera position for each shot.

Points to Remember

- Take preliminary photographs early on, before the scene is altered in any way, or further altered.
- Take a complete set of pictures—overall, midrange and close-up.
- Use fresh film and keep it away from heat.
- Keep pictures sharp. Use a tripod if necessary, and focus carefully.
- Use a clean lens.
- Be conscious of lighting.
- Avoid backlighting.
- Use side lighting where necessary.
- Front lighting is usually the best.
- Use a flash where necessary.
- Create a documentation book. Record camera settings, filters, angle, and other aspects for each frame. Notes should include scene number, date, time, and a brief description of the item photographed.

Digital Photography

Concerns about admissibility are raised when discussing digital imagery. Today, almost any photo or image can be retouched to produce any kind of result desired. Federal Rules of Evidence allow for digital imagery, and a digital photograph stored in a computer is considered an original. Reference Article X, Rules 101(1), 101(3), 101(4), and Rule 103. Most states have laws or rules of evidence that apply to digital evidence.

Digital images should be preserved in their original file formats. If an image is analyzed or enhanced, the new image files created should be saved as new file names. The original file must not be replaced (overwritten) with a new file.

One of the advantages in using a digital camera at a scene is the preview screens, which help prevent errors, as you can view the image at the time it was taken. A concern, however, is whether an image is a true and accurate depiction of the scene. Another concern is that most digital cameras cannot perform as well as traditional film cameras. Because digital pictures have the equivalent of only 0.3 percent as many pixels as 35 mm film, traditional cameras generally produce sharper, crisper, and clearer images.

Another consideration is that when a digital photograph is enlarged, the quality of the image gets reduced considerably. Lack of camera accessories is another major drawback to digital photography.

> *Every piece of evidence must be photographed by the scene photographer and cataloged before it is removed from the scene for forensic examination or it is destroyed by development or by natural catastrophes.*

Consider All Aspects of a Scene

In viewing, and considering, a parcel of land, consider everything that is attached to it, since everything is a part of the whole. Consider all the title documents, deeds, mortgages, probate records, plans, and sketches. Also consider all of the improvements, buildings, wells, fences, burial sites, ditches, and evidence of land use, such as farming, and man-made changes. Any or all of them may provide clues to something, possibly something very important, or related to the title and boundary location.

In viewing a document, consider all documents that relate to it. Deeds often have related documents or other instruments. Wills and other probate items often appear with estate inventories, appraisals, and partitions, although even related partitions sometimes can be found in separate files, particularly if processed later by the court.

Case #6

NO DEED, BUT MORTGAGE PROVIDED THE DESCRIPTION

A chain of title was missing the original deed, the creating document for the parcel in question. Following the deed was a recorded mortgage with (1) a metes-and-bounds description that identified the premises and (2) a reference to the deed that identified the parties. Secondary evidence can suffice in the absence of primary evidence.

CONSIDERING THE LANDSCAPE AS A SEPARATE ENTITY

> Reading the landscape is like solving mysteries akin to reading the Adventures of Sherlock Holmes.
>
> —*Tom Wessels*, Reading the Forested Landscape

One of the rules of construction is that a deed should be construed with reference to the actual state of the land at the time of its execution.[21] In addition, courts have self-imposed the rule that descriptions must be construed in accordance wth the surrounding circumstances.

With an ancient description, regardless of its source, the surrounding circumstances, conditions, and state of the land must be considered. The character of the majority of land parcels has changed over time, and the longer the period of time between the original description and today's document, the more changes that are likely to have taken place. Woodland may no longer be woodland, pastures seed in and grow to forested land, flooded land becomes dry land, dry land becomes flooded, roads and rivers change course. There are clues on the landscape as to how it used to appear, and recognizing these clues for what they are is a science in itself.

Major disturbances to the landscape over time include:

Fire

Flooding, natural and man-made

Wind damage

Pasturing

Logging

Mining

Tree and plant disease

Animal activity, such as beaver or grazing stock

Human activity: filling, dredging, grading, construction, rechanneling rivers and streams, and otherwise changing the character of the landscape

RECONSTRUCTING A SCENE

Normal Disturbance

All sites change over time, some dramatically, others subtly. Normal erosion and severe weather are natural occurrences, and there are innumerable possibilities for man-made disturbances, with dredging, filling, and movement of earth being among the most notable. It is important to picture yourself back in time, to the point where the boundary lines were created, to the point where the original survey was performed, or to some other beginning point. Then, and only then, can you evaluate the surrounding circumstances and relate them to the contemporary description. The difficulty arises when changes are very subtle or have occurred over a long period of time.

[21] A deed of land conveys the property described in its existing state; and is to be construed in all its parts with reference to the actual, rightful state of the property conveyed, at the time of the conveyance, unless some other time is expressly referred to. *Dunklee v. Wilton R. Co.,* 24 NH 489 (1852).

Case # 7

SITE DISTURBANCE OVER TIME

A site was created in the early eighteenth century for which there was a recorded plan dated some years later. In the mid- to late 1800s a railroad line was put through and altered the site considerably, mostly through filling and extending the abutting shoreline outward. Reconstructing the site through original descriptions and survey, railroad surveys (preliminary and as-built) along with field notes, and a recent survey, the progression of changes could be documented from beginning to end. This resulted in being able to identify the location(s) of the original title lines and original parcel boundaries even though the site bore little resemblance to what it was when created.

The older the scene, the more it will have been changed. Just because the site is heavily wooded today does not mean that it was not completely cleared 100 years ago. And just because a site is cleared today does not mean that it was not wooded when the original survey, or even the last survey, was done. Fire, people's activities,

Figure 3.6   A view along a boundary, now deep in the woods, but once separating two fields. There is a stone wall in the foreground connecting with a balk, or turnrow, in the background. The feature is higher than the land on either side, since when plowing was done, the last row piled up along the perimeter. Explanations of the use of the land from history books, former owners, or words in a land description may provide clues as to what features are likely to be found on the landscape.

catastrophic weather, and other influences continually change the landscape—and the evidence of past events.

Mapping and Sketching the Scene

Sketches should be made whenever appropriate, of either an entire site or a very small part. Boundary corners, buildings, and other improvements or any noteworthy items can be sketched in detail as well as photographed. Sketches can highlight or emphasize details that photographs are sometimes unable to. Detailed survey maps and plans should always be considered and used whenever necessary to show spatial relationships between features and items of physical evidence.

CHAPTER 4

INVESTIGATION: THE SEARCH FOR EVIDENCE

> Before the investigator can successfully conduct a search
> for evidence, he must know what he is searching for.
>
> —*Luke S. May*, Scientific Murder Investigation

> It is a capital mistake to theorize before one has data.
> Insensibly one begins to twist facts to suit theories, instead
> of theories to suit facts.
>
> , —*Sherlock Holmes*, "A Scandal in Bohemia"

> Put another way, it is a capital mistake to theorize before
> you have all the evidence. It biases the judgment.
>
> —*Sherlock Holmes*, A Study in Scarlet

***Collect anything and everything you think might possibly relate to the
case. It may be gone when you return.***

Investigation. The following up or making research by patient inquiry and observation and examination of facts.[22]

When is an investigation necessary?
   Whenever something is missing
      Unrecorded document
      Wood evidence deteriorated
      Marker destroyed

---

[22] *Webster's Dictionary.*

Whenever something has been disturbed
Moved or otherwise altered

Whenever there is a conflict
Two or more items disagree with one another
More than one possible conclusion from the
evidence (rare—perhaps more investigation is in order)

Evidence Collection and Preservation

While paint fragments, spent cartridges, and other types of physical evidence may be collected from a crime scene for later comparison and analysis, or even for presentation in court, the retracement surveyor does not have such luxury. Corner trees, yards of fencing material, and physical corner monumentation cannot be removed from their locations. Sometimes samples may be procured, but for the most part, evidence must be left as found, and intact. However, all such evidence can be located through the survey measurement process and photographed and later depicted on a diagram generally known as a survey plan or plat. And hopefully you can return to the scene at a later date to find all, or at least some, of that evidence still in place. There is usually no such thing with a crime scene, as it must be cleaned up, and most items of evidence either will be removed or at least filed elsewhere for future reference.

Trace Evidence

We hear the term "trace evidence" and immediately think of gunshot residue, wood splinters, glass fragments, paint chips, hairs, and fibers. Retracement surveyors often are looking for wood fragments to identify the remains of a wooden stake or post, stump holes where a corner tree or a bearing tree once stood, or soil discoloration from rusted metal objects or rotted-away wood. Remains of fences long fallen down and deteriorated are frequently important pieces of evidence in the location of property boundaries. Paint flecks and remnants of flagging tape may indicate the presence of a marker or the place where one once stood.

Photographs

Few crime scenes are ever investigated without an abundance of photographs being taken to preserve the condition of the scene and its contents. Out of necessity, bodies must be moved within a short time, and it does not take long after the initial investigation for a scene to look entirely different from when it was discovered.

Retracement people should consider that their scene may change as well. People have been known to alter or destroy evidence, severe weather and other earth processes some times change the character of a site, and the development process for the installation of improvements. Soil testing and other types of construction can alter a site to make it unrecognizable from what the observer initially saw. If the court were to visit a site long after the retracement was done or the survey was finished, the scene might look entirely different from the way the surveyor described it. Photographs will aid in fixing a scene at a point in time.

Scientific Analysis

Much of the evidence in a criminal investigation becomes the subject of testing. Fingerprint and DNA comparisons, ballistics testing, and chemical tests all comprise a series of processes necessary to arrive at proper conclusions. Boundary evidence also may be subject to tests for wood fragment identification, aging of trees and other wooden evidence, soil testing, fence wire identification and comparison, and mathematical analyses of the current survey and past measurements.

RULES FOR INVESTIGATIONS

Searching for boundary evidence is an investigation. It is an investigation into a scene—not a crime scene, but a scene nonetheless—where the investigator is searching for evidence and for clues when evidence is absent or not readily visible. Crime scene investigators are highly trained for their tasks, and they usually employ sophisticated scientific techniques. Other investigations, if taken to the same level and applying appropriate techniques, also can be very successful in locating valuable evidence.

Many of the same techniques are employed in both types of investigation, even if the evidence sought and the tools employed might be quite different. One thing does not change: the thought process and the scientific process of reasoning. Sherlock Holmes made a habit of explaining his reasoning throughout his stories. Today's sleuths have attained higher levels, and reference materials demonstrating and analyzing methods of reasoning are readily available. Learning this part of the investigative process is like learning the multiplication tables for the first time. There did not seem to be any immediate practical value to the fact that 2 times 2 is equal to 4, but it did make sense that it might be useful at some time in the future. Such it is with the science of reasoning, especially when translated into practical rules, and more so when illustrated with examples.

Good investigators know that lists of questions from officially issued procedure manuals have limited use. Reading the signs and asking questions at a site does not involve completing a form or responding to circumstances by following preestablished rules. Each site, perfectly preserved or irrevocably compromised, has unique elements that modify the questions and define the playing rules for that particular site. Asking the right questions, of oneself or of others, depends on identifying the rules of each new challenge.

Reasoning backward analytically at a scene involves discovering the rules while playing the game. Sherlock spoke of that in *A Study in Scarlet*: "In solving a problem of this sort, the grand thing is to be able to reason backwards. That is a very useful accomplishment, and a very easy one, but people do not practise [*sic*] it much. In the everyday affairs of life it is more useful to reason forward, and so the other comes to be neglected. There are fifty who can reason synthetically for one who can reason analytically."

There is no place for guesswork in an investigation; it is much too serious for that. Thinking logically does not involve guessing. Guessing is blind and riddled with doubt. Guessing is merely desperate, and is not necessary where there are ordinary facts, as facts raise no doubts.

> *I never guess. It is a shocking habit—destructive to the logical faculty.*
>
> —**Sherlock Holmes,** *The Sign of Four*

Yesterday Sherlock Holmes and today scientific reasoners employ the art of *abductive reasoning.* "Abduction" is the process of finding a best explanation for a set of observations, and it leads to subtle implications for evidence evaluation. It is about certainty and the logico-computational foundations of knowledge. "Abduction" can be described as "inference to the best explanation," which includes the generation, criticism, and possible acceptance of explanatory hypotheses. What makes one explanatory hypothesis better than another are such considerations as explanatory power, plausibility, parsimony, and internal consistency. *In general, a hypothesis should be accepted only if it surpasses other explanations for the same data by a distinct margin and only if a thorough search was conducted for other plausible explanations.*

Ask any forensic investigator to name the biggest problem encountered on the job and you will consistently hear the same response: crime scene contamination by others. Surveyors encounter that on almost every scene, and the older the scene, the more likely the contamination, or compromise. Developers will not even hire a surveyor until the soil testing is completed. Backhoes have an uncanny way of seeking out the corner evidence and running over it. *Rule Number 1: Protect the scene.* Once evidence is lost, opportunities are lost. And you may never know what was lost when a scene is not controlled. State guides for police practice on crime scenes state: "Once the scene has changed, you cannot change it back."

Most investigators will not visit a scene alone. It is always a good idea to take someone on an investigation with you. Another person, or preferably more than one, will most likely see something that you may not. It is always good to have independent corroboration of a scene.

A good investigator will keep his or her perceptions clear. If on the scene for a while, take something to eat and drink with you. Avoid anything that could impair your senses, such as alcohol.

Most investigators will do their research first, trying to find out as much about the site as possible. Without research, you cannot know what you should be looking for, nor can you know what you have when you do find something.

Some investigators make it a practice to arrive at the scene with skepticism. While you should always maintain an open mind, remember that there just may not be anything out there. By doing your homework first, you get an idea as to what to expect.

Beware of false readings. Measurements, mathematical closures, magnetic attraction, errors in reported information can all lead to false conclusions or provide false leads. Make sure that your equipment is working properly, that the operator knows what he or she is doing, and that you are on the right parcel of land, not the neighbor's land or someplace totally irrelevant.

Most investigators will take lots of photographs, digital or otherwise. Make certain you have plenty of film and you know how to take good pictures, with or without a flash. If you are not a good photographer, take along someone who is. The next time you visit the site the conditions may have changed dramatically, or the evidence may have been totally obliterated.

In taking parol testimony, be prepared to write things down, have the witness sign the sheet(s), and have it witnessed. If one person is a notary public or a justice of the peace, so much the better. When you are taping an individual's statement, be certain that the person knows he or she is being taped and has given consent.

These rules, at the very least, should be second nature to any successful investigator. Sometimes it is easy to find and locate the evidence, but explaining procedures or a lack of success to a judge or jury may be entirely another matter. People watch television, and they watch shows like *CSI*, and have come to expect from practitioners what they see and hear on television. Make certain that your good and careful work, successful or otherwise, is not compromised or discounted by those who have a different expectation.

THE INVESTIGATION

Methods of search should be customized to suit the scene or the site. Each one is different. However, under no circumstances should untrained individuals be allowed to search in a haphazard manner, picking up wood fragments, digging up iron pipes, and tipping over stones.

The land description in use is someone else's narrative or rendition of the scene. Valid questions to consider are:

- Who were they?
- Could they communicate

- Did they have a survey or a map?
- Had they seen the [entire] property?

A successful property retracement and determination of how original boundaries were established is dependent on the ability of the investigators and the surveyors. Observations must be made using logical approaches, evidence must be examined scientifically, opinions must be formulated as to where the original surveyor went and what he or she marked, and finally a conclusion must be made as to the location of all the original lines and corners.

Examining evidence scientifically means checking or testing everything. Do not take anything for granted. Do not assume that the chain is 66 feet long; they used a 16.5 foot rod. They may have paced the line and converted it to feet.

Searching the Scene

Methods of search are customized to suit the scene. A large, open area such as a field might be searched one way, while an alley would likely be searched in a different manner. There are five basic patterns for conducting a detailed search:

1. Circle or spiral search
2. Strip and line search
3. Grid search
4. Zone search
5. Point-to-point search

These techniques are described in a number of books on crime scene and other methods. As defined, they suggest a two-dimensional process. A scene is three-dimensional, and the investigator must look up and down as well as in a horizontal plane. Key evidence may be found underground, underwater, or under pavement. If a line of trees was marked during the wintertime by a person on snowshoes, depending on the depth of the snow, the marks could be considerably higher up the trees than normal.

The traditional method of searching for a boundary corner is with the circle, or spiral, search, which has been recommended by (government) for some time. The only fallacy in this method is that it presupposes that you are in the right vicinity to begin with. Generally, getting to that point has been/is by following field note calls, following a corrected bearing, or searching an area indicated by another who claims to know where the corner is or was.

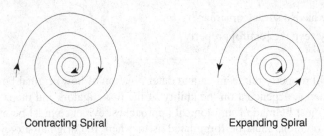

Contracting Spiral   Expanding Spiral

Figure 4.1 Circle, or spiral, search. *Source*: Henry C. Lee, Timothy Palmbach, and Marilyn T. Miller, *Henry Lee's Crime Scene Handbook*. Elsevier Academic Press, 2001.

Circle or Spiral Search (Individual).   With this technique, the searcher begins on the outside of the area and, moving in a slow circle, searches in a spiral pattern inward. The width of the swath or area evaluated with each spiral is dependent on the nature of the scene. The circle search is also done in reverse, by moving outward from a primary focal point. If, by circling one way, there is no success, it is worth repeating the opposite way as sometimes evidence shows itself by looking at it from the opposite direction. The critical consideration of the circle search is the management of the pace of the search. As the circle closes, the searcher often begins moving along the path of the spiral at a faster pace. Therefore, the searcher must consciously slow the speed as the area to be examined narrows. Either way, the searcher must maintain a pace that allows a full evaluation of the area being searched. This is an effective way for the individual to search.

Strip and Line Search (Group).   The strip search is used most often when a large area is being evaluated. A strip or swath size is designated, one that can be functionally evaluated by a single searcher, then the area is divided into similar strips. Very large areas can be divided with flagging tape or string, preventing searchers from wandering from a given lane. The searcher begins on an outer strip and moves down the strip, then reverses direction and searches the next adjacent strip. The process is repeated until the entire area has been evaluated.

 The line search is a variation that requires more searchers than the strip search. The number of searchers is scene dependent. This method is particularly useful

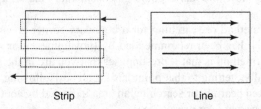

Strip   Line

Figure 4.2 Strip and line searches. *Source*: Henry C. Lee, Timothy Palmbach, and Marilyn T. Miller, *Henry Lee's Crime Scene Handbook*. Elsevier Academic Press, 2001.

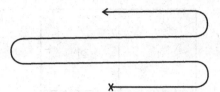

Figure 4.3   Sidehill one-person search.

when dealing with searches over uneven terrain. Searchers should move as a group, remaining in line and flagging items as they are found. Scene processors later locate and photograph each item. Where terrain is particularly rough, this type of search may be the only effective method to search the area.

When doing an individual search, especially on a slope, the sidehill one-person technique is recommended.

Grid Search (Group).   The grid search is a variation of the strip search. Instead of limiting the search to a single direction, an area is divided into two sets of strips or lanes at right angles to each other. Once a searcher has reached the far side of an area, a second search is begun using the strips at right angles to those just completed. This allows a second search over the same area, being more effective, since a given searcher is approaching from two different directions.

Zone Search.   The zone search is used mostly for searching small, sometimes confined, areas. For instance, if we know a corner is within the confines of three (found) bearing trees, we can concentrate all searches for corner evidence and location within that relatively small area. Depending on how many searchers are involved, such an area can be divided as one would slice a pie and assign each searcher one or more sectors.

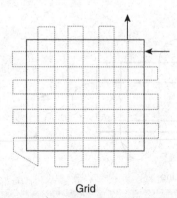

Grid

Figure 4.4   Grid search. *Source*: Henry C. Lee, Timothy Palmbach, and Marilyn T. Miller, *Henry Lee's Crime Scene Handbook*. Elsevier Academic Press, 2001.

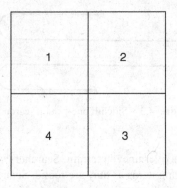

Zone

Figure 4.5   Zone search. *Source*: Henry C. Lee, Timothy Palmbach, and Marilyn T. Miller, *Henry Lee's Crime Scene Handbook*. Elsevier Academic Press, 2001.

Another variation is to divide an entire property, or large section of property, into smaller sections, assigning them to different individuals or teams. A more formal approach to dividing up an area is to physically grid off the zones with tape or string.

When searching for a corner marker, it is helpful to break up the site into quadrants, giving a thorough inspection.

Variations.   There are several variations to the zone search. If one fails to produce results, it is recommended that another technique be tried.

The line variation can be employed when there is a group effort, while the wheel/ray approach can be used by an individual. Mark the point from which to seach, then methodically fan out in various directions, at approximately 45-degree angles, until the area is covered.

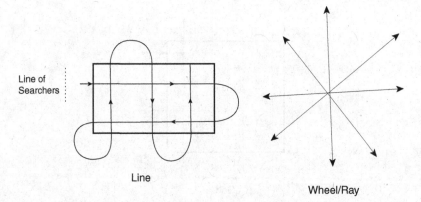

Line

Wheel/Ray

Figure 4.6   Two variations on the zone search. *Source*: Henry C. Lee, Timothy Palmbach, and Marilyn T. Miller, *Henry Lee's Crime Scene Handbook*. Elsevier Academic Press, 2001.

Different Lighting Conditions

If searchers are on the scene in the early morning or late afternoon when the light is low and are unsuccessful, they should try again at midday, when the light is overhead. The reverse can also be true.

Approach from a Different Direction

If unsuccessful coming into a scene from one direction, try coming in from a different direction. Sometimes it is necessary to view a scene from a different angle, such as vertical instead of horizontal. Aerial views offer a different perspective and allow observers to view the subject property in relation to its surroundings.

Maintain an Open Mind

One of the biggest reasons why people fail is because they get locked into a preconceived notion of what they are looking for. When you look at records and previous survey maps beforehand and listen to people supposedly knowledgeable about the scene, you naturally formulate a mental picture. When conditions don't fit the picture, some people give up, get frustrated, sometimes don't believe what they see or miss the obvious. There are an infinite number of possibilities that can be misleading. Some common ones are:

- Being told by the client or landowner what things should look like (frequently people say what they want it to be, not what it actually is [e.g., the amount of frontage or acreage based on someone's estimate or faulty data])
- The land description contains mistakes or improper references (e.g., improper tree identification)
- Previous surveyors did not do their work correctly, even though they generally do excellent work and usually get it right (humans make mistakes, and no professional gets it right every time, for a variety of possible reasons)
- The overall scene looks different than previously thought or remembered because of either natural or man-made changes (construction or natural catastrophes)
- A previous owner gives facts or descriptions of property or evidence based on faulty memory (see Chapter 17)

Simply having a wealth of data will not resolve the questions and problems of a scene. You can possess all the facts in the world, but unless you can correlate those facts in a logical and appropriate fashion, their true meaning will remain elusive. In the words of Dr. Zakaria Erzinçlioglu, "a microscope and a computer no more make a forensic scientist than a brush and paints make an artist."

> *Search and search well. If it is there, find it. If it is not, be able to say with certainty that it isn't there.*
>
> —**Rule Number Three,** *Carlisle Madson's Compilation of Rules for Land Surveyors*

CHAPTER 5

THE THOUGHT PROCESS

> The ideal reasoner would when he has once been shown a
> single fact in all its bearing, deduce from it not only all the
> chain of events which led up to it, but also all the results
> which would follow from it.
>
> —*Sherlock Holmes*, "The Five Orange Pips"

> We balance probabilities and choose the most likely. It is
> the scientific use of the imagination.
>
> —*Sherlock Holmes*, The Hound of the Baskervilles

> Crime is common. Logic is rare. Therefore it is upon the
> logic rather than upon the crime that you should dwell.
>
> —*Sherlock Holmes*, "The Copper Beeches"

Logic and the Art of Reasoning

Thinking logically means in addition to following the rules of scientific logic adhering to these five rules:

1. Do not jump to conclusions.
2. Do not make assumptions.
3. Do not skip necessary steps.
4. Do not try to put square pegs into round holes.
5. Do not mix apples and oranges.

RULES ACCORDING TO SHERLOCK AND HIS COUNTERPARTS

Sherlock Holmes, the world's first consulting detective (and every forensic individual's secret hero), said "the game's afoot" whenever he began his pursuit of a mystery. Surveyors and title investigators, in tracking property descriptions or searching for corner evidence, undertake a forensic approach in solving a mystery. The work is not unlike a game, charades in a sense perhaps, and certainly similar to a scavenger hunt. The differences are that in surveying there is either no list of items or at best a partial list, and there is a lot more at stake than a simple prize. The prize is the satisfaction of filling out the list without knowing what the list is beforehand and doing what the rules require, thus separating the mere contestant from the true professional. More than that perhaps, a survey is not a contest, and there should be no losers, only winners. The saving grace is: The rules are available—in abundance.

It may come as no surprise that Sherlock Holmes had rules for his practice. These were rules that worked for him, or maxims he accumulated along the way. A study, or a collection, of such rules would benefit any forensic surveyor, and bedside reading of Sherlock Holmes stories probably should be a requirement for any retracement artist. In fact, it is likely that the most successful of retracement people, in the records, on the ground, or both, have read much of Sir Arthur Conan Doyle and view forensic television programs at every opportunity.

Sherlock dealt with evidence and clues. Surveyors deal with evidence and clues. Sherlock perfected the art of *abductive reasoning,* that is, the process of finding a best explanation for a set of observations. Probably his best-known rule is the oft-quoted (and most appropriate) "When you have eliminated the impossible, whatever remains, however improbable, must be the truth" from *The Sign of Four.* This rule was important enough that it was stressed in the writings by being twice repeated, in "The Blanched Soldier" and "The Bruce-Partington Plans." Sherlock went on to dwell on what is improbable by saying in "Silver Blaze": "Improbable as it is, all other explanations are more improbable still."

A rule that courts have stressed perhaps more than any other procedure, and one which surveyors wrestle with and argue over probably more than any other, is *following the footsteps of the original surveyor.* Study that rule and follow it through its course in the various decisions where it is discussed in order to fully grasp its meaning. It is not enough to know the rule; it is essential that you understand it and the reasons for it. Sherlock Holmes said in *A Study in Scarlet*, "There is no branch of detective science which is so important and so much neglected as the art of tracing footsteps."

Going back to the scavenger hunt briefly, and "the list of items," Sherlock addressed this as well, by saying in "A Scandal in Bohemia": "It is a capital mistake to theorize before one has data. Insensibly one begins to twist facts to suit theories, instead of theories to suit facts." Holmes repeated this thought in *A Study in Scarlet,* saying: "Stated another way, it is a capital mistake to theorize before you have all the evidence. It biases the judgment." The New Hampshire Court addressed that pitfall in 1926 by discussing a common mistake people (mostly landowners, but some others) make: "The issue is what the deed means. It is an application of the deed to the land, and not of the land to the deed." *Smart v. Huckins*, 82 N.H. 342 (1926) (See Case #11.)

Yesterday's Sherlock Holmes and cronies are today's teams viewed on *CSI*, *Cold Case, CSI: Miami, NCIS, CSI: NY, Forensic Files, Law & Order*, and several others. Holmes had his team as well, his trusted friend and colleague Dr. Watson and his brother, Mycroft (when he needed insight), who was brilliant in observation and deduction but so lazy that he seldom moved from his accustomed cycle: his rooms, his office in a government building, and the Diogenes Club. And today's advice is the same as yesterday's advice, only stated a little differently and with a slightly different application. Much of that advice can be translated into useful rules. Yesterday it was called *detective work*; today it is called *forensics*.

Gil Grisson stated in the first *CSI* episode, "There's always a clue." Translated, that might mean the rule *Don't give up*, or perhaps more appropriate, *If you are convinced it exists, don't quit until you find it.* This could apply to an unrecorded or "lost" deed or a nonvisible "lost" monument or marker. Both the Montana court in *Myrick v. Peet* (180 P. 574 [1919]) and the Federal Court in the case of *U.S. v. Doyle* (468 F.2d 633 [1972]) stated, "Before courses and distances can be used, all means for ascertaining the location of lost monuments must first be exhausted." That can be viewed as a strict standard, for there is no interpretation to the word "all," and its meaning is clear. Finding what is sought is success; not finding it is failure.

Grissom is also quoted as saying in the pilot episode: "Concentrate on what cannot lie—the evidence." Witnesses, in court and outside of court, don't always tell the precise truth, sometimes deliberately, sometimes inadvertently based on what they believe or remember. Evidence provides clues and, by itself, is an indication of something.

Mac Taylor stated in the many advertisements promoting CSI: NY, "This evidence all ties together. We just have to figure out how." Record evidence, physical evidence, and parol evidence all tie together in a land investigation. The job of the investigator, surveyor, or title person is to figure out how.

THE THOUGHT PROCESS

> Being a good investigator means using one's head. It also means being creative
> and taking advantage of the evidence and whatever tools are at one's disposal.
>
> —*Sound Reasoning*: The Forensic Mind

The practice of forensic science is not simply the application of a set of laboratory techniques; it is an attitude of mind, a tendency to think in a particular way. It is the acquisition of the habit of starting with a doubt, of being eager and willing to question the unquestioned. It is the cultivation of a suspicious mind.

The mixture of suspicion and reason is the forensic scientist's forte. It is not merely an interesting optional extra; it is essential. Without it, forensic science is reduced to the routine application of scientific recipes.

What seems reasonable or obvious may yet be totally wrong. Remember, the techniques of forensic science are the techniques of reconstructing the past, whether that past is of legal interest or not.

HOW TO THINK

Like all other arts, the science of deduction and analysis is one that can be acquired only by long and patient study, and life is not long enough to allow a person to attain the highest possible perfection in it.

In logic, reference is often made to the two broad methods of reasoning: the *deductive* and *inductive* approaches. *Deductive reasoning* works from the more general to the more specific. Sometimes this is informally called a top-down approach. Beginning with thinking up a *theory* about our topic of interest, we then narrow that down into more specific *hypotheses* that we can test. We narrow that down even further by collecting *observations* to address the hypotheses. This ultimately leads us to be able to test the hypotheses with specific data, resulting in a *confirmation* (or not) of our original theories.

A common example is where an individual claims a particular parcel through a certain instrument and wants someone to "prove it." Researching the document, collecting addition documentary evidence, and verifying calls in the various descriptions may prove or disprove the person's ownership.

Inductive reasoning works the other way, moving from specific observations to broader generalizations and theories. Informally, we sometimes call this a bottom-up approach. In inductive reasoning, we begin with specific observations and measures, begin to detect patterns and regularities, formulate some tentative hypotheses that we can explore, and finally end up developing some general conclusions or theories.

Again, a common example is where an individual is given a description and shown its accompanying physical evidence and, through analysis, conclude what the individual owns or has rights or an interest in.

These two methods of reasoning have a very different "feel" to them when you are conducting research. Inductive reasoning, by its very nature, is more open-ended and exploratory, especially at the beginning. Deductive reasoning is more narrow in nature and is concerned with testing or confirming hypotheses. Even though a particular study may look like it is purely deductive (e.g., an experiment designed to test the hypothesized effects of some treatment on some outcome), most social research involves *both* inductive and deductive reasoning processes at some time in the project. Even in the most constrained environment, researchers may observe patterns in the data that lead them to develop new theories. This frequently happens with property research, both on the ground and in the records.

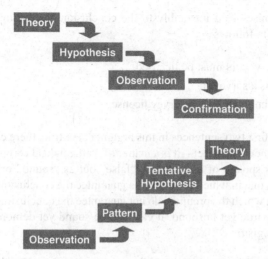

Figure 5.1   Diagram of the comparison of inductive (bottom-up approach) and deductive (top-down approach) reasoning.

Either of these reasoning methods can be used in the investigative process for land boundaries or a title examination. It may be a matter of choice as to which may be the most efficient analysis, or there may not be a choice; the very nature of the problem may place the investigator in one procedure or the other. Possible procedures include:

- Starting with the current deed and tracing it backward to a resting deed
- Starting with an early deed and tracing it forward to determine the current owner of record
- Taking a description and applying it to the ground to see what it covers
- Taking what is believed to be the property and searching for descriptions and sources of title to cover it
- Selecting any part of documentary evidence or a piece of physical evidence and researching all related evidence to arrive at an explanation or supporting data

DEDUCTIVE AND INDUCTIVE ARGUMENTS

Characteristics of Deductive Reasoning

A deductive argument offers two or more assertions that lead automatically to a conclusion. Though they are not always phrased in syllogistic form, deductive arguments usually can be phrased as *syllogisms*, or brief, mathematical statements in which the premises lead inevitably to the conclusion. An example of a sound deductive syllogism follows:

> Premise: All surveyors must be licensed.
> Premise: John is a surveyor.
> Conclusion: John must have a survey license.

As long as the first two sentences in this argument are true, there can be no doubt that the final statement is correct—it is a matter of mathematical certainty. Deductive arguments are not spoken of as "true" or "false" but as "sound" or "unsound." A sound argument is one in which the premises guarantee the conclusions. An unsound argument is one in which the premises do not guarantee the conclusions. A deduction can be completely true yet unsound. It can also be sound yet demonstrably untrue. Consider this syllogism:

> Premise: All surveyors must be licensed.
> Premise: John has a survey license.
> Conclusion: John is a surveyor.

Because all that a deductive analysis can do is measure soundness, or validity, it is a limited tool. Nonetheless, understanding the structure of syllogisms can be extremely important when making, or critiquing, arguments that depend on deductive reasoning. Basically, a deductive argument has three parts: a major premise, a minor premise, and a conclusion. These different parts can, theoretically, come in any order.

The *major premise* is a statement of general truth dealing with categories rather than individual examples:

> Charlie Brown always set crimped iron pipes at his corners.
> There is a crimped iron pipe here.
> Therefore, it must be a corner set by Charlie Brown.

> All crimped iron pipes in this area were set by Charlie Brown.
> Here is a crimped iron pipe.
> It must have been set by Charlie Brown.

Sound Deductive Arguments

In simple syllogisms, the minor premise can relate in four different ways to the major premise. Two of these ways produce sound arguments: *affirming the antecedent* and *denying the consequent*.

To *affirm the antecedent,* a minor premise must assert that a particular instance equates to the antecedent of the major premise:

All surveyors set corners.
John is a surveyor.
John sets corners.

To *deny the consequent,* a minor premise must assert that a particular instance does not equate to the consequent.

All surveyors set corners.
John does not set corners.
John is not a surveyor.

DEDUCTIVE FALLACIES

Fallacy #1: Affirming the Consequent

The fallacy of *affirming the consequent* is committed when a minor premise equates a specific instance to a consequent:

Surveyors set corners.
John sets corners.
John is a surveyor.

Fallacy #2: Denying the Antecedent

The fallacy of *denying the antecedent* occurs when the minor premises asserts that a specific instance is not an instance of the antecedent:

Surveyors set corners.
John does not set corners.
John is not a surveyor.

Fallacy #3: Equivocation

The fallacy of *equivocation (ambiguity)* occurs when a word is used in a difference sense in the minor premise than in the major premise:

This deed calls for maple trees as the corners.
There are maple trees on the property.
They must be corners.

Fallacy #4: Division

The fallacy of *division* occurs when the major premise deals with attributes that apply only to a group collectively and cannot be divided to apply to individual cases:

All deeds contain descriptions.
This instrument contains a description.
This instrument must be a deed.

GENERALIZATION

The most basic kind of inductive reasoning is called *induction by enumeration* or, more commonly, *generalization*. You generalize whenever you make a general statement (all surveyors charge too much) based on observations with specific members of that group (the last surveyor I hired was expensive). You also generalize when you make an observation about a specific thing based on other specific things that belong to the same group (my cousin Fred is a surveyor, so he probably makes lots of money). When a specific observation is used as the basis of a general conclusion, it is called *making an inductive leap*.

Fallacy #1: Hasty Generalization

Unlike deductive fallacies, which are easy to point to, inductive fallacies tend to be judgment calls. Different people have different opinions about the line between correct and incorrect induction. The fallacy most often associated with generalization is *hasty generalization,* which you commit when you make an inductive leap that is not based on sufficient information.

Investigators who spend insufficient time collecting and analyzing evidence tend to make hasty generalizations, such as: "There are no monuments in the vicinity," "You'll never find it, it's too old," "There is no deed indexed under that name; therefore, it is not on record." In the first instance, perhaps there are monuments in the vicinity, but (1) not where the searcher thought they should be, or (2) they are underground and not readily visible. Even if a metal detector was used, the monument may exist but not be metal. In the second instance, things that are very old and therefore should likely have disappeared sometimes still exist. Even though they have disappeared, their location may have been perpetuated, that is, a corner could have been re-monumented. In the third instance, a deed could be improperly indexed, recorded much later than the transaction date, or perhaps the source of title is not a deed but some other means, such as an inheritance. In any case, it may be on record. Occasionally a document is in fact on record but does not appear in the index. It appears to be not of record since most documents are in fact indexed, but a document certainly can be on record and not indexed or not indexed properly. Older documents were often difficult to read because of the handwriting, and the name in the index is not always exactly the same as the name on the document.

Generally speaking, the amount of support needed to justify an inductive link is inversely related to two other factors: the *plausibility* of the generalization and the *risk factor* involved in rejecting a generalization.

Implausible inductive leaps require more evidence than plausible ones do. It requires more evidence to support the notion that a property was surveyed by a professional surveyor than whether it was indeed surveyed. The evidentiary requirements are greater for the first assumption simply because induction requires us to combine what we observe with what we already know, and most of us realize that many properties have been surveyed but not necessarily by a professional surveyor.

Generalizations require less support when there are tremendous negative costs involved with rejecting them. Consider these statements:

Getting a subdivision approval requires having a survey.
I don't have a survey; therefore, I probably will not get my subdivision approved.

Getting a loan requires having a mortgage inspection done.
I bought my house without a mortgage inspection, which means I probably won't get a loan.

Technically, the amount of evidence for these two arguments is the same. However, most people would take the second argument much more seriously, simply because the consequences for not doing so are more disastrous.

Fallacy #2: Exclusion

A second fallacy that is often associated with generalization is the fallacy of *exclusion*. Put in simple terms, "exclusion" occurs when you omit an important piece of evidence from the inductive chain used as the basis for the conclusion. Frequently encountered is the situation where one piece of evidence is discounted because by using it, the result does not fit the forgone conclusion. Human nature causes people to draw conclusions first, then select evidence that supports the conclusions and discount evidence that does not. A common example is the interpretation of land descriptions. People believe they already know what they own; therefore, regardless of the wording in the description, the conclusion must be what they meant when they wrote it, despite what they actually said. As the court stated in the case of *Smart v. Huckins,*[23] "it is a matter of fitting the deed to the land, not the land to the deed."

Analogy   To make an induction based on an analogy is to draw a conclusion about one thing based on its similarities to another thing. Consider, for example, this argument:

In the 1960s, surveyor licensure was instituted.

Your survey was done in the 1940s, so since it was not done by a licensed land surveyor, it probably isn't any good.

Fallacy #3: False Analogy

The *false analogy* argument enumerates the similarities between one event and another event and argues that these similarities will produce a similar result. While arguments by analogy tend to be very persuasive, they can very easily fall into the trap of the false analogy, which is the major fallacy associated with this kind of reasoning. Both valid and false analogies compare similar things; false analogies, however, use hasty generalizations as the grounds for comparison. Consider this example:

Charlie Brown was not a careful surveyor; his measurements were sloppy and his traverses did not close mathematically.

Bill Black worked for Charlie Brown until he got his surveying license. Don't hire Bill, because he doesn't know how to do careful work.

[23]82 NH 342 (1926).

	Inductive Reasoning	Deductive Reasoning
Argument begins . . .	with specific evidence	with a general claim
Argument concludes . . .	with a general claim	with a specific statement
Conclusion is . . .	reliable or unreliable	true or false
Reasoning is used..	to discover something new	to apply what is already known

Table 5.1   Comparison of inductive and deductive thinking.

STATISTICAL INFERENCE

A third variety of inductive reasoning is *statistical inference*. We make statistical inferences whenever we assume that something is true of a population as a whole because it is true of a certain portion of the population. Inductions based on statistics have proven to be extremely accurate as long as the sample sizes are large enough to avoid huge margins of error. However, when amateurs attempt to use statistics as the basis for inductive leaps (and as evidence for arguments), they often end up committing the fallacy of unrepresentative sample.

Fallacy #4: Unrepresentative Sample

An *unrepresentative sample* is a statistical group that does not adequately represent the larger group that it is considered a part of. Sampling survey equipment in one county and finding that all the surveyors using a total station does not insure that all surveyors in the state use the same equipment. The sample is by far too small and therefore unrepresentative.

Summary of Inductive Reasoning

- Inductive reasoning moves from the *specific* to the *general*. Beginning with the evidence of *specific facts, observations, or experiences,* it moves to a *general conclusion.*
- Inductive conclusions are considered either *reliable* or *unreliable* instead of true or false. An inductive conclusion indicates *probability*, the degree to which the conclusion is likely to be true. Inductive reasoning is based on a *sampling of facts*.
- An inductive conclusion is held to be reliable or unreliable in relation to the *quantity* and the *quality* of the evidence supporting it.
- Induction leads to *new truths* and can support statements about the unknown on the basis of what is known.

Summary of Deductive Reasoning

- Deductive reasoning moves from the *general* to the *specific*. The three-part structure that makes up a deductive argument is called a *syllogism*. A syllogism has at least two premises and a conclusion that comes naturally from them.

- A deductive argument is *valid* if the conclusion logically follows from the premises.
- A deductive conclusion may be judged *true* or *false.* If all the premises are true and the argument is *valid,* then the conclusion must be true.
- If any of the premises are subjective, they must be supported by data, the words of persons in authority, an example, or a separate inductive argument.
- Deductive reasoning applies what the observer already knows. Though it does not yield anything new, it builds stronger arguments than does inductive reasoning because it offers the certainty of a conclusion being either true or false.

Laws of justification are not identical to laws of thought. Logic provides us with laws for justifying the truth of conclusions rather than laws for arriving at conclusions.

Creating an Argument with Solid Premises

To create an argument with solid premises, first review the evidence to insure that it is fair, objective, and complete. Then consider these points:

Suppression of facts.
   Any contrary arguments must also be considered and not ignored.
   Don't pretend they don't exist. Use them to strengthen your own argument.
Manipulation of facts.
   When the support for the argument is weak, don't try to stretch it.
   State it as it is, and let others decide how weak it is, if at all.
Is there enough evidence?
   Strengthen where necessary.
Is there too much evidence?
   Don't let supporting data outweigh the discussion.
Is the evidence current and reputable?
   Exercise care not to ignore current information that could conflict with or alter
      your point of view.
Avoid logical fallacies.

Logical fallacies are mistakes in reasoning. They may be intentional or unintentional, but in either case they undermine the strength of any argument. They include:

Hasty generalization—Based on too little information or is biased.
Either/or—Only two possibilities presented when in fact several exist.
Non sequitur—Conclusion does not follow logically from the premise.
Ad hominen—Arguing against the person instead of against the issue.

Red herring—Distracting by drawing attention to an irrelevant issue.

Circular reasoning—Asserting a point that has just been made.

False analogy—Assuming that because two things are alike in some ways, they must be alike in all ways.

Post Hoc, Ergo Propter Hoc (False Cause)—Assuming that, because event *a* is followed by event *b*, event *a* caused event *b*.

Equivocation—Falsely equating two meanings of the same word.

ABDUCTIVE REASONING

Abduction is inference to the best explanation, a pattern of reasoning that occurs in such diverse places as medical diagnostics, scientific theory formulation, accident investigation, language understanding, and jury deliberation.

Abduction, "the logic of Sherlock Holmes," leads to subtle implications for evidence evaluation in areas such as accident investigation, confirmation of scientific theories, law, diagnosis, and financial auditing. It is about certainty and the logico-computational foundations of knowledge, about inference in perception, reasoning strategies, and building expert systems.

> *Abductive reasoning is the process of finding a best explanation for a set of observations.*

"Abduction" can be described as "inference to the best explanation," which includes the generation, criticism, and possible acceptance of explanatory hypotheses. What makes one explanatory hypothesis better than another are such considerations as explanatory power, plausibility, parsimony, and internal consistency. *In general, a hypothesis should be accepted only if it surpasses other explanations for the same data by a distinct margin and only if a thorough search was conducted for other plausible explanations.*

SEPARATING FACT FROM HYPOTHESIS OR WANT

> *Desire can be our greatest enemy.*

Bias, or how we can be misled:

By the owner, client or other, such as a neighbor

By the record (poor, erroneous, or ambiguous description)

By the evidence—inconsistency, premature conclusions, generalizations

By ourselves—formulating an erroneous mental picture, either from land description(s) or someone's verbal account

Horatio Caine, the team leader in *CSI: Miami* remarked in the episode "Double Cap": "We are being detoured into the land of make-believe." It is very easy to be misled by land descriptions, changes in the landscape, and the words of people who claim to know. Horatio also stated about a suspect in one episode, "He is a liar. I just don't know what the lie is yet." It is not unusual to have part of the answer without the basis for it. And, like his counterpart Gil Grissom, recognized in one instance that "The only thing that matters is the evidence."

Mind-set

One of the biggest enemies of investigators is having a foregone conclusion, or believing they know the outcome and setting out to prove it.

The difference between the person taking a position and the objective investigator is that the former knows what he or she wants; the investigator does not know what he or she wants and is willing to accept whatever is the truth. This is where the client and surveyor sometimes differ. For example, the client wants the investigator to prove the fence is the boundary, yet all of the evidence indicates it is not. Sometimes the client cannot accept that and will disagree with the investigator.

Case #8

THE STONE POST THAT WASN'T

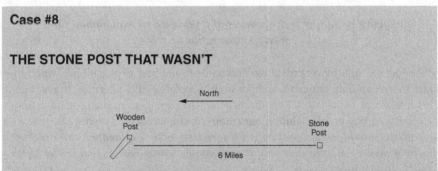

Figure 5.2 A surveyor was once attempting to locate the position of a municipal boundary. Obtaining the description of the line and correcting the bearings for the change in declination, he proceeded to the southerly corner. Here he found a sizable stone post, standing about three and half feet high, with the initials of the appropriate towns chiseled into the faces of the monument. Following the corrected bearing for the six-mile length of the line, the surveyor spent the next three days searching for another stone like the one found at the other end. He had the appropriate mind-set in that he believed the marker should be still there, and undisturbed, so he persevered. Finally it was suggested to him that perhaps the northerly corner was not monumented the same as the southerly corner. He returned with an open mind, and within 15 minutes found a wooden post that had fallen over and was lying on the ground, appropriately marked. In fact, on one of his previous visits he had sat down on that post to take a brief rest but did not recognize what it was. He had it in his mind that the corner was marked by a stone, and naturally, he could not find that it was.

Case #9

BEWARE THAT WHICH IS ABSURD ·

Bradford v. Pitts

2 Mills. Const. Rep. 115 (South Carolina, 1818)

Rules as to the locality of lands, and ascertaining boundaries, considered: all liable to a variety of exceptions, arising out of different matters of evidence, and particularity of circumstances.

The court stated in this case:

> Suppose a surveyor had run, as in this case, one line, and part of another, and distinctly marked them, and had, without closing the lines, called for a watercourse as his western boundary, when, in fact, the watercourse lay at a distance to the east, would it be permitted to the grantee in opposition to conclusive evidence, that the land was located by the marked lines, to shift it from this position to the watercourse, for the purpose of preserving that boundary: Surely not—if it were, our freeholds would soon become as ambulatory as their owners. This, however, may be said to be an extreme case, but it answers the purpose of showing, that a blind devotion to this rule would lead to infinite error; and, as applied to this case, it is erroneous in the extreme; every surveyor knows, that small watercourses are designated in the plats annexed to many of the old, and even of the recent grants, without regard to any rule of location; and it would be monstrous to suffer them to control the principle so long and so uniformly acted on.

Blind devotion to a rule may lead to infinite error.

Case #10

ONLY GETTING WHAT YOU ARE ENTITLED TO

Riley v. Jameson

14 Am. Dec. 325, 3 N.H. 23 (1823)

Rights to real estate acquired by adverse possession under claim of right, but without title or color of title, are limited to the lands actually occupied.

The court stated in this case:

> Naked possession must be actual possession. The constructive possession, which the law annexes to a title, cannot exist in such a case. There can be no constructive naked possession. An entry upon part of a lot of land, claiming the whole, without color of title, gives no possession of any part, except that upon which entry is made. If it could, an entry upon part of a lot might give actual possession of all the land in a town or county.

Case #11

A CASE FULL OF IMPORTANT PRINCIPLES, MISINTERPRETED BY MANY

Smart v. Huckins

82 N.H. 342 (1926)

This case is one of most importance to the investigator, whether a surveyor or a title examiner. It deals with a single description that is misleading is several aspects. The court, in its wisdom, set down some very useful guidelines, and as a result, this case is frequently quoted in its jurisdiction.

*It is a matter of fitting the deed to the land, not the land to the deed.
It is not a boundary dispute, it is a question of what the deed means.*

The plaintiffs in this action owned a parcel of land with this description:

> Beginning at the westerly corner of the door yard of John Leavitt by the road and running thence South about 52° West eight rods and twenty links to an old stump . . . thence South 30° West thirty-two rods twenty links to a stake and stones . . . Thence North 84° West . . . to the check line of lot number Seven, thence South about 40° East to the corner of the lot one hundred eleven one half rods thence Northwesterly [northeasterly] on the range line to the road Thence by the road to the bounds commenced at, containing forty acres and seventy-eight square rods . . . Also that part of lot No. 5 . . . lying between the Northerly and Southerly lines of the lot of land above described extending in a straight course to the Westerly check line of said lot No. 5.

A diagram would look about like this (see opposite page).

The forty-acre tract is in lot 7, the westerly line of which is the easterly line of lot 5. At the time of Leavitt's deed he owned all but five acres in lot 5, and the defendant now owns the part not owned by the plaintiffs.

In the description of the area in lot 7, the court's N. 84° W. and the range line, which are respectively the northerly, or a part of the northerly, and the southerly lines of that area, are not parallel. The range line is at right angles to the easterly and westerly lines of both lots, and in its continuance becomes the southerly range line of lot 5. The course N. 84° W. if continued from its intersection with the easterly line of lot 5 across the lot gives the plaintiffs from thirty to thirty-five acres more than a line from such intersection across the lot drawn at right angles to the check lines and parallel with the southerly or range line of the lot. It is this thirty or more acre area that is in controversy.

Lot 5 contains in all two hundred acres or slightly more, and the part of it conceded to belong to the plaintiffs contains seventy acres or slightly more.

The court remarked that the case involves the construction to be given the words "extending in a straight course" as meaning at right angles to the check

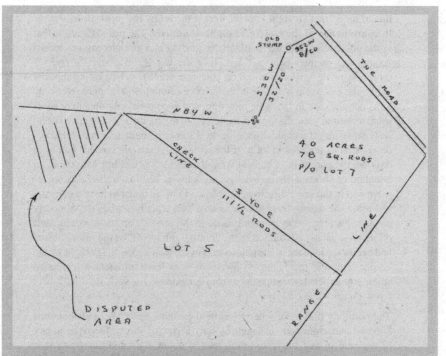

Figure 5.3 Sketch of the description in *Smart v. Huckins*.

lines of lot 5 or as meaning in the same direction as the course N. 84° W., which forms either all or a part of the northerly boundary line of the land in lot 7 conveyed by the deed.

This deed itself seems equally susceptible of either of the constructions claimed. [Emphasis added.] *This is where a lot of people have trouble, trying to decide which of two or more possibilities is the correct one.* While the defendant rightly contends that a straight line is the shortest distance between two points, a straight course does not require the meaning of the shortest distance between two lines. It may be straight although diagonal.

A straight line is the shortest distance between two points.

A straight course is not necessarily the shortest distance between two lines.

Maintaining the same direction throughout its course, it is a straight course. The boundary line claimed by the plaintiffs is just as straight as that claimed by the defendant. But because of the ordinary meaning of straight as most direct or shortest, the use of the word in the deed gives it a doubtful meaning. The southerly range line of both lots is admittedly continuous; **there is nothing in the deed, however, to show whether the line in lot 7 described by the course N. 84° W. is to be extended from its intersection with the easterly** line of lot 5 "in a straight course" across the lot by the most direct way or in continuance of the course. [Emphasis added.] The plaintiffs argue that in the absence of altering language, "extending in a straight course" means "continuing in their courses," but since the words "in a straight course" ordinarily relate to shortest distance, language which may be altering appears. While the extension of a line in the science of mathematics may refer only to its continuance in the same direction, and even conceding this may be its usual meaning, yet a change in direction is not in general meaning inconsistent with extension. If technically so, the deed shows no undertaking of technical description of the area in lot 5. If the words "in a straight course" had been omitted, the plaintiffs' argument would require adoption, but with them, an intention of shortest distance across the lot is at least as probable as one of extension in the same direction. Words are to be assumed to have a purpose, and since the words "in a straight course" are equally applicable to modify or to support the theory of continuance in the same direction, the deed itself offers no satisfactory solution of the doubt. If the words "in the same course" had been used in place of the phrase employed, their supporting weight would be decisive; and the failure to use them is at least sufficient to create an ambiguity called for competent extrinsic evidence for such light as it may throw on the situation.

The northerly line of lot 7 described by the course N. 84° W. and about one hundred and eight rods in length is only a part of the northerly boundary of the land in that lot. Two other lines of about eight and thirty-two rods in respective length make up the rest of such boundary. If the reference in the deed to the northerly line of the land in lot 7 were construed to include its entire distance of these three lines, as opposite to the southerly line, a line across lot 5 parallel with its southerly line would require adoption, since the continuation of a line made up of three sections, each having a different direction, would be impossible of application, and extension by making the northerly line across lot 5 parallel with its southerly line would be the only alternative. But there is at best nothing in the deed from which it may be inferred that the entire northerly boundary was meant rather than the part of it which is greatest in distance and which runs to lot 7. There is hence again an ambiguity which the deed on its face does not resolve.

The extrinsic evidence which the record presents points out nothing of value to show what the deed means, unless as to one item. The deed shows that the land in lot 7 had been in part at least improved. It mentions buildings on the land, the boundaries are clearly defined, and the acreage is definitely stated. Lot 5 appears never to have been improved and never to have had much value except for the growth on such parts of it as are capable of sustaining growth. Much of it was heath land, and swampy except in dry periods. Where the land was elevated above the swamp level, the growth had value. A part of the disputed area is thus elevated, but this fact is indeterminative of a purpose to include it in, or exclude it from, the land conveyed. The evidence is not satisfactory to show that any lines were run or bounds set by the parties, or that any agreement was made between them, by or from which the northerly line may be established.

The only evidence of any significance is the statement of acreage in the Leavitt mortgage, which is equivalent to a claim nearly contemporaneous with the deed of 1869 that his southerly boundary ran as the defendant now claims. To give him one hundred and twenty-five acres would limit the plaintiff's acreage to about seventy, whereas to give the plaintiffs the disputed area would cut down the one hundred and twenty-five acres to ninety-five. The use of the words "more or less" does not militate against this claim. As meaning about or approximately, they denote Leavitt's understanding and claim of substantially the acreage stated. The character of the lot, which to this day has made walls or fences on it impractical, and the lack of a survey and defined bounds help to give the statement of variation of acreage its ordinary narrowed scope of limited rather than general indefiniteness. See 3 Words & Phrases, 2d ed., 446, 447. The statement of acreage would have an empty meaning if the qualification of "more or less" were construed to admit a variation of thirty acres either way from the stated number of one hundred and twenty-five. The variation is too disproportionate to be reasonable. When it is considered that lots in Effingham usually overrun their assumed acreage, the argument for a smaller acreage than that stated in the deed becomes still further weakened, since the approximating phrase in the light of that circumstance would more especially refer to increased rather than lessened acreage.

The issue here is not a disputed boundary. It is what the deed means. It is an application of the deed to the land, and not of the land to the deed. While the object of the inquiry is to determine where the boundary is, the inquiry itself what the requirements of the deed are rather than what the situation of the land is. [Emphasis added.] The line as claimed by Leavitt is the true line if the deed is to be construed as so giving it. Otherwise it is not. There is no difficulty in locating and establishing the line when it is ascertained how the deed requires it to be run. **The boundary is doubtful only because the meaning of the language of the deed is doubtful, and the problem is not how or where to establish bounds answering the calls of the deed but to say what the calls of the deed are. [Emphasis added.]** "What the monument is is determined by the deed, but where it is is a question of fact to be determined by the jury." *Coburn v. Coxeter*, 51 N.H. 158, 162. That the evidence would be admissible to show where bounds are is not a reason to make it competent to show where they should be.

The issue here is not a disputed boundary.

It is what the deed means.

It is an application of the deed to the land, and not of the land to the deed.

This is a <u>most</u> important concept. Too often people try to make the land fit the deed. That is a backwards analysis. The deed defines the title, the ownership, and the boundaries of it, and it is to be applied to the land. The tendency is often to see it as one wants it, then try to make the deed fit what is wanted. That is the wrong approach.

While in a case of ambiguity "... it is the duty of the court to place itself as nearly as possible in the situation of the parties at the time the instrument was made, that it may gather their intention from the language used, view in the light of the surrounding circumstances" (*Weed v. Woods*, 71 N.H. 581, 583), yet the rule "that the construction of a written document is the ascertainment of the intention of the parties ... does not authorize the use of evidence of mater not proper for consideration in the interpretation of a writing." *Lancaster &c. Company v. Jones*, 75 N.H. 172, 174.

In a case of ambiguity, it is the duty of the court to place itself us nearly as possible in the situation of the parties at the time the instrument was made, that it may gather their intention from the language use, view in the light of the surrounding circumstances

What is to be determined is the meaning of the deed, and not the parties' understanding of its meaning. [Emphasis added.] What they in fact intended cannot control or affect the language used when its meaning is ascertained. **The process of construction builds upon the language to develop the intention, and not upon the intention to interpret the language. [Emphasis supplied.]** The meaning of the language being established, an actual intention of a different meaning may not be shown except in an effort to reform the instrument. "The belief of the parties, as to the effect of the deed, could neither add to nor diminish its force." Furbush v. Goodwin, 25 N.H. 425, 456. Whatever may be the present state of the law in admitting evidence of the common intention of both parties to aid in arriving at the meaning of doubtful language, neither reason nor authority countenances the use of evidence of the intention of only one of the parties for such purpose. The objectionable character of the evidence to show intention seems to rest both on its doubtful relevancy and its inexpediency. Its incompetency is perhaps also fortified by the analogous rule by which, even in the case of unilateral instruments like wills, evidence bearing on the construction of a patent ambiguity is limited to acts and statements of the maker from which intention may be inferred, to the exclusion of direct declaration of intention.

What [parties] in fact intended [in a deed] cannot control or affect the language used when its meaning is ascertained.

The evidence of the Leavitt mortgage being accordingly incompetent in construing the deed, and the record showing no other extrinsic evidence satisfactory to explain the ambiguity of the language of the deed, the rights of the parties must be settled by resort to the rule of *contra proferentem*.[24] *Canning v. Pinkham*, 1 N.H. 353; *Bullen v. Runnells*, 2 N.H. 255; *Tenney v. Beard*, 5 N.H. 58; *Cocheco &c. Company v. Whittier*, 10 N.H. 305; *Clough v. Bowman*, 15 N.H. 504; *Nutting v. Herbert*, 35 N.H. 120, 125; *Richardson v. Palmer*, 38 N.H. 212, 218; *Sanborn v. Clough*, 40 N.H. 316, 330.

[24] Against the party who proffers or puts forward a thing. *Black's Law Dictionary*.

"'Where all other rules of exposition fail' is a description (less appropriate now than formerly) of the situation of a case in which there is no preponderance of evidence in favor of either party." *Smith v. Forbish*, 68 N.H. 123. Such is the situation here. Broadened liberality in the use of parol evidence cannot be invoked when such evidence is lacking and there is nothing to show what the intention is as a matter of probability. The case therefore calls for a construction of the deed to the disadvantage of the grantor. The construction is a reasonable one, and must prevail in the absence of evidence of a different one.

This is a wonderful example of the court's analysis of a troublesome problem, and its attempt at the application of rules to achieve a result of which rule solves the problem. Several of the rules are not helpful, in that they support either argument equally, while a rule of last resort, construing in favor of the grantee, finally brings about a solution. This is a case where other rules fail, so construing the description against the grantor, the very person who chose the wording and signed the deed, is applicable.

The process of construction builds upon the language to develop the intention, and not upon the intention to interpret the language.

Once again, this is one of the most important principles there is concerning the application of a description to the surface of the earth. The parties' intention being the controlling factor above all, which is to be interpreted by what they said, and not what they meant to say, or how they wish to interpret it according to what they want the result to be.

Forensic scientists have, for the most part, treated induction and deduction rather casually. They have failed to recognize that induction, not deduction, is the counterpart of hypothesis and theory revision. Too often a hypothesis is declared as a deductive conclusion, when in fact it is a statement awaiting verification through testing.[25] Two glaring examples of this is surveying the current deed, thereby indicating that is what the client owns, and treating a fence line as locating the boundary, with no knowledge of the origin or history of the fence.

In summary. The Honorable Ruggero J. Aldisert[26] once put forth these statements, which should be kept in mind by all reasoners and investigators:

A professional in forensic science can use logic and still be wrong.

Bertrand Russell once stated, "Even when the experts all agree, they may well be mistaken."[27]

A professional who does not use logic cannot be right.

[25] John I. Thornton, "The General Assumptions and Rationale of Forensic Identification" in David L. Faigman, David H. Kaye, Michael J. Saks, and Joseph Sanders, Editors. *Modern Scientific Evidence: The Law and Science of Expert Testimony* (St. Paul, MN: West Publishing Co., 1997), vol. 2.

[26] Senior Circuit Court Judge, U.S. Court of Appeals for the 3rd Circuit, *Logic for Lawyers: A Guide to Clear Thinking* (Notre Dame, IN: National Institute of Trial Advocacy, 1997).

[27] "On the Value of Scepticism," from *The Will to Doubt*.

This point is proven continually by those who are misled by outside influences or are not conscious of their thought patterns.

Most professionals use inductive and deductive reasoning all the time, without realizing they are applying principles of logic.

However, they do not always have the correct premise(s). Applying abductive reasoning will often yield sounder, more strongly supported, results.

Every now and then we have to break the rules. Start with the conclusion and work your way backwards.

—**Gil Grissom,** *CSI: Crime Scene Investigation, "Overload"*

The ideal reasoner would when he has once been shown a single fact in all its bearing, deduce from it not only all the chain of events which led up to it, but also all the results which would follow from it.

—**Sherlock Holmes,** *The Five Orange Pips*

CHAPTER 6

STUDYING A LAND PARCEL

The World is full of obvious things which nobody by any
chance ever observes.

—*Sherlock Holmes*, The Hound of the Baskervilles

Detection is, or ought to be, an exact science and should be
treated in the same cold and unemotional manner.

—*Sherlock Holmes*, The Sign of Four

Forensic science, unlike much other science, is not
concerned with predicting the future, but reconstructing the
past.

—*Zakaria Erzinçlioglu*

It has long been an axiom of mine that the little things are
infinitely the most important.

—*Sherlock Holmes*, "A Case of Identity"

There is far more to a parcel of land than its deed, its title, its boundaries, its corners, or its improvements, even though these are the most often considered and the most commonly dealt with. They are all part and parcel of an entity sometimes termed "a bundle of rights," sometimes called the "land and its improvements" or other characterizations. Inherent in whatever it is termed are elements and principles of law and evidence, abutting parcels, rights of others including the public, and a host of other considerations. Any of these "other considerations" may provide valuable clues or may become a factor requiring independent research. In researching a land parcel, in the records or on the ground, the investigator should consider the influence or assistance of peripheral concerns and what effect they might have on the investigation or the ultimate project or use of the parcel.

Focus may ultimately be not on the entire parcel itself but on one line, one corner, one improvement, the outhouse, the garden, a fence, a burial ground, or anything attached to the parcel itself—in any way, shape, or manner.

Evidence involving the use of high-tech apparatus is not always the most revealing; *very often the simplest of simple clues can be the most revealing.* The forensic scientist (investigator) must look for clues wherever they may be found. It is the way forensic scientists think that matters, not so much the techniques at their disposal. A microscope and a computer no more make a forensic scientist than a compass and chain make a surveyor.

No evidence is infallible. Arriving at the truth in a forensic investigation is a matter of attacking the problem from many different angles. If the story that emerges is supported by all the strands of evidence, one can then have good reason to believe that it is the truth.

There is a general assumption that the more scientifically complex a forensic investigation is, the more impressive is the evidential value of the results. In fact, the opposite is often true; a simple, straightforward clue, correctly interpreted, may shed a good deal more light on a case than can a clue that requires investigation using sophisticated and expensive equipment.

DETAILS

This involved dealing with those little pieces of evidence that keep cropping up in forensic cases—the sort of evidence with which Sherlock Holmes would have felt comfortable. If the type of barbed wire is consistent on all sides of a land parcel, it was likely placed by the owner of the parcel or its abutting tract. If the monumentation at all corners of a land parcel is consistent in physical character, likely it was all placed at one time, by same person, to mark that particular tract of land. There are numerous other examples.

> *You know my method. It is founded upon the observation*
> *of trifles.*
>
> —**Sherlock Holmes**, *"The Boscombe Valley Mystery"*
>
> *Trifles make perfection—and perfection is no trifle.*
>
> —**Michelangelo**

Any forensic case is made up of little details. The task of the scientists involved is to try to reconstruct events; and if all the evidence seems to point one way, you have a scenario that is compelling. A story supported by many different lines of evidence will have a strength and integrity that is difficult to shake. However, you

Figures 6.1a and 6.1b   Figure 6.1a is the same hemlock stump as Figure 2.1. When it was sampled, naturally the most solid sample would be the best, so a piece was taken from the above. When analyzed, the wood was found to be white pine (*Pinus strobus*). However, the saving grace was that with the wood sample were samples of the bark, which proved to be hemlock. Knowing that white pine does not have hemlock bark, even though they often appear similar, further observation and study was necessary. It was found that a small pine had grown out of a rotted hemlock stump, thus misleading the untrained observer.

must be careful not to interpret the evidence in such a way that it is made to point one way only; often evidence can point many different ways. The probability of your interpretation of the evidence—all of it—must be assessed. This is not always done, and miscarriages of justice are known to happen as a result.

IN THE RECORDS

Mental pictures are formed when descriptions are read. One must be constantly on guard not to get locked in to a preconceived notion without having all of the facts. People's narratives can be very misleading, even totally wrong. When property on the ground does not resemble what the documents lead you to believe, don't assume (1) you are at the wrong property, or (2) the description is either the wrong one or was compiled in error. Both conclusions could be correct, or both incorrect, they must be tested.

ON THE GROUND

A good investigator will take in every detail and rely on notes and photographs to refresh the memory. Memories fade, and probably the last thing you would want to rely on as a substitute would be a land description or someone else's survey plan. Neither will provide an independent view, and either or both could be incorrect or biased.

COLLECTING AND PRESERVING THE EVIDENCE

Surveys and photography play an important role in collecting and preserving the evidence. Since property line and corner evidence cannot be taken back to the lab, photographs and samples must be used. What to take as a sample is of utmost concern.

CHAPTER 7

INTERPRETING THE EVIDENCE

> Any procedure in scientific testing is only as good as the
> scientist's ability to perform such a test. A standard
> procedure will not guarantee the reliability of the testing
> results.
>
> —*Dr. Henry C. Lee*, Cracking Cases

The value of a piece of evidence is not based solely on its mere presence at the scene. A scene has sometimes been characterized as a jigsaw puzzle. Some would say that they are puzzles with all of the pieces being about the same size and shape and turned upside down so that the details of the resulting picture are not visible. Unfortunately, we never have all of the pieces of the puzzle. Each artifact is like a piece of the puzzle, and the investigator must determine whether they fit together, and if so, how. Chisum and Rynearson commented on the interpretive value of evidence by noting that "the full meaning of evidence is a function of time and the item's surroundings."[28]

Stated another way, the value of evidence is more of an issue of context than it is of content. Merely having the pieces of the puzzle is not enough. The investigator must be able to place the pieces in the overall picture or offer an explanation why one or more pieces do not belong.[29] As an example, say an iron pipe is found at or near the position of a property corner. Only by thorough investigation and study will a conclusion result about whether it was placed by the original surveyor, and if not, what its placement by someone else, perhaps a following surveyor, is based on. It is not sufficient, or even proper, to accept it as marking the position of the corner just because it is in *about* the right location, without corroborative evidence. Iron pipes

[28] W. J. Chisum and J. M. Rynearson, *Evidence and Crime Scene Reconstruction*. Rynearson, J.M., Ed., Shingletown, CA. 1989.
[29] Ross M. Gardner, *Practical Crime Scene Processing and Investigation* (Boca Raton, FL: CRC Press. 2005).

have, through investigation, found to be unrelated to property corners because they were placed as supports for mailboxes, playing horseshoes, and other uses.

Some artifacts are evidence of something, but not necessarily something the investigator cares about.

In considering the context of the evidence, Chisum and Rynearson suggested that such context may manifest itself in several ways, and classified them as:

1. Predictable effects
2. Unpredictable effects
3. Transitory effects
4. Relational details
5. Functional details

Predictable effects are those changes to the scene or the evidence that occur with some rhythm or regularity. Based on this regularity, such evidence provides the investigator a factual reference or, at the very least, an inference as to other information. A wooden stake set in 1850 has more than likely disappeared, although, under the right circumstances, remains or traces might still be found. A fence called for in an 1800 description would not be made of wire, because wire fences were invented much later. Any wire fence found at the scene would therefore, not be original evidence, even though it might be good evidence under the right circumstances.

Unpredictable effects are changes that occur in an unexpected or random fashion. Unpredictable effects alter the original scene and the evidence. If unrecognized, such effects can cause significant misinterpretation of the scene. Widespread flooding may cause corner evidence to be inundated whereas it might otherwise be visible. Under the right conditions, movement of the water may destroy or otherwise alter original evidence. Significant earth movement in the form of slippage, horizontal or vertical, can shift an entire parcel of land from one location to another while leaving its boundary and corner evidence intact.

Transitory effects manifest themselves at a scene in a number of ways. Ultimately, time and environment will destroy any transitory effects present at the scene. Temporary stakes, marks on trees, and paint will disappear in time, erasing much of the original evidence of survey and boundary marking. Records in the form of deeds, sketches, survey plans, notes, and the like become more difficult to locate as time passes. Many are destroyed if not in some type of protective custody.

Relational details manifest themselves through the investigator's ability to physically place items at the scene. Most corners were once marked or locatable in some fashion. Today, when some or all have disappeared, a competent investigator should be able speculate where they once existed, often leading to traces of where they were or the collection of related evidence (such as from one or more abutting land parcels) resulting in conclusions as to where markers once existed. Significant in the understanding of relational details is the belief that the item was actually at a given location at a point in time. Many searchers give up too easily or too quickly, after convincing themselves that the evidence has disappeared. Persistence and perseverance often prove that the evidence remains, having been found after most have given up the search. Believing that an item still exists produces a positive mind-set that often results in success rather than failure.

Activity on a site can easily alter the relational information. Natural processes such as wind, rain, and water flow can have a dramatic effect, especially on smaller items. Original sketches and measurements, where available, will aid in documenting the original scene. This is one reason that measurements and fixing methods should be applied to all items of evidence, not just those thought to be important at the time. Although the measurements are often seen as laborious and unnecessary, but without them, information to prove or disprove a relational detail may be lost. Many a corner has been recovered through the use of accessories, such as bearing trees and objects, which otherwise might have been lost.

Functional details manifest themselves in the operating condition of items in the scene. Knowing when the survey work was done may explain the accuracy of the measurements and the tolerances to which original evidence may have been set or located. Knowing something about the original surveyor and the equipment owned or used would explain the parameters within which he or she operated. For example, an original survey made with a compass and chain could not result in bearings to seconds (even minutes) and distances to decimals of a foot without some manipulation of the numbers (such as with adjustment routines or even fudging to make the result come out right). Recent descriptions may lead one to believe greater precision was produced, because someone converted a crude measurement into something it is not (such as a conversion of rods and links to feet and hundredths of a foot).

Each of these tells the investigator something about what was possible or impossible given that specific item. Functional details assist the investigation in a variety of ways. They can disprove specific allegations (a description reciting directions to seconds and distance to hundreds of a foot is likely based on a relatively recent survey, *not* an older one) or help define when the procedure occurred (a description in rods and links must be based on an older survey; a description by abutters reciting an area in acres and [square] rods must have been based on survey data, even though it is not mentioned in the particular description).

Using Chisum and Rynearson's five effects is a functional and appropriate way to evaluate information and evidence present in an investigation. Bevel and Gardner offer an additional approach to evidence observations when discussing scene reconstruction.[30] When considering each item observed at a scene, the investigator should

---

[30]T. Bevel and R. M. Gardner, *Bloodstain Pattern Analysis: With an Introduction to Crime Scene Reconstruction*, 2nd ed. (Boca Raton, FL: CRC Press, 2001).

ask three questions about that evidence:

1. What is it and what function did it serve?
2. What relationship (if any) does it have to any other items of evidence or to the scene itself?
3. What does it indicate about timing and sequencing aspects?

Every action taken at a scene has some level of destructive effect on the scene. Frequently development projects will dispatch soil-testing equipment, such as a backhoe, to a scene to examine the suitability of the soils for the intended project. All too often this equipment unwittingly destroys and displaces valuable evidence. Some activity cannot be prevented; however, some can, or the damage can be minimized. Scene degradation begins the moment after the first event has taken place and continues until the last person evaluates the scene and its remaining evidence. The older the site, the more degradation and contamination that is likely. Regarding boundary evidence, trees die, blow down, or are harvested; fences deteriorate; corner markers become lost. All evidence is subject to change or loss from the moment it is created.

Knowing that scene alteration is inevitable, the basic goal of any processing is to limit the damage and recover as much evidence and context from the scene as possible. There is no "right" way to process a scene since each one is different, but proper methodology will demand five ingredients of good scene examination practice:

1. Knowledge
2. Skills and tools
3. A methodical approach
4. Flexibility
5. A coordinated effort

The scene processor is more than just a collector of things; he or she must seek the interpretive value of the evidence.

The investigator at a scene is more than just a collector of "things."
This is not a scavenger hunt.

Knowing what he or she is looking for (description calls and other likely evidence) and what can be done with evidence (identification, interpretation, and aging), the investigator will not overlook information and evidence. In addition to the knowledge, the processor must have the skills and tools to recover the evidence. Camera equipment, metal detectors, and increment borers are all part of an investigator's toolbox. Since land parcel creation and resulting land use can provide almost any imaginable type of scene, the investigator needs to be prepared for a wide variety of conditions and be equipped accordingly. Because of this, the investigator must be flexible. Each new scene presents its own challenges.

Since the investigator's goal and responsibility is to collect evidence and scene context in as pristine condition as possible, the five ingredients of good scene examination practice must be combined with an understanding of basic scene integrity concerns. The scene processor's methods must consider three specific scene integrity issues:

1. Addition of material to the scene
2. Destruction of material in the scene
3. Movement of material in the scene

Addition of material results from using the scene in any way since it was created. A common example is the existence of surveys subsequent to the original, adding to or changing original evidence. Sometimes this can be helpful in that perpetuates the existence and location of original evidence. A case in point is *U.S. v. Champion Papers, Inc.*[31] After the passage of 135 years, the original trees marked as corners and witnesses had disappeared, but subsequent resurveys and surveys of abutting lands had located their remains, therefore perpetuating the corners. A frequent occurrence is the removal of iron markers from lawns since they interfere with mowing. If the investigator is fortunate, they are pushed below the surface and are no longer visible but still can be found. Most metal material can be located with a metal detector, not so with wood and stone. Evidence gets buried with normal fill to make a site ready for improvements; some is inundated with water in normal flowage situations; some is removed by excavation in mining activity. In some situations, especially with commercial sites, markers are covered with pavement and often difficult to recover.

Destruction of material results in the loss of an item's evidentiary value. Documentary evidence in particular is subject to fading, alteration, and so on. Physical evidence can be mishandled or damaged. One item of information that is difficult to deal with because it is always suspect is a call for an iron pipe in a description yet there is an iron pin at corner. Was someone mistaken with the call, or was there once a pipe, which has since been replaced with a pin? And if the latter is true, who did it, and what was it based on?

Movement of material is sometimes the result of investigative processing techniques, or can be by others at some time in the past. When such movement is unchecked and unrecognized, it is deceiving to the investigator and can result in erroneous conclusions. Such movement can significantly alter the relational aspects of the scene. Natural earth movements due to landslides, slippage, subsidence, earthquakes, and the like will likely alter the spatial relationships of items at a scene.

---

[31]361 F.Supp. 481 (Tex., 1973).

CHAPTER 8

LAND RECORDS

> Never trust to general impressions, but concentrate yourself
> upon details.
>
> —*Sherlock Holmes*, "A Case of Identity"

Land record problems are many, and fall into several categories: missing heirs, missing documents, misspellings, confusing language, conflicting information—almost anything you can think of. They lead to two types of problems: title problems (ownership) and survey problems (location). Many of them become the subject of litigation, and a few of them can be solved only through the court system. Most of them, however, have been solved before, and can be resolved without involving the court system, because the rules, techniques, and tools are available. Some of them involve forensic procedures by experts from other fields.

Tool Kit

Working with land records requires drawing pictures and sketches of descriptions. It is generally a mistake to try to compare descriptions, either those within a chain of title or between chains of title, without sketches. Subtle differences and changes are difficult to detect, and can easily be overlooked.

A straightedge and a protractor are indispensable tools, as is a scale (not a ruler). An engineer's scale is usually more helpful than an architect's scale, as the former is divided into decimal equivalents rather than feet and inches or some other units. It will suffice for scaling from most survey plans, provided they are to scale. Sketches should be done reasonably to scale, even though they are only sketches and may have been done by eye.

A calculator is a necessary tool for converting units, working between descriptions and plans, and computing the area (acreage) of a figure. Colored pencils and markers are useful for color-coding different parcels, calling attention to certain features, or highlighting important wording in documents. Be selective with colors if an image is to be photocopied, since many colors will obscure the image so that it does not copy, and using a red pen to underline will show up as an obvious black line. Yellow is usually the best for highlighting.

There is no substitute for a thorough search of the pertinent records.
Boundaries locate the title on the surface of the earth and define its
extent.

Case #12

WHICH ONE OF US OWNS THE PARCEL?

Ski Roundtop, Inc. v. Wagerman, et al.

79 Md. App. 357; 556 A.2d 1144 (1988)

This dispute involved the ownership of a parcel of land that was claimed by the two parties. One of the parties' deed description included the tract in question, and that had been consistent in the chain of title since 1812. The patents creating the titles were based on surveys created nearly 200 years prior.

The court found that even though one chain of title included the tract in question, the title was not valid since there never was a patent issued from the state.

The appeals court stated:

> It appears that a requisite for valid title to real property is an original conveyance of public land by the State. Absent such a conveyance, one purporting to transfer an ownership interest in such property transfers nothing, and no quantity of successive transfers by deed nor the mere passage of time will metamorphose good title from void title.

This court referred back to an earlier decision of *Maryland Coal and Realty Co. v. Edkhart*, 25 Md.App. 605, 337 A.2d 150 (1975). In that case the court said, "An albatross in the wake of every title searcher is the ominous question of whether he has gone back far enough in the chain of title." To be certain, to be on the safe side, a description should be searched back to the origin of its description.

Since neither party could show a valid title to the parcel, the court could not find that either one was the owner. This case illustrates a very important consideration that is often ignored. Given two choices, many want to know which one is the correct one, that is, which is right and which is wrong. It is a rare situation in which both would be right (probably for different reasons); rarely, also, both could be wrong. The proper question should then be: "*If* one of the two choices is correct, which one?"

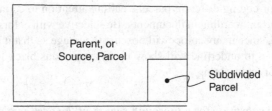

Figure 8.1   A subdivided parcel consisting of two existing lines (the side line and the road) and two new lines (the lines interior to the parent parcel). The deed creating the parcel is the creating document for the two interior lines, but the creating document for the parent tract lies farther back and may consist of more than just one document.

REASONS FOR SEARCHING A DESCRIPTION OR TITLE BACK TO ITS ORIGIN

There are a number of reasons for taking a description or a title source back to its origin.[32] Origin of description is one thing; origin of title is another. A title may be taken back to where it originated, but some of its boundaries may be coincident with boundaries that were created earlier, such as those of a parent parcel.

Without taking a description back in time, there are seven things that cannot be known about a tract of land, any of which may be very significant. They are:

1. Comparison of the current description with the original
2. Interpretation under the surrounding circumstances
3. Order of conveyancing
4. Declination conversion
5. Early surveys and plans
6. Easements and other encumbrances
7. Agreements and other instruments affecting the boundaries or the title

Each of these areas is discussed next.

Comparison of the Current Description with the Original

Comparison of the current description with the original will provide clues or information as to whether there have been any changes to the description over time.

---

[32] Procedures useful to assist in tracing backward are elaborated on in *Interpreting Land Records* (Hoboken, NJ: John Wiley & Sons, Inc., 2006) and are not repeated here.

Changes may be due to several factors: whether part of the original tract had been conveyed away, whether land was added to the original tract by one or more means, or whether mistakes and transcription errors have occurred. The latter tends to be far more frequent in the records than we would like, to the point where most things become untrustworthy. Any changes should be documented, unless they are plain mistakes in description.

Interpretation under the Surrounding Circumstances

Given a description in deed that has been carried over for many years and through many transfers, it is a rule which has been recited by numerous courts that it must be interpreted according to the surrounding circumstances.[33] Without knowing the time frame of the creation of the description and the conditions at the time, it is very easy to be misled, especially if we assume that the conditions and circumstances have not changed over time.

Order of Conveyancing

Without a time frame and a comparison of a chain of title (or a deed) with other tracts from the same source, it is impossible to determine whether a parcel enjoys senior standing, is a remainder, or was created at the same time as adjoining or nearby tracts. This issue becomes critical when (1) either excess or deficiency is encountered in the surveying (measuring) process, or (2) it is necessary to analyze evidence found on adjoining and nearby tracts from earlier surveys that treated them as stand-alone parcels.

Declination Conversion

When it becomes necessary to convert magnetic bearings for changes in magnetic declination, it is crucial to determine when the original observations were made.[34] While the date of a deed may not be the date of the observation, it will likely be close enough to accomplish the purpose for which the corrected magnetic bearing is

---

[33] Fundamental rule of construction is that purpose or intent of a written instrument is to be determined from language usd in light of circumstances under which it was written. *Phipps v. Leftwich*, 222 S.E.2d 536 (Va. 1976).

[34] *Greer v. Hayes*, 216 N.C. 396 (1939).

needed. Retracement is one thing, as the change is intended only to get the investigator in the general vicinity, but setting a corner demands refinement to the extent possible, as the requirement is that the marker be set in the *exact* position in which it was created.

Early Surveys and Plans

Especially in offices (courts and otherwise) that record or index surveys by name, it is necessary to have the name of the owner at a point in time to find or access a survey made for him or her. Without a full search of all the names in a complete chain of title, it is easy to overlook a survey or plan that could be helpful or provide an answer.

Easements and Other Encumbrances

Without a search to the beginning of title, it is impossible to determine whether there are easements that will either benefit or burden the property. Easements by implication, for instance, can be identified only by relating a tract to its source parcel. Besides, frequently a right-of-way to a parcel, either in the creating document or in a separate document, is created at the time of the parcel creation and not recited subsequently. The determination of such important access rights, and perhaps other easements, can be found only by a complete search of the pertinent records.

Agreements and Other Instruments Affecting the Boundaries or the Title

Numerous instruments affecting either or both the boundaries or the title to a tract may be filed subsequent to its creation. Without a complete search from the creating document forward, these may go unnoticed but still be a very important part of the title elements of the property.

TECHNIQUES FOR GETTING BACK IN THE CHAIN OF TITLE

Several techniques are available to the researcher for searching backward in the records, particularly if the usual checking of the grantor's name in the grantee index,

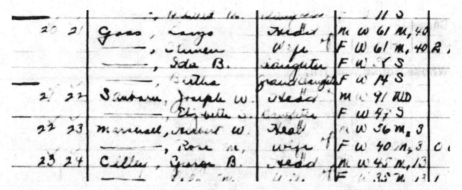

Figure 8.2 A project was under way where an important parcel was identified by an abutting tract as "heirs of Joseph W. Sanborn." An attempt to find him in the records was unsuccessful, and there was no recorded death (probate) records for him, so a search was made in the related records, including census records. There he appeared with his daughter Elizabeth, who was searched as his (possible) heir. The necessary records were found. The difficulty arose in that when his name was transcribed, it was written as Joseph W. Saubarn instead of Sanborn.

to see where he or she may have acquired the property, does not yield the desired information. These techniques presuppose that the acquisition was by deed, which is not always the case.

Additional information is in great abundance in most cases. Town line descriptions, road layouts, town histories, genealogies, local (family) burial grounds, railroads, utility lines, and other sources frequently provide a researcher with possible names. Sometimes this information has to be verified for accuracy, but ordinarily in a thorough research project, proof becomes self-evident.

Optional Spellings of Surname

In the event there is more than one spelling of the surname (one instance produced 28 aliases) or there is a misspelling, one must think in terms of options.

Early Maps and Atlases

Early maps and atlases with names can be found in most libraries and historical societies. If there is more than one generation of maps, sometimes names of several owners can be followed for a location, providing a skeleton chain of ownership that can later be verified in the official records.

Case #13

Whose House Was This?

Figure 8.3   (a) bricks

Figure 8.4   (b) remains of foundation

Figures 8.3–8.7   These photographs were taken of a site for which ownership was unknown. Noting that it once had a dwelling (b) and a well (c) indicating habitation, as well as being on a road (d), led to the examination of early maps for possible names. The map (e) showed one dwelling with the name G.D. Hanson, which was searched in the deeds office, producing a description of the parcel and, tracing forward, the current owner.

Figure 8.5   (c) remains of rock-lined well

Figure 8.6   (d) line of stones along edge of road

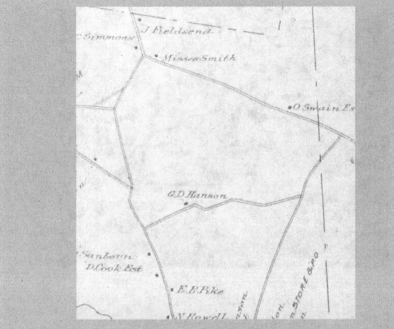

Figure 8.7 (e) early map showing one house and the name G.D. Hanson

Case #14

STREET LAYOUT PROVIDES NAMES AND NUMBERS

> Beginning at a stake in the highway near the barn of George L.Brown in said Hampton Falls, thence S.74 deg.40' E.through land of George L.Brown 87 rods and 11 links to a stake; thence on the same course 7 rods and 12 links through land of James Janvrin in said Hampton Falls to a stake; thence on the same course through land of Jonathan Janvrin of Amesbury, Mass.27 rods and 14 links to a stake; thence on the same course through land of James Janvrin in said Hampton Falls 21 rods and 8 links to a stake; thence on the same course through land of Jonathan Janvrin of Amesbury,Mass.21 rods and 8 links to a stake; thence crossing the Eastern Railroad 4 rods on the same course; thence on the same course through land of Charles F.Chase and George L.Brown in said Hampton Falls 26 rods to a stake; thence S.53 deg.E.through land of Charles F.Chase and George L.Brown in said Hampton Falls 36 rods and 13 links to a stake and stones at the south easterly corner of Charles F.Chase's land.
>
> The line above described is to be the middle of the highway and the highway is to be TWO AND ONE HALF RODS WIDE.

Figure 8.8 A boundary dispute between two parcels ensued due to a recent survey, neither of the parcels contained measurements in its deeds or in any predecessor deeds. In fact, none of the tracts along the road had measurements in its descriptions, even though each was searched back to the parent parcel. Once back to the parent tract, however, it was apparent that there was no acknowledgment of the street. Therefore, it was concluded it had been laid out after the parcels had been sold from the parent tract. Once the various parcels had been created, the street had been laid out through them, providing distances for all of the parcels that it traversed.

Case #15

USING GENEALOGICAL INFORMATION

A title was traced back to a will of Mary E. Giddings in 1910, which contained an "18 acre woodlot," but further search failed to uncover earlier records. Through the use of calls in descriptions of abutting tracts, it was determined that Zebulon Wiggin had owned this tract and had sold it to a Mary E. Scammon in 1857. It then was a matter of determining whether these are two persons or only one who married and changed her name. The answer was found in the family genealogy, as shown in Figure 8.9.

> **84 John Lyford** (*Thomas*,[30] *Thomas*,[11] *Thomas*,[2] *Francis*[1]) born 1 Mar., 1777; died 1803; published 30 Aug., 1799, to Anna Hilton of Kingston (born 4 Sept., 1776), daughter of Andrew Hilton of Exeter (son of Benjamin Hilton) and Jemimah Prescott (born 23 Oct.,
>
> 1742). Letters of administration on the estate were granted 23 Jan., 1804. The widow married for her second husband, Kinsley Lyford[71], and she died 7 Mar., 1865, in Exeter.
>
> Children, all born in Exeter:
>
> 198. A CHILD.
> 199. JAMES, b. 1800; d. at sea while still a young man.
> 200. ANN, b. 4 June, 1803; d. 4 Mar., 1857, in Stratham; m. 6 June, 1828, Ira Scammon of Stratham. He was b. 11 June, 1803, and d. 14 Jan., 1852. Their dau. Mary Ellen, b. 1 Mar., 1837; m. 22 Oct., 1882, John Colcord Giddings.

Figure 8.9 Excerpt from Francis Lyford, of Boston, and Exeter, and some of his descendants.

This information was helpful in two ways. In addition to the foregoing, the tract in question was made up of two parcels, one of which was sold to "Ellen" Giddings. It could only be assumed that the "E." in Mary E. Giddings stood for Ellen, but this information confirmed it. The two parcels had more detailed descriptions in the earlier records, which aided in locating the ownership on the ground.

Family Genealogies

Family histories and relationships can be very useful in making connections when compiling chains of title. Whenever possible, it is important to have a continuous

chain from the originating document, whether for title purposes or boundary description, to the present title document. Often that is easier said than done, and the
older the source, the more difficult having a continuous chain sometimes becomes.
In particularly difficult situations, it may be necessary to spend as much or more time
with family records as with the title documents. Family records, besides providing
links between individuals, may lead to other clues or provide a source for finding
additional land records.

Many published genealogies are available, and there is a wealth of information on
selected families in town histories and other sources. There are also several sources
of family records on the Internet. A few free sites provide with limited information,
a few compilations on CD are available for purchase, and several sites with an
abundance of information are available on a subscription basis.

Whatever source you rely on, consider the information as a clue or a possible
solution, but satisfy yourself that whomever compiled it was correct. Make your own
verifications. Just because it is in print or appears in the right form on the Internet at
a reliable site does not insure its accuracy.

Probates for Last 50 Years

Checking all the probate records of the surname ended with sometimes uncovers an
estate that lists the last grantor as receiving property or title. Family records will also
give relationships so that you can sometimes get an idea regarding from whom the last
grantor may have inherited or been left property. People do not always inherit from
the father. Inheritance or title by will may come from anyone, including a mother who
may have remarried, an uncle or an aunt with different surnames, another relative, or
even a friend. Sometimes further investigation beyond that of just checking probates
with the same name is necessary.

Cemetery on Site or Nearby

See Figure 8.10.

Cemetery Records

From the town or from others who have accumulated such records, located either
in libraries, historical societies, or in personal files. Many gravestone records are
a part of town histories and in published family histories. Information such as this
can be invaluable in compiling chains of title and tracing records back to find title
information or boundary description.

Figure 8.10 Stone from the Wiggin burial ground located at the site of former Wiggin property. Some of the family relationships are shown on this stone. Others can be found on nearby stones.

In the absence of collected information, the investigator may wish to visit the family burial ground to seek his or her own information. Many times, especially in more rural communities, a small family burial site will be found on the property being investigated. This is important to determine for other reasons as well. Other people may own the burial area or, at the least, may have a right of burial or access. In some

Figure 8.11 Small family burial ground located on the subject property. This one is not well maintained, which is sometimes the case. Great care must be taken in searching records and examining the site to find such locations. Frequently others have a right of access to the graves of family members or have the right of burial. Occasionally the burial site is a separate ownership, having either been sold from or excepted out of the chain of title some years ago.

cases, the burial spot itself may be an independent title, or an abutting property to the subject parcel.

Topographic Maps and Related Information

Topographic maps often depict cemeteries and burial grounds, sometimes with names. An index by state to topographic maps, entitled *Geographical Names: Alphabetical Finding List,* is available. This official list contains are all the major features

Sanborn Brook	stream	33013	431639N0712157W		432134N0712340W	0139 C
Sanborn Brook	stream	33013	425817N0711055W			0184
Sanborn Cemetery	cem	33015	425900N0710359W			0185
Sanborn Cemetery	cem	33015	425606N0705330W	120		0186
Sanborn Corners	ppl	33015	432135N0712131W			0139
Sanborn Hill	summit	33001	432016N0712320W	670		0138
Sanborn Pond	lake	33013	425620N0710317W			0185
Sanborn School	school	33015	432921N0713458W	822		0124

Figure 8.12 A page from the *Geographic Names* list, indicating on which maps Sanborn cemeteries may be found.

Figure 8.13 Sanborn Cemetery shown on a topographic map. Most likely it carries that name because Sanborn is the predominant name in the cemetery, a stone or sign states it is the Sanborn Cemetery, or there is an obvious stone that identifies it as having Sanborns buried there.

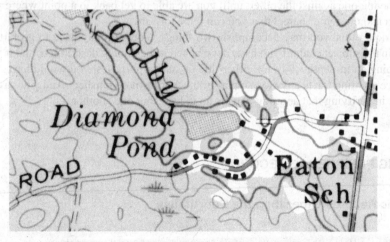

Figure 8.14 Colby Brook is a stream on which people by the name of Colby had a mill; Diamond (*sic*) Pond is a body of water mostly surrounded by land of Dimond; and Eaton School likely was on land owned by Eaton.

labeled on these maps. Much this information is now on the Internet, and a search can be made that way. The investigator can find such things as bays, bridges, cemeteries, islands, lakes, streams, hills, mountains, and so on, any of which may provide one or more clues as to where people may have settled. Information also can be found in a multitude of other records, such as town and church records and the like.

Check Grantor Indexes

After you have thoroughly exhausted the grantee index without results, check the grantor indexes. Sometimes the grantor sells other parcels and recites the source(s) or title. And sometimes these recitation(s) are, or include, your parcel or a clue as to where your parcel may have come from.

Use of Abutting Tracts

When it becomes impossible to trace backward, it is usually possible to search an abutting tract back to a point where it calls for an abutter (your tract), which is a previous owner in your chain of title. In doing such a search, always start with what appears to be the easiest parcel to search or the most unusual one, if there is a choice. Also, work with the most unusual name or the name of a person who did not buy many parcels of land. Otherwise your search may become very lengthy. If you reach a dead end in that chain, work with its abutter, and if necessary, its abutter and so on, playing one against the other, until you are able to get back to a point where you have a name for your parcel that you can trace forward.

In doing this, you may accomplish two things. You still may have to go farther back in the subject chain if you have not reached its beginning description, and if you are doing an in-depth study for either title or survey purposes, you would have had to trace abutting tracts anyway, so part of the work is now finished. And you have a check in verifying that abutting parcels do indeed abut one another.

USING PERIPHERAL RECORDS

Public Records Other than the Courthouse

Usually a title examination is confined to the records at the courthouse or similar repository of deed and probate information. Many, *many* important records that have a significant effect on the property are not found at the courthouse but are a matter of public record elsewhere.[35] These include, but are by no means limited to:

- International boundary information.
- State and county boundary descriptions and surveys.[36]
- Town line descriptions, perambulations, and surveys.[37]
- Highway and other road layouts and surveys.[38]
- Vital statistics: state, town and church records of births, deaths, marriages, adoptions, and so on.
- Tax records.
- Court records, including not only decisions and testimony, but also exhibits that may not be found elsewhere.[39] These may be the source of descriptions of and actual condemnations.
- Records of state, county, and municipal lands.[40]

---

[35] Generally, relevant documentary evidence is admissible, whereas irrelevant and immaterial evidence is not.

For the purpose of identifying the premises sought to be described reference may be had to other conveyances, plats, or records, well known in the neighborhood, or on file in public offices. *Pittsburgh, C., C. & St. L. Ry. Co. v. Beck*, 53 N.E. 439, 152 Ind. 421 (1899).

Proper aids for identifying land described in deeds include judgements, certificates, patents, deeds, leases and maps. *Sorsby v. State*, 624 S.W.2d 227 (1981).

[36] Courts are bound to take cognizance of the boundaries in fact claimed by the state, and will exercise jurisdiction accordingly. *State v. Dunwell*, 3 R.I. 127 (1855).

The court will take judicial notice of the lines of counties and the towns enclosed therein. *Ham v. Ham*, 39 Me. 263 (1855); *State v. Jackson*, 39 Me. 291 (1855).

[37] Perambulations of the line between two towns, made by the selectmen of such towns, are competent evidence of the true line in a suit between individuals owning land on opposite sides of the line. *Adams v. Stanyan*, 24 N.H. 405, 4 Fost. 405 (1852). *Lawrence v. Haynes*, 20 Am. Dec. 5511, 5 N.H. 33 (1829).

On a dispute between private parties as to the division line between their private lands, which was also the boundary line between two towns, a "field book" of one of them is admissible as an "ancient record," when it came into possession of the town clerk a "large number of years ago," and deeds of land in the town, executed over 45 years previous to the trial, referred to it, and it was in the town clerk's custody at the time of the trial. *Aldrich v. Griffith*, 29 A. 376, 66 Vt. 390 (1893).

[38] Where a deed describes property as being bounded on a public highway, the return of the surveyors laying out such highway is competent evidence to establish the boundary, without showing their appointment or whether the highway was legally laid out. *Haring v. Van Houten*, 22 N.J. Law, 61 (N.J. Sup. 1849).

[39] In establishment of ancient boundaries, judgments of trial court were held admissible, notwithstanding parties were not the same. *Blaffer v. State*, 31 S.W.2d 172 (Tex. Civ. App. 1930).

[40] In trespass to determine the location of a boundary line between certain lots, an old plat of the town obtained from a volume of state papers, showing the division of the lots, etc., was admissible, in the absence of objection as to its authority. *Twombly v. Lord*, 66 A. 486, 74 N.H. 211 (1907).

Ancient surveys of a city, showing streets and lots appearing on the county records, are presumed to have been recorded by authority, though not formally certified for record. *City of Lexington v. Hoskins*, 50 So. 561, 96 Miss. 163 (1900).

- Records of state, county, and municipal easements,[41]
- Public utility easements and surveys,
- Federal lands and surveys, including field notes,
- Records from other jurisdictions,
- Political boundaries. These change, but records, particularly true of county and town records, sometimes remain within the original jurisdiction.
- Railroad records: lands and surveys.

Document Not on Record

Sometimes a deed or other document does not get recorded. However, beware of the pitfall that it may indeed be on record but not indexed properly or indexed at all. The record is what puts the public on constructive notice, not the index; the index is generally there for convenience.

Even when a deed actually never did get recorded, it may still exist. Checking with family members who have old papers stashed away, law firms that represented the parties or their families, real estate firms that may have handled a sale, and other obscure possibilities may produce the deed. Don't give up until at least all of the obvious sources have been checked.

Missing Link Not Found

Some courts and jurisdictions have accepted one or more documents as missing from a chain of title if their existence can be shown by parol evidence, or if it can be shown that it is more probable that the document was executed than that it was not.[42]

In *Pratt v. Townsend*, the court said:

> The doctrine of the presumption of execution of a deed, or, more properly, the proof of its execution, by circumstances when no better evidence is obtainable, is well established, and especially so by the decisions in this state, and the disposition has been to extend rather than to limit it. . . . The destruction of the records in so many of the counties of this state, coupled with the carelessness, well-nigh universal at an earlier period of our history, on the part of the people in keeping the original deeds after they had been recorded, and the disposition now so prevalent to uncover, by reason of carefully prepared abstracts of title, the absence of such written muniments of title as are necessary to make a complete chain, require, in our opinion, a liberal application of this rule for the protection of titles long relied upon in good faith, the evidence of which has been lost through carelessness or accident, the destruction of records, and the death of

---

[41] In boundary dispute, copy of profile of street sewer showing boundary line and fence, dated October, 1892, and produced by city engineer's office, held competent. *Hews v. Troiani*, 179 N.E. 622, 278 Mass. 224 (1932).

[42] 67 ALR 1333, Presumption or circumstantial evidence to establish missing link in chain of title.

all persons originally connected therewith, or likely to know anything about the facts. And especially is this true when such claim of title is accompanied by an entire absence of any assertion or claim of right or title inconsistent with the claim under the deed so sought to be established. It was clearly established in this case that diligent but unsuccessful search had been made for the originals of the deeds sought to be established by circumstances.[43]

Problems within the Records

Handwriting can frequently be a problem. In fact, some people shy away from early records, especially those older than typewritten copies, because of the difficulty in reading them. *Interpreting Land Records* provides clues on how to deal with this problem, as well as a number of sources identified in that book. Handwriting experts are available, as are others who are used to various old styles of writing. Anytime difficulty arises in reading any record, take the known and proceed to the unknown. Often by finding words you know, you can figure out the words you are having trouble with. Remember, all words in a document must be given meaning, so it is never a matter of picking and choosing the words you understand, or like.

Deciphering Handwriting

Many times documents found in public repositories are not originals but have been copied from the originals. This is true for deeds and related instruments, some probate records, town records, and the like. When this is the case, the key to being able to read the records is understanding a person's handwriting. Often the same handwriting will appear for several pages in succession, even for several successive years, so there is generally an abundance of work to compare.

Comparison is one of the methods that proves to be successful. If a word cannot be read, searching for similar words in the same document or in other documents written by the same hand may help to decipher it. Sometimes it is necessary to break the word down into individual letters and apply the same technique. Comparison with known words to determine what the individual letters are often reveals the ultimate word.

Early records were transcribed by hand, so familiarity with handwriting of the period or the particular writer is important. Later records were transcribed by typewriter, and, in most cases, today's records are copied through photographic processes, either with photocopy machines or microfilm, or both. Today's process are theoretically more accurate transcriptions since direct copies are made. Early documents were subject to interpretation, and a person copying could only write what he or she could read. The potential for a mistake was always present. Sometimes it is necessary to

---

[43] 125 S.W. 111 (Tex Civ. App. 1910)..

view the original document, rather than the copy, if possible. *The problem arises, however, in that the later documents were produced from the earlier ones—the very records where the problems originated.*

In genealogical work, researchers insist on the original document whenever available. Any copy, unless it is a direct reproduction, is suspect. Mistakes can occur the first time a transcription is made, and they will likely be carried forward forever. One method of checking is to *check the source whenever anything appears to disagree* or is not in complete harmony.

Today's records owe their origin and basis to earlier records. If the early record was faulty, it is probable that the current record is faulty as well. Never forget that the value of *any record is only as good as its basis.*

Beware of Pitfalls

Many things have changed over the years, including the calendar, the way we number the months, the choice of words and labels, the meanings of words, and the units of measurement common to the time a parcel was described. Don't assume. Make sure any choices can either be verified or supported with other information.

Deciphering Words

People not only wrote differently in the past, but they also talked differently, and they used words not in common use today. Therefore, it is sometimes necessary to consult early dictionaries and encyclopedias to locate and define early words. Colonial documents are a good example of this type of problem.

A word of caution should be inserted here, however. Just because your practice is in the twenty-first entury or because your work is limited to areas outside of the Colonial states does not exempt you from these problems. Early settlers left colonial states for other areas and took their handwriting, language, and other habits with them. Also, today's records are based on earlier records, and frequently the current record is the same as the original, which may have been written a century or more ago, barring any transcription errors. Never lose sight of the fact that *the more times a writing is transcribed, the more likely it is that the record may contain errors or mistakes.* That is one good reason why it is imperative to trace a description back to its origin to compare with the present description for such errors or differences. If you routinely do this, you will be absolutely astounded by the frequency of errors.[44]

---

[44] A trail of descriptions is like the game *telephone* played by children. The first person relates a story or a phrase to the second person who retells it to a third person and so on around the room. By the time the last person recites to the original teller, the story has changed. It may even be a new story, bearing no resemblance to the original version.

Early Writing and Punctuation

Knowledge of the history of writing is helpful in understanding how problems or conflicts may have originated and developed. There have been marked improvements in writing and punctuation since land records began, so it is easy to be lulled into thinking that things were always as they are now. Not only is that incorrect, but we must always keep in mind that *today's records are based on, and are sometimes the same as, those of long ago*. It is those original mistakes, or departures, that we must constantly be on guard against.

Early on, and until the middle of the sixteenth century, most public documents were written by scribes, secretaries, or priests. At that time writing was considered very much an art form. Writing during the seventeenth and most of the eighteenth century was influenced by European standards. It wasn't until 1713 that English law provided that clerks and scribes use common English and legible handwriting. Even this law wasn't foolproof, and many old habits remained.

Latin abbreviations continued. In the English colonies some old short forms persisted that remain today. These include the ampersand for the word *and* (&), the sign for the word *at* (@), and the abbreviation of the Latin *et cetera* to *etc.*

One of the biggest problems in the American colonies is that not all scribes were English or used English words and translations. There were scribes of Dutch, French, German, Jewish, Spanish, Swedish, and many other national origins. Each person brought his own language, interpretations, and slang. In addition, they all brought their own terminology relating to land and had their own ideas of land ownership, transfer, and tenure.

At first there were no pencils, and documents were written with quill pen. This method was popular in the American colonies until about 1830, when the modern steel nib pen was first introduced.

Spelling

People generally received a bare minimum of formal education until about 1850. Consequently, misspelled words abound, since much of the writing or spelling, especially of surnames, was phonetic. For instance, there are numerous examples where the town of Epping, New Hampshire, is written *Eppin*, just as it was pronounced.

Early Terminology

Terminology has changed significantly over the years. Some terms will be found to be local or colloquial.

DOCUMENT EXAMINATION

Document examination is the forensic science concerned with the study and investigation of documents, usually what is called the area of *questioned documents*. It includes the examination and analysis of all parts of a document, including signatures or other handwriting, typewriting, inks, papers, and anything else making up a document. Ordinarily, examiners are concerned with possible forgeries, alterations, and dates, so they may also investigate possible insertions or obliterations. While most of the application of document examination is in solving crimes and in proving guilt, many of the techniques can be used in solving routine problems likely to be encountered in the reading and interpretation of ordinary records.

The analysis of a person's handwriting is a part of forensic document examination known as the science of *graphology*.[45] It is a behavioral science that identifies personality characteristics and expected behavior from the study of a person's handwriting.

Forensic document examiners, in general, use a five-point scale in determining whether a particular individual was the author of a document or not. The scale follows.[46]

1. *Common authorship.* One hundred percent positive, written by the same person.
2. *High probability.* Very strong positive evidence; very unlikely to have been written by a different person.
3. *Probably,* or could well have been written by the same person.
4. *Inconclusive.*
5. *No evidence.* (This does not mean that the subject did not write the document, merely that there is no evidence that he or she did.)

DOCUMENT EXAMINATION

You can sometimes use the office copying machine to enhance a document that is difficult, or impossible, to read. This is particularly true of faded lettering or anything produced with a pencil. Early blueprints, particularly if they have faded with age, frequently present a problem.

Placing a sheet of yellow acetate film between the document to be copied and the glass will usually result in a copy that is better than the original.

This technique cannot be used, however, where pages are fed *through* the machine, such as between rollers.

---

[45] The graphologist analyzes handwriting to determine the personality of the writer. The document examiner analyzes handwriting to identify the writer or detect a forgery. Dorothy V. Lehman. *Questioned Document Examination—Identification of Handwriting on Document.* 15 POF 3, p. 595.

[46] Zakaria.Erzinçlioglu, *Forensics.*

Excellent results have been obtained using this technique in order to be able to read and decipher early records written in ink where the ink has bled through and stained the reverse side of each page, which also contains writing. In addition, the author once was able to obtain only a copy of a blueprint instead of the original, and the print itself was difficult to read. Copying the copy using yellow acetate resulted in a second-generation copy that was more readable than the blueprint I received.

Images can be enhanced by photographic means in countless ways. The use of special films, filters and developing techniques, along with combinations thereof, can sometimes produce excellent results.

PHOTOGRAPHIC TECHNIQUES

Use of High-Contrast Film

Photographs that are old, faded, or stained can be copied with high-contrast film, producing a photograph that is far superior to the original, even though there may be a slight loss of quality. The key to a satisfactory result is often the correct choice of film, filter, and developer. The yellow color can be removed through the use of a blue filter; reddish stains can be removed by using a red filter. Expert assistance is frequently necessary, since the selection of photographic paper may also be important. With severely faded or deteriorated images, several trials and combinations may be necessary to produce the best, or a satisfactory, result.

Ultraviolet Photography

The purpose of ultraviolet photography is to provide information about an object or material that cannot be obtained by other photographic methods. There are three types, depending on the techniques used: (1) straight ultraviolet photography, (2) ultraviolet luminescence photography, and (3) fluorescent photography.

Straight Ultraviolet Photography.   This is the method of taking a picture whereby a camera is used to record the differences in a subject's reflection, transmission, or absorption of ultraviolet rays in the same manner as visible light photography. To accomplish this, an ultraviolet light source is used and an ultraviolet filter is placed on the camera lens to eliminate all visible light and transmit only ultraviolet.

Ultraviolet Luminescence Photography.   This method is the process of taking a picture of a subject that is emitting invisible radiation while it is being exposed to ultraviolet rays. As an example, a document in a completely dark room can be exposed to (short-wave) ultraviolet rays. Certain areas of the document, such as ink lines or paper fibers, may emit (long-wave) ultraviolet radiations that cannot be seen but can be photographed. By employing this technique, the range of human vision is extended and facts that cannot be learned any other way may be discovered.

Fluorescence Photography.   This technique is similar to the familiar use of ultraviolet light to "black light" in a dark room whereby objects give off a visible glow called *fluorescence*. Fluorescence is merely *visible* luminescence. Fluorescence photography is visible light photography of the image produced from exposing the subject to ultraviolet light. This is what is usually thought of or spoken of as *ultraviolet photography*.

Document Photography

The preceding techniques are used routinely in document examination, perhaps more widely than in any other type of forensic examination. Fluorescence photography can demonstrate that two sheets of paper are not the same.

Ultraviolet rays are especially helpful in bringing out erasures. In some cases, not only the fact that there was an erasure can be shown, but also the wording of the erased matter is found. Fluorescence photography usually results in the erased writing showing as white lines on a dark background, while straight ultraviolet photography shows the lines more naturally, as dark lines on a lighter background.

Infrared Photography

Infrared photography deals with recording images that are invisible but which evidence themselves as heat. Recording images involving infrared rays are complex and include four distinct areas, three of which are of interest here: (1) straight infrared photography, (2) indirect infrared photography, (3) infrared luminescence photography, and (4) thermography.

Straight Infrared Photography.   This method involves the use of infrared film and an infrared filter over the lens. Otherwise the procedure is the same as for ordinary black-and-white photography. It is useful since many subjects reflect, transmit, or absorb infrared radiation such that they produce results other than those obtained by means of ordinary visible light.

Indirect Infrared Photography. This technique involves the conversion of the infrared images on a fluorescent screen which can be observed visually and photographed directly with fast panchromatic film. The results are very similar to those obtained by straight infrared photography.

Infrared Luminescence Photography. When an object is illuminated, it emits not only ultraviolet, but also infrared, radiation, which is not visible to the naked eye. By keeping an object from being irradiated by infrared rays, visible light will cause the object to luminesce, thereby allowing an image to be recorded on infrared film.

Infrared photography is one of the most important and widely used techniques for examining documents. It can demonstrate differences in inks and thereby allow deciphering of overwriting and other obscuring factors. Sometimes an infrared photograph can make a charred or stained document readable.

Ordinarily, infrared film and a filter, such as a Kodak Wratten 87 filter, are employed. If stronger effect or penetration is found to be necessary, the No. 87C filter can be used.

Infrared photography sometimes overcomes these problems with illegible documents:

- Illegibility due to charring
- Deterioration because of age or dirt
- Obliteration by application of ink
- Invisible inks
- Chemical bleaching
- Mechanical erasure with subsequent overwriting

COMPARISION OF PHOTOGRAPHS TAKEN YEARS APART

Aerial photographs often are used to in the comparison of two or more photos of the same area taken at different times. Existing photos may show changes in conditions or changes in land use over time and are very useful in cases of boundaries affected by accretion or avulsion, erosion, timber and gravel trespass, hazardous waste dumping, and the extent of, or change in, development.

Terrestrial photographs may be used in the same manner, if taken from the same viewpoint. Early photos may be found in archives, libraries, historical societies, and private collections and on postal cards. Arcadia Publishing Company produces several series of titles containing historical photographs. They contain mostly items

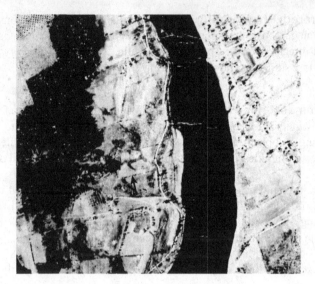

Figure 8.15

Figure 8.16

Figures 8.15–8.17   These three photos were taken 1938, 1952, and 1970 respectively. Note the changes on the two adjacent parcels. The tree line between them in the earliest photo was subsequently demonstrated to be a fence line and line of occupation, which the court found to be the best evidence of the location of the dividing line between the two ownerships. The tree line is visible in all three photographs.

Figure 8.17

of historical interest, many of which no longer exist. A picture is worth, according to the Chinese proverb, "more than ten thousand words."*

USE OF POSTAL CARDS

An often-overlooked resource is the simple postal card. As a Chinese proverb says, "A picture is worth more than ten thousand words." Paper shows and postcard shows carry literally thousands of cards on a wide variety of subjects. Some collectors and dealers will do a search or keep a searcher's name on file for a particular image or photograph. a partial list of subjects that can be very helpful to a researcher or investigator follows.

*The text material for much of the foregoing was excerpted from Donald A. Wilson, *Interpreting Land Records* (Hoboken, NJ: John Wiley & Sons, Inc., 2006).

Railroad Stations and Crossings

Figure 8.18 Railroad station and post office on a trolley or short line railroad line.

Toll Gates

Figure 8.19 An old toll gate on a turnpike. This photograph shows the size and loction of the building and the gate and the character of the road. A utility pole line can be seen to the left. Searching the records for the utility pole may also provide information.

Dams

Figure 8.20   A rock wall dam. Many dams like this no longer exist, but maps may either show where they were located or provide clues to their location. Photographs such as this provide information as to the character of the structure, particularly its height, providing clues to the extent of flowage that could be made upstream.

Road Intersection

Figure 8.21   Road intersection as it was when the railroad existed. Today the tracks are gone, but the railroad right-of-way exists as a trail. Fences mark the edges of the highway at that point in time.

Other Features

Old roads and streets, buildings, mills—almost anything imaginable can be found in picture format, and many as picture postcards, both color and black and white. Many times the subject of the photograph can be referenced to some other feature shown. If one feature is gone and another remains, at least an approximate reconstruction may be possible. Model builders are particularly adept in reconstructing scenes from a minimum of data.

FOR FURTHER REFERENCE

American Jurisprudence. Proof of Facts. *Ancient Documents.*

31 ALR 1431	*Use of photographs in examination and comparison of handwriting or typewriting.*
106 ALR 721	*Determination of date of document from inspection and examination of typewriting.*
28 ALR2d 1443	*Mutilations, alterations, and deletions as affecting admissibility in evidence of public record.*
41 ALR2d 575	*Genuineness of handwriting offered as standard or exemplar for comparison with disputed writing or signature, or shown by age of writing.*
65 ALR2d 342	*Carbon copies of letters or other written instruments as evidence.*
11 ALR3d 1015	*Admissibility of expert evidence to decipher illegible document.*
36 ALR4th 598	*Admissibility of evidence as to linguistics or typing style (forensic linguistics) as basis of identification of typist or author.*

CHAPTER 9

DEALING WITH WORDS, PUNCTUATION, AND NUMBERS

Each word in deed is presumed to have been used for some
purpose and to have some effect.

—Wittmeir v. Leonard
122 So. 330, 219 Ala. 314 (1929)

It cannot be presumed that words or terms in a conveyance
were used without a meaning or having some effect given
them, and a construction will be adopted, if not contrary to
law, giving effect to the deed and every word and term
employed.

—Rhomberg v. Texas Co.
40 N.E.2d 526, 379 Ill. 430 (1942)

DEALING WITH WORDS

There are many examples of more serious instances whereby words are very significant to the understanding of a case. Since one rule is that *all words in a description are presumed to have been inserted for a purpose,* followed by another rule, *a description must be honored in its entirety if it is possible to do so,* it is imperative that readers determine meanings when something is unknown.

Ordinary Words

Descriptions are to be construed according to their plain terms, and words are to be given their ordinary meaning.[47]

Technical Words

Technical words should be construed according to their technical meaning. Frequently, on commercial sites or on lands other than regular house lots, unusual and little-known words and terms may be used. It is often necessary to consult technical sources, early dictionaries and encyclopedias, and other, perhaps unusual, sources to determine the meanings of these words.

Legal Words

Words that have definite legal significance must be given their legal effect.[48]

Case #16

MEANING OF A PHRASE

Wilson v. DeGenaro

415 A.2d 1334, 36 Conn. Sup. 200 (1979)

Courts look to the usual meaning of words in construing a deed.

In respect to 1927 conveyance whereby defendant's predecessor in title conveyed property by deed but reserved a right-of-way "as now laid out," the quoted words, given their ordinary meaning, clearly referred to the "location" of the right-of-way when it was reserved in the 1927 deed; it would be illogical to assume that the grantor intended to specify the right-of-way's

---

[47]Words in a deed must be given their ordinary and popular meaning, unless used in a technical sense, or the context shows that they are used in a different sense. *Wood v. Mandrilla*, 140 P. 279, 167 C. 607 (1914); Ordinary meaning must be given words where meaning is not doubtful. *Raritan State Bank v. Huston*, 161 N.E. 141, 329 Ill. 604 (1928).

[48]Grantor is presumed to know the legal meaning of the terms which he employs in a deed. *Holloway's Unknown Heirs v. Whatley*, 104 S.W.2d 646, affirmed 131 S.W.2d 89, 133 Tex. 608, 123 ALR 843 (Tex. Civ. App. 1937).

location (boundaries)—but without so stating, not specify its width—which was inevitably part of those boundaries.

Location, by definition, signifies and implies the determining of applicable boundaries.

This case involved the meaning of wording in a deed, "Reserving. . .the drift-way[49] *as now laid out.*" One party stated that the phrase was ambiguous. In resolving the phrase, in that context, the court relied on *Ballentine's Law Dictionary* and *The American Heritage Dictionary* and explored the phrases "lay out," "laying out," and the words, "locate," and "location." The court's interpretation in this instance was that the words "as now laid out" were not only analogous to the phrase "as presently used," but they speak even more clearly to the subject of location. The court further stated, "Giving the words their ordinary meaning, 'as now laid out' clearly referred to the 'location' of the right-of-way when it was reserved [in the deed that created it]."

The court found that the phrase was not ambiguous and offered some words of guidance:

When a deed is ambiguous: there should be considered, when necessary and proper, the force of the language used, the ordinary meaning of words, the meaning of specific words, the context, the recitals, the subject matter, the object, purpose, and nature of the reservation. . .and the attendant facts and surrounding circumstances at the time of the making of the deed.

Descriptions are made up of words, each with a specific meaning and importance in its own right. This case serves to emphasize the court's view on the importance of even a single word.

Courts have frequently relied on three sources for the determination of meanings of words and/or phrases: *Black's Law Dictionary* or the equivalent; *Webster's Dictionary* or the equivalent; and the 100-volume set published by the West Publishing Company entitled *Words and Phrases*. In addition to *Black's*, there are several other important legal dictionaries, and not all editions of *Black's* contain the same assemblage of words and definitions. There are many dictionaries besides *Webster's*, and it is sometimes important to consult more than one of them. *Words and Phrases* contains excerpts from several thousand decisions wherein a word or a phrase was defined or given meaning by a court.

Ordinary words are to be given ordinary meaning, technical words are to be given technical meaning, and legal words are to be given legal meaning. A grantor, or a

[49] A road or way over which cattle are driven. *Smith v. Ladd,* 41 Me. 314. A byroad is one which inhabitants who live some distance from a public road in a newly settle country make in going to the nearest public road—that is, it is a road which is used by the inhabitants, but is not laid out—and very often it is called a "driftway," and is a road of necessity in newly settled countries. *Van Blarcom v. Frike, 29 N.J.L. (5 Dutch.) 516.*

scrivener, selects words to describe or define a particular thing at a particular point in time. The resulting compilation is a narrative of how that individual viewed the situation according to his or her recollection, understanding, education, experience, and background. What was clear to these people may not necessarily be clear to today's reader, especially if considerable time has passed and conditions have changed. The large collection of cases dealing with such instances is a testament to how troublesome the meaning of a single word can be.

Many land descriptions were/are written by agents of the grantor. They write from a survey (someone else's interpretation of the property), a verbal description (someone else's mental picture), an assessor's plat (someone else's compilation of parcels assessment purposes, not conveyancing), or an previous description (usually without verifying its reliability or accuracy). Seldom does the scrivener factor in all the elements necessary to produce a description that will stand alone. When reading land descriptions, generally several things cannot be known:

- Whether scriveners ever viewed the premises being described
- Whether they have a grasp of mathematics and words
- Whether they understand the meaning of legal terminology
- Whether the premises have ever been measured, or surveyed

> A deed must speak for itself, and its obnoxious provisions cannot be aided, modified, or explained by extrinsic evidence.
>
> —*Parran v. Wilson*
> 154 A. 449, 160 Md. 604 (1931)

Ancient or Obsolete Words and Terms

When you encounter older terms and words, an early dictionary, encyclopedia, or other reference will often aid in explaining their meanings. Language and terminology changes over time, and words can be particularly troublesome in very early documents.

Case #17

"As Many Tenter Bars as Is Wanted"

The deed language on a particular mill site was "to extend south of the river to where the south end of the old clothes shop formerly stood and running westerly such a course as will be a sufficient room on the up land south of the river for as many tenter bars as may be wanted. . . ." An old dictionary defined a "Tenter-bar" as a device for stretching cloth.[50]

[50] Edward H. Knight, Knight's American Mechanical Dictionary (1880).

DEALING WITH PUNCTUATION

The rule of construction is that punctuation is ordinarily given slight consideration in construing a deed. Words will control punctuation rather than the other way around; however, punctuation will be resorted to in order to settle the meaning of an instrument, after all other means fail.[51]

The court in *Holmes v. Phenix Insurance Company* stated: "words control punctuation marks, and not the punctuation marks the words."[52] In *Ewing v. Burnet*, the federal court held: "Punctuation is a most fallible standard by which to interpret a writing; it may be resorted to when all other means fail; but the Court will first take the instrument by its four corners, in order to ascertain its true meaning: if that is apparent on judicially inspecting the whole, the punctuation will not be suffered to change it."[53]

Case #18

WHEN PUNCTUATION CAN BE AN IMPORTANT FACTOR

North Hampton School District v. North Hampton Congregational Society

84 A.2d 833, 97 N.H. 219 (1951)

In 1875 the defendant conveyed by warranty deed to plaintiff's predecessor, a parcel of land. Following the habendum clause were the usual covenants of warranty except that the last of them read that the grantor "will warrant and defend the same to the said grantees, their heirs, successors and assigns against the lawful claims and demands of any person or persons whomsoever, so long as said lot shall be used as a lot for a school house lot as aforesaid."

The school was built on the land in 1876, and used as a schoolhouse until 1950, when a new school was constructed at another location. At the adjourned annual school meeting held in 1950, it was voted to sell the school building to be moved away or torn down and to turn over the land to the town for a public common.

The plaintiff brought suit and asked the court to decree that it was the owner of said land and building free and clear of any all claims of the defendant. The defendant in its answer sought a decree that said land and building were owned by it in fee simple free from all claims of the plaintiff of the town.

The court stated that the main issue to be decided is the nature of the estate held by the plaintiff by virtue of the deed. The defendant maintained that the provision

---

[51]See *Berry v. Hiawatha Oil & Gas Co.*, 108 S.W.2d 497 (Ky. 1946); *Franklin Fluorspar Co. v. Hosick*, 39 S.W.2d 665, 239 Ky. 454 (1931); *Teachers' Retirement Fund Ass'n of School Dist. No. 1 v. Pirie*, 34 P.2d 660, 147 Or. 629 (1934); *Vinson v. Vinson*, 4 Ill. App. 138 (1879).

[52]98 F. 240, 47 L.E.A. 308 (1899).

[53]11 Pet. 41, 9 L.Ed. 624 (1837).

"so long as said lot shall be used as a lot for a school house lot as aforesaid" applies to the grant, with the result that the plaintiff acquired a determinable or qualified fee. The plaintiff, however,contended that the provision applied only to the covenant of warranty and that consequently it acquired by said deed a fee simple, the provision at most impressing a trust thereon, the purpose of which is set out by it.

The court went on to cite an important rule of construction for its interpretation: "It is the duty of the court to place itself as nearly as possible in the situation of the parties at the time the instrument was made, that it may gather their intention from the language use, viewed in the light of the surrounding circumstances." The words "so long as," "while," "until," and "during" are the usual and apt words to create a limited estate, such as a determinable fee. An earlier case,[54] decided just eight months prior to the execution of this deed, held that a proviso substantially in the same language and placed in the identical part of the deed was not restricted to the covenant of warranty but must be applied to the grant, thus creating a qualified fee.

Continuing, the court found important in this case,

> as indicative of the intention of the draftsman to create an estate similar to that in the previous case, is the fact that there had been some discussion about the fact that the limiting proviso, although it followed a period after the last covenant of warranty, nevertheless began with a small letter instead of a capital as would be the usual custom. The deed in this case contains a comma instead of a period after the last covenant of warranty thereby making the proviso which begins with a small letter more in conformity with the usual use of the English language.

The court concluded therefore that the plaintiff's predecessor in title by the deed it received from defendant obtained a determinable fee that terminated when the plaintiff decided not to use said lot as a schoolhouse lot. When that event occurred, the fee reverted to the defendant, who holds title to it free from any claim of the plaintiff.[55] In addition, interestingly, the court stated that the building was erected while the plaintiff was the owner of the land, and since the record revealed nothing that would prevent it from becoming part of the real estate, it also became the property of the defendant free from all claims of the plaintiff when it ceased to use said lot as a schoolhouse lot.

This case demonstrates how important specific words and the placement of punctuation can be in the interpretation of the grant and in giving meaning to the intent of the parties creating the estate.

---

[54] *Reed v Hatch*, 55 N.H. 327 (1975).
[55] Citing *Lyford v. Laconia*, 75 N.H. 220, 72 A. 1085 (1909).

DEALING WITH NUMBERS

Researchers must always be on the lookout for the transposition of numbers in distances, bearings, acreages, and the like when descriptions are traced back in time. A deed once gave a direction as *North 97° East*. We might be inclined to think that the two numbers were transposed and the direction should have been North 79° East; however, the next line in the description read *North 97° West*.

Consider the use of the rod as a unit of distance. The common belief is that a rod measures $16\frac{1}{2}$ feet. However, that is not always true. The rod in England varied from 12 to 22 feet, depending on the quality of land being measured. The longer rod was used for poorer land, so to compensate, the recipient got more acreage, since there are always 160 square rods in an acre. More than one definition of the rod was brought to America. In areas where the Scotch-Irish settled, the 18-foot rod is common. Incorrectly substituting one measure for the other can yield all sorts of bizarre results when dealing with retracement or, even worse, senior-junior considerations.

Using the conversion of 160 square rods to the acre, there is also the area measurement of the rood, which is equivalent to 40 square rods, or $\frac{1}{4}$ acre. Finding inadvertent changes in the records of roods to rods also presents difficulties.

Occasionally we encounter measurements in chains. Are they Gunter's chains of 66 feet, engineer's chains of 100 feet, or perhaps some other definition? And of course there are the 2-pole chains and those called 4-pole chains, again 33 and 66 feet, respectively (provided the rod used is $16\frac{1}{2}$ feet and not 18, or some other length).

The mile can also present difficulty. The statute mile contains 5,280 feet. However, it was not uncommon in coastal communities, particularly when laying out original lots, to use a nautical, or geographical, mile of 6,080 feet.

The Spanish *vara* has numerous definitions that are not consistent from state to state. *Land Survey Descriptions* by William Wattles lists that the *vara* in Arizona, California, and New Mexico is 33 inches, with the *vara* measuring 33.3333 inches in Texas and 33.372 inches in Florida. The book also lists 59 definitions for the *vara*, depending on which Spanish-settled area it is used.

Arpents, the French unit of length measurement (but also of area like the English rod), can be confusing as well. Used in northern Maine and other border states, the arpent is likely a spill-over from Quebec, where it is equivalent to 191.835 feet. But in Louisiana, which was originally settled by the French from Canada, the arpent is likely to be equivalent to 191.994 feet. The same is true in Alabama, Mississippi, and Florida, although 192.50 feet is the measure of an arpent in Arkansas and Missouri.

It is generally accepted that the acre contains 43,560 square feet, but then there is the builder's acre and the Block Island Acre, both of 40,000 square feet, as well as others.

Another misleading item in the system of land records that we so rely on is the date. While much of the country is not old enough to experience the change of the calendar, eastern states can be very much affected. In English countries, the long-used Julian calendar was tossed out in 1752, because it was not accurate, and the

Gregorian calendar was substituted. The Julian calendar was off by 11 days by 1752, and the year began on March 25. That is why September, now the ninth month, was originally the seventh month, October was the eighth month, November the ninth month, and December the tenth month.

A document dated the seventeenth day of the third month was transacted when? Well, it depends on the year: Prior to 1752, it would have been May 17; after 1752, March 17. In converting from one calendar to the other, it is necessary to allow for the 11-day error. George Washington was born on February 11, 1731, but when the new calendar was adopted, he changed the day he celebrated his birthday to February 22 by adding 11 days, so as to be in conformance with the new system.

Another consideration is that not all cultures follow the same standard in expressing dates. Depending on who is writing a document, the date of January 10, 2000, could be expressed as 01-10-2000 or as 10-01-2000. The latter might be confused to mean October 01. Some scientific communities would express that date as 10-I-2000; with Roman numerals expressing the month, there can be no confusion regardless of the order of the elements of expression. Dates not fully expressed can also be troublesome. For example, 01-25-53 contains four possibilities: January 25, 1653, 1753, 1853, or 1953.

CHAPTER 10

FOLLOWING FOOTSTEPS

There is no branch of detective science which is so important and so much neglected as the art of tracing footsteps.

—*Sherlock Holmes*
A Study in Scarlet

The foundation of land surveying is the retracement and recovery of evidence of the original surveys and the perpetuation of this evidence by substantial monumentation and by documentation of corner perpetuation for the public records.

—*Donald D. Lappala*
Retracement and Evidence of Public Land Surveys

Titles to land are not to fail merely because old markers may have disappeared or because it may be difficult to trace footsteps of the surveyor.

In suit involving boundary question, search must be made for the footsteps of the original surveyor and, when found, the case is solved.

—*Hart v. Greis et al.*
155 S.W.2d 997 (Tex. 1941)

Two decisions addressing the philosophy of "following the footsteps" and the reasoning behind it are noteworthy. In *Stanolind Oil & Gas Co. v. Wheeler*, the Texas court stated: "In determining location of boundary line it is not where surveyor intended

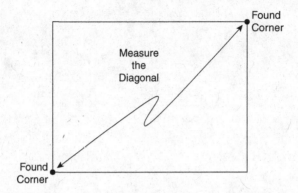

Figure 10.1   Test the accuracy of the survey by measuring between two known, or verified, points.

to run boundary or should have run it but it is where boundary line was actually run which controls"[56] And just a few years later, in another jurisdiction, the court addressed the duty of the retracement surveyor this way:

> In cases deciding the boundary between two parcels of land, the law is settled that it is the duty of the surveyors to follow the original survey lines under which the property and neighboring properties are held notwithstanding inaccuracies or mistakes in the original survey. The purpose of this rule of law is that stability of boundary lines is more important than minor inaccuracies or mistakes.[57]

For many years, the various courts have recited the phrase "following the footsteps" of the original surveyor. Since today the footsteps have now faded, and even though we do not know exactly what the original surveyors did and how they did it, we can derive some insight into what an individual surveyor probably did, and how. By knowing something about how surveyors were educated and trained and the type of equipment they used, we can develop and test hypotheses. Once a corner or two has been found, and located, we can draw certain conclusions about where and how to find other corners. This applies to a site in question or to other sites the same surveyor may have surveyed. Patterns may emerge.

As an example, the test shown in Figure 10.1 is but one of several that could be performed to gain insight as to what a surveyor may have done.

> It has been declared that all the rules of law adopted for guidance in locating boundary lines have been to the end that the steps of the surveyor who originally projected the lines on the ground may be retraced as nearly as possible; furthermore, that in determining the location of a survey, the fundamental principle is that it is to be located where the surveyor ran it.

[56]247 S.W.2d 187 (Tex. 1952).
[57]*Froscher v. Fuchs*, 130 S.2d 300 (Fla. 1961).

To "survey a parcel of land" means to locate it on the surface of the earth. When the surveyor is retained to locate a boundary line wthat has heretofore been established, he or she "traces the footsteps" of the "original surveyor" in locating existing boundaries. *The sole duty, function and power of the retracement surveyor is to locate on the ground the boundaries, corners and boundary lines established by the original survey.* The following surveyor, rather than being the creator of the boundary line, is only its discoverer and is only that when he/she correctly locates it.

—Rivers v. Lozeau
Fla. App. 5 Dist., 539 So. 2d 1147 (Fla. 1989)

Locating previously established boundaries and corners sometimes can be a formidable and frustrating task, particularly if they are ancient and their evidence and remains have disappeared. Knowing how surveyors did their work and the inherent errors in their results often is the key to discovering where they went and to finding traces of evidence they left behind. Often an individual project is a forensic study of the highest order, demanding expertise, time, and patience to achieve the goal of locating corners and boundaries in accordance with rules of law: where they were originally established.

The Ohio court recognized this in the case of *Sanders v. Webb*, wherein it was stated:

> The law provides that the original corners established during the progress of the survey shall forever remain fixed in position, and that even evident errors in the execution of the survey must be disregarded where these errors were undetected prior to the sale of the lands. The original monuments thus assume extreme importance in the location of land boundaries. Unfortunately, most of the public lands were surveyed before the present day regulations relative to the character of monuments went into effect, and as a consequence most of the monuments used were of a very perishable nature. Their disappearance or destruction has rendered the relocation of old lines a very difficult task.[58]

This court went on to say, adding perspective:

> In the original survey it was presumed that permanent monuments were being carefully established and witnessed, so that long lines of monuments would be perfect guides to the place of any one that chanced to be missing. Unfortunately, the "monuments" were often nothing but green sticks driven into the ground, lines were carelessly run, monuments were inaccurately place, witnesses were wanting in permanence, and recorded courses and lengths were incorrect. As the early settlers made little effort to perpetuate either the corner monuments or the witnesses, the task of the present-day surveyor in reconciling much conflicting evidence (or in even finding any kind of evidence) is often an extremely difficult task. It must be remembered that no matter how erroneously the original work is done, lines and monuments that can be identified still govern the land boundaries.

[58] 85 Ohio App.3d 674, 621 N.E.2d 420 (1993).

Surveyors would be advised to take note of what the courts have had to say about negligence and failure to fulfill a duty. In addition, contracting to do a survey, and failing to do exactly that, can result in a breach of the contract. A concise summary of the duty of the retracement surveyor is provided next.

SURVEYORS' DUTY AND RESPONSIBILITY

Resurveys

> In surveying a tract of land according to a former plat or survey, the surveyor's duty is to relocate, upon the best evidence obtainable, the courses and lines at the same place where originally located by the first surveyor on the ground. In making the resurvey, he has the right to use the field notes of the original survey. The object of a resurvey is to furnish proof of the location of the lost lines or monuments, not to dispute the correctness of or to control the original survey. The original survey in all cases must, whenever possible, be retraced, since it cannot be disregarded or needlessly altered after property rights have been acquired in reliance upon it. On a resurvey to establish lost boundaries, if the original corners can be found, the places where they were originally established are conclusive without regard to whether they were in fact correctly located, in this respect it has been stated that the rule is based on the premise that the stability of boundary lines is more important than minor inaccuracies or mistakes. But it has also been said that great caution must be used in reference to resurveys, since surveys made by different surveyors seldom wholly agree. A resurvey not shown to have been based upon the original survey is inconclusive in determining boundaries and will ordinarily yield to a resurvey based upon known monuments and boundaries of the original survey.
>
> —12 Am. Jur. 2d Boundaries, § 61

"The object of a resurvey is to furnish proof of the location of the lost lines or monuments, not to dispute the correctness of or to control the original survey. The original survey in all cases must, whenever possible, be retraced, since it cannot be disregarded or needlessly altered after property rights have been acquired in reliance upon it." It is generally held, therefore, that a resurvey that changes lines and distances and purports to correct inaccuracies or mistakes in an old plat is not competent evidence of the true line fixed by the original plat.[59]

Several courts have not only emphasized the philosophy, or directive, but have made it clear that anything other than locating an existing line placed by a surveyor

---

[59] *Winters v. State*, 500 So.2d 303 (Fla. App. 1 Dist., 1986), quoting in part from 8 Am. Jur. Boundaries, § 102.

(whether marked or not) is outside of the surveyor's responsibility when surveying (locating) a boundary. In fact, the court stated in *Pereles v. Gross* that anything other than locating the original, existing, line may be unlawful:

> In resurveying a tract of land according to a former plat or survey, the surveyor's only function or right is to relocate, upon the best evidence obtainable, the corners and lines at the same places where originally located by the first surveyor on the ground. Any departure from such purpose and effort is unprofessional, and, so far as any effect is claimed for it, unlawful.

> To fix lines variant from the originals and according merely to his notion of a desirable arrangement of lots and streets leads naturally to confusion of claims among lot owners, and, when done by a city surveyor as a basis for occupation of land for streets, is attempted confiscation.[60]

In the case of *Johnson v. Westrick*, a city surveyor attempted to locate a right-of-way line according to the intended layout and in accordance with the stated width of the street. In doing this, there was found a conflict with an abutting owner's improvement, and the case went to litigation. The court held that "the east line of the street was where the original surveyor placed it, not where it should be according to resurveys or subsequent surveys; that subsequent surveys are worse than useless; they only serve to confuse, unless they agree with the original survey."[61]

Retracement is the process of uncovering physical evidence of monuments and corners by intelligent search on the ground for the calls of the description and field notes of the original survey, guided by the controlling influence of known points. It should proceed from a known location to hypothecate the unknown. In public land states—those in which the land was once under federal domain—the known starting point is almost invariably some section or quarter section corner, accepted and recognized as having been established by the original government survey. When such starting point called for by the legal description is missing, it must first be reestablished in its original position before the retracement may proceed. Such location, too, is governed by the rules of retracement.[62]

The retracement, commencing at some known point which was recognized and accepted by the original survey, is run in accordance with the plan of the original survey to ascertain the probable position of each succeeding point. An intensified search for evidence of the original location of each succeeding point is made in the vicinity hypothecated by retracement. The search may uncover the actual monument in its undisturbed position, which may be identified by its conformity in character to that described in the record (legal description, field notes, and plat or map) and by its physical appearance as mellowed by age and elements. Or the search may uncover

[60] 126 Wis. 217 (1905).

[61] 200 Wis. 405 (1930).

[62] Robert J. Griffin, "Retracement and Apportionment as Surveying Methods for Re-establishing Property Corners," *Marquette Law Review* 43 (1960): 484–510.

record accessories or witness marks, which, if not greatly at variance with record ties, may satisfactorily establish the exact location of the original monument. Accessories or witness marks to section and quarter section corners may be such things as bearing trees (trees blazed and marked), bearing objects, mounds of stone, or pits dug in the sod or soil. Accessories to private survey corners may be such things, as disks set in trees or poles, offset pipes or stakes driven into the ground, nails, tacks or cross-cuts in pavement or sidewalk, or the corner of some permanent object such as a house or other building. The record may disclose distances between such accessories and the missing survey corner. The missing corner may be reestablished by intersecting two or more such known distances.[63]

RETRACEMENT AND RESURVEY

The location upon the ground of property lines is determined by resurvey. One of the methods of resurvey is *retracement*, which is a surveying method for resurrecting evidence of the location of a once- established property corner. Its aim is to follow, as closely as possible, in the footsteps of the original surveyor and reestablish property corners in the exact position in which they were originally placed.

Rules and Principles of Retracement

The purpose of a resurvey of land is to locate and mark upon the ground the boundaries of the parcel of land evidenced by the description in a particular deed. The legal description fails as a complete protection of the boundaries, however, because it merely describes what they are and how they are to be determined.[64] The extent of the parcel actually transferred by the deed is resolved by the intention of the grantor, so far as that intention is effectively expressed in the deed interpreted in the light of the then existing conditions and surrounding circumstances.[65] The expression of the intention in the deed may be incomplete and ambiguous; but, regardless of how bunglingly expressed, there is a strong presumption that the grantor intended a certain encompassing boundary to define the lands granted. The fact that uncertainties or ambiguities appear in the legal description does not nullify the rule of intention, but rather makes its application even more pertinent.[66] The intention of the

---

[63] Griffin, "Retracement and Apportionment as Surveying Methods for Re-establishing Property Corners."

[64] Skelton, *The Legal Elements of Boundaries and Adjacent Properties*, § 2, 26 (1930).

[65] *Perry v. Buswell*, 113 Me. 399, 94 A. 483 (1915). The cardinal rule for the interpretation of deeds and other written instruments is the expressed intention of the parties gathered from all parts of the instrument, giving each word its due force, and read in the light of existing conditions and circumstances. It is the intention effectually expressed, and merely surmised. This rule controls all others.

[66] Griffin, "Retracement and Apportionment as Surveying Methods for Re-establishing Property Corners."

o*riginal grantor* (author's emphasis) as expressed and is inferred from the deed, is the paramount consideration in determining the location of property lines and corners.[67]

When a deed is interpreted in the light of the then existing conditions and surrounding circumstances, the interpreter considers the original survey which marked the boundaries.[68] The highest and best proof of intention lies not in the words of expression, but in the work performed on the ground itself. Lines actually run and corners actually established on the ground *prior to the conveyance* are the most certain evidence of intention. It is by the work as executed upon the ground, not as projected before execution or represented on a plan afterward, that actual boundaries are determined.[69] When the monuments or marks of the original survey are found and they lie wholly within the original grantor's ownership, the resurvey is complete and the boundaries conclusively established.[70]

Case #19

I OWN THE REMAINDER, NO YOU DON'T, *I* DO
Wells v. Lagorio
112 Va. 522 (1911)

> The owner of a tract of land intending to sell the whole tract, and believing it to contain sixty acres, sold the western half thereof to one purchaser and the eastern half to another, describing each half as containing thirty acres. In each case, the grantor pointed out to the purchaser the lines of the property sold, and they received the identical pieces of land that they purchased, bounded by the very lines they had contemplated and had been pointed out to them, but the acreage of each piece was seventeen and one-fourth acres, instead of as was supposed. Subsequently it was ascertained that the grantor was mistaken as to the location of his western boundary and that the true line was further west than he had supposed, and different from the line he had shown to the purchasers, such that the area of the difference between the two lines was 14.2 acres. This area the purchasers claimed to have passed to them under their deeds.

[67]Numerous sources. See Griffin, "Retracement and Apportionment as Surveying Methods for Re-establishing Property Corners."

[68]*Wells v. Lagorio*, 112 Va. 522, 71 SE 713 (1859); *Wisconsin Realty v. Lull*, 177 Wis. 53, 187 N.W. 978 (1922).

[69]*Oven v. Davidson*, 10 U.C.C.P. 302 (1859). Exceptions to the rule as quoted: It has no application where: 1. The lines were never located and definitely fixed upon the ground. *Nissley v. Moeslein*, 23 Pa. Super. Ct. 119 (1903); 2. The monuments or stakes of the original survey are not referred to in the deed or on the plat. *Warren Powers v. Henry Jackson*, 50 Cal. 429 (1875).

[70]Griffin, "Retracement and Apportionment as Surveying Methods for Re-establishing Property Corners."

The court held that the 14.2 acres did not pass by the grantor's deeds and is still his property. The purchasers got the exact lands they proposed to purchase, though of less acreage than was supposed, and it is not to be presumed that the grantor intended to sell that which he did not know that he owned.

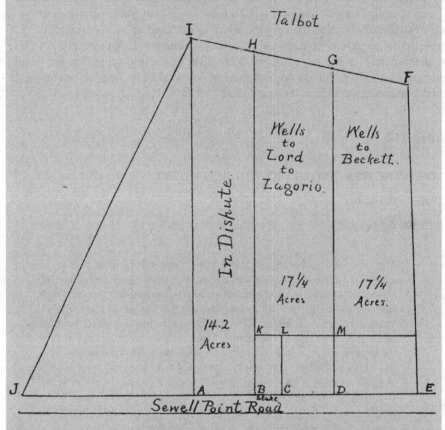

Figure 10.2   Diagram from the case of *Wells v. Lagorio*.

The controversy in this case arose with respect to the western line of the original tract owned by Wells, his deed to Claude Lord called on the west for "lands occupied by Jackson Denby." That line is shown on the diagram as B-H. The land sold to Lagorio is B-H-G-D; the land sold to Beckett is D-G-F-E. At that time Jackson Denby's wife held title to triangle A-I-J, but her husband seems to have extended occupancy beyond the limits of her title, so as to embrace A-I-H-B, which is the land in controversy.

Lagorio was put in possession of the western half of the tract owned by Wells, which was supposed to contain 30 acres, but upon survey was found to contain only 17 QF acres. Some years later the Denbys conveyed their land to one Edwards, who upon investigating his title and having a survey of the premises made found that his line on the east was I-A, and not H-B, as supposed. He disclaimed all interest in that quadrilateral containing 14.2 acres and informed Wells, who thereupon made claim to it. The land is in a wild state, grown up in scrubby woods and bushes—practically in a state of nature.

The controversy in this case is as to the ownership of the quadrilateral of 14.2 acres. Lagorio and those interested with him claimed that the land he purchased was bounded on the west by land occupied by Jackson Denby, that he was entitled by the terms of his deed to the line I-A as his western boundary, and that as he only bought one-half the entire tract, that would put his eastern line farther to the west; and the same with the line of Beckett; so that each having purchased half of the original tract, the purchase of each would be increased by one-half of the 14.2 acres—at least that would be the logical result of their claims. Certainly, if all the facts had been known at the date of the deed from Wells to Lagorio, and he had sold one-half of the quadrilateral A-I-F-E, the dividing line G-D on the sketch would have fallen farther to the west, and would have left in the half that Beckett afterward bought 7.1 acres now embraced in the quadrilateral B-H-G-D. They agreed among themselves, however, that instead of readjusting their rights in that way, they would consider themselves co-owners of the land in dispute, and that it be sold and the proceeds divided among them.

The court stated that it appeared from the evidence that the line B-H was shown to Lagorio and accepted by him as his western boundary at the date of his deed, and so the line G-D was the western boundary of Beckett. Lagorio and Beckett were both placed in possession of what they bought, or what they believed at the time of the transaction they were buying, and they and their assigns are still in the enjoyment of it. As the facts have since developed, it appears that the western line of Wells's original is farther west than either he or his grantees at the time supposed. They were under the impression that his western line was H-B; it turns out now, in the light of subsequent developments, that his western line was I-A, and at the time of making the deeds to Lagorio and Beckett, he was the actual owner, though ignorant of the fact.

As was said by the commissioner to whom the cause was referred:

> It is true that Wells thought he was selling all of the land that he owned, but his possessions were greater than he knew, and it cannot be reasoned that he intended to sell that which he did not know that he owned. It is true that Wells thought he was selling thirty acres to each of the parties, Lord and Beckett, and they thought they were getting approximately that amount; and it is further true that they did not get approximately thirty acres apiece, but they did get a tract of land, the exact boundaries of which they knew, and had been pointed out to them. They got the identical piece of land that they purchased, bounded by the very lines which they had contemplated and which had been pointed out to them—the exact lands they had proposed to purchase, although their lands contained less acreage than supposed.

The lesson to be learned from this case is that without the extensive research and knowledge of conveyancing law, two parties claimed the same parcel, for two different reasons, yet were willing to work it out between themselves. Regardless, the court ruled that they did not have that option, since they did not have title to the parcel in question.

Again, the question was which of the two grantees was entitled to the property. The answer is neither one. The basic premise is wrong: One of the two must be the rightful owner, ignoring the possibility that someone else might be.

GRIFFIN'S RULE #1

Location of a Boundary Line is Determined as of the Time of Its Creation[71]

A boundary line once established should remain fixed in its original position through any series of mesne conveyances.[72] A grantee who purchases the entire extent of particular lands owned by the grantor determines the boundaries of his purchase as of the time that the particular parcel was carved out of some larger tract. He takes to the bounds of the estate of his grantor, who in turn took to the limits of his grantor's estate, etc., to the time of creation of the boundary. A grantee purchasing only a part of the lands of his grantor will determine the common boundaries as of the time of the conveyance, while he while he will determine boundaries on the perimeter of the grantor's original tract with reference to the time that they were created. *Each line of the same parcel must be considered separately and a determination of the proper surveying method to be used must be made with respect to each line of the parcel.* [Emphasis added.].

The time of creation principle is closely related to the original government survey of public lands, since private boundary lines often run along lines established by this first survey or are described and located in relation to the original survey corners.

In an article entitled "Deeds/Senior Rights v. Junior Rights," James J. Demma, L.S., Esq. analyzed the case of *Millar v. Bowie*, 115 Md. App. 682 (1997). His closing statement was: "In synthesizing all of the case law quoted in *Millar*, it becomes clear that when the language of a senior deed is sufficiently clear and definite to convey an exact parcel of land, without reference to other evidence, later deeds are irrelevant to the establishment of the boundary."[73]

---

[71] Griffin, "Retracement and Apportionment as Surveying Methods for Re-establishing Property Corners."
[72] *Ibid.*
[73] James L. Demma, "Deeds/Senior Rights v. Junior Rights," Legal Notes, *Professional Surveyor* 18, no. 1 (January/February 1998).

The Florida case of *Rivers v. Lozeau* is a classic in identifying the role of the survey with regards to responsibility, and to the sanctity of original boundaries and titles:

> Although title attorneys and others who regularly work with them develop expertise as to land descriptions, *the only professional authorized to locate land lines on the ground is a registered land surveyor. In fact, the definition of a legally sufficient real property description is one that can be located on the ground by a surveyor.* However, in the absence of statute, a surveyor is not an official and has no authority to establish boundaries; like an attorney speaking on a legal question, he can only state or express his professional opinion as to surveying questions. In working for a client, a surveyor basically performs two distinctly different roles or functions:
>
> FIRST, the surveyor can, in the first instance, lay out or establish boundary lines within an original division of a tract of land which has theretofore existed as one unit or parcel. In performing this function, he is known as the "original surveyor" and *when his survey results in a property description used by the owner to transfer title to property** that survey has a certain special authority in that the monuments set by the original surveyor on the ground control over discrepancies within the total parcel description and, more importantly, control over all subsequent surveys attempting to locate the same line.
>
> SECOND, a surveyor can be retained to locate on the ground a boundary line which has theretofore been established. When he does this, he "traces the footsteps" of the "original surveyor" in locating existing boundaries. Correctly stated, this is a "retracement" survey, not a resurvey, and in performing this function, the second and each succeeding surveyor is a "following" or "tracing" surveyor and his sole duty, function and power is to locate on the ground the boundaries corners and boundary line or lines established by the original survey; he cannot establish a new corner or new line terminal point, nor may he correct errors of the original surveyor. *The following surveyor. rather than being the creator of the boundary line. is only its discoverer and is only that when he correctly locates it.*[74] [Emphasis added.]

*This is a most important qualification.

As in this case, sometimes professionals are mistaken as to their responsibility, duty, and the scope of their agreement. In this case, the surveyor was subdividing out a small parcel from the parent tract with a right of way to it. Two of the lines of the new parcel were new lines, the other two coincident with existing title lines of the parent tract. In the court's view, the surveyor treated all four lines the same, believing that in part he was an original surveyor and in part he was a retracement surveyor. It is vitally important to recognize this distinction, as the rules for survey are different. In addition, as stressed elsewhere, in retracement it is critical to be certain one is in the right time frame.

[74] 539 So.2d 1147 (Fla., 1989).

GRIFFIN'S RULE #2

Retracement Should Proceed from a Known Location to Hypothecate the Unknown

Retracement is the process of uncovering physical evidence of monuments and corners by intelligent search on the ground for the calls of the description and field notes of the original survey, guided by the controlling influence of known points. It should proceed from a known location to hypothecate the unknown. In public land states—those in which the land was once under federal domain—the known starting point is almost invariably some section or quarter section corner, accepted and recognized as having been established by the original government survey. When such starting point called for by the legal description is missing, it must first be reestablished in its original position before the retracement may proceed. Such location, too, is governed by the rules of retracement.[75] The publication *Restoration of Lost and Obliterated Corners & Subdivision of Sections—A Guide for Surveyors* is designed to provide guidance on how to retrace original government surveys and how to resurrect corners.[76] In fact, several courts have recognized this book and the *Bureau of Land Management Manual of Surveying Instructions* as official guides.[77]

The retracement, commencing at some known point which was recognized and accepted by the original survey, is run in accordance with the plan of the original survey to ascertain the probable position of each succeeding point. An intensified search for evidence of the original location of each succeeding point is made in the vicinity hypothecated by retracement. The search may uncover the actual monument in its undisturbed position, which may be identified by its conformity in character to that described in the record (legal description, field notes, and plat or map) and by its physical appearance as mellowed by age and elements. Or the search may uncover record accessories or witness marks that, if not greatly at variance with record ties, may satisfactorily establish the exact location of the original monument. Accessories or witness marks to section and quarter section corners may be such things as bearing trees (trees blazed and marked), bearing objects, mounds of stone, or pits dug in the sod or soil. Accessories to private survey corners may be such things as disks set in trees or poles, offset pipes or stakes driven into the ground, nails, tacks or cross-cuts in pavement or sidewalk, or the corner of some permanent object, such as a house or other building. The record may disclose distances between such accessories and the missing survey corner. The missing corner may be reestablished by intersecting two or more such known distances.[78]

---

[75] Griffin, "Retracement and Apportionment as Surveying Methods for Re-establishing Property Corners."

[76] Bureau of Land Management, U.S. Government Printing Office, 1975.

[77] *United States v. Doyle*, 468 F.2d 633 (Colo. 1972); *Reel v. Walter*, 131 Mont. 382, 309 P.2d 1027.

[78] Griffin, "Retracement and Apportionment as Surveying Methods for Re-establishing Property Corners."

GRIFFIN'S RULE #3

Retracement Should Apply Rules of Construction to Contradictory Evidences of Intention

> "Intention" as applied to the construction of a deed is a term of art, and signifies a meaning of the writing.
>
> —*26 C.J.S.*, § 83

The main object in construing a deed is to ascertain the intention of the parties from the language used and to effectuate such intention where not inconsistent with any rule of law.[79] "Intention" as applied to the construction of a deed is a term of art, and signifies a meaning of the writing.[80] The intent of the grantor as spelled out in the deed itself must be interpreted, not the grantor's intent in general, or even what he *may* have intended (*Wilson v. DeGenaro*).[81] In ascertaining the intent of the grantor in a deed, the court will not attempt to ascertain and declare what the grantor meant to say, but only the meaning of what the grantor did say (*Holloway's Unknown Heirs v. Whatley*).[82] Stated similarly, the Texas court said a year later that "generally, secret intentions of the grantor not communicated to the grantee are not binding on the grantee" (*Cawthon v. Cochell*).[83]

The Colorado court in *Tilbury v. Osmundson* stated it this way: "A deed conveys the land actually described, regardless of the mistake of the parties. Land cannot be transferred by the intent of the parties alone, especially when the specific words used state less than what was intended."[84]

The cardinal rule for construction of deeds is that the intention of the parties is to be ascertained and given effect, as gathered from the entire instrument, together with the surrounding circumstances, unless such intention is in conflict with some unbending canon of construction or settled rule of property, or is repugnant to the terms of the grant.[85] The only legitimate or permissible object of interpreting written contracts and conveyances is to determine the meaning of what the parties have said in them.[86]

---

[79] *Iselin v. C.W. Hunter Co.*, C.A.La., 173 F.2d 388 (1949).

Grantor's intention controls unless in conflict with some positive rule of law. *Leeper v. Leeper*, 147 S.W.2d 660, 347 Mo. 442 (1941), see 133 ALR 586.

[80] *U.S. v. 15,883.5 acres of land in Spartanburg County*, S.C., D.C.S.C., 54 F.Supp. 849 (1944).

[81] 415 A.2d 1334, 36 Conn. Sup. 200, 1979).

[82] 104 S.W.2d 646 (Tex. 1937).

[83] 121 S.W.2d 414 (1938).

[84] 352 P.2d 102 (1960).

[85] *Hedick v. Lone Star Steel Co.*, Civ. App., 277 S.W.2d 925 (1955).

[86] *Yukon Pocahontas Coal Co. v. Ratliff*, 24 S.E.2d 559, 181 Va. 195 (1943).

It is especially important to ascertain the intention of the grantor, for it is the grantor who made and signed the instrument.[87] Such intention is to be ascertained from the words that have been used in connection with the subject matter and the surrounding circumstances. The presumption is that the intention of the parties is expressed by the language of the deed and the grantor is presumed to intend what his words communicate. It is the intent existing at the time the deed was made.[88]

The intent of the parties or, more specifically, the intent of the grantor, when ascertained, is controlling and is to be given effect as long as it is not repugnant to any rule of law or inconsistent with settled rules of law or some principle of law, or in violation thereof, in violation of some rule of property, or when there are no expressions in the deed that positively forbid it or render it impossible.[89]

The intention sought in the construction of a deed is that expressed in the deed, and not some secret, unexpressed intention, even though such secret intention be that actually in mind at the time of execution.[90]

It is not what the parties meant to say, but the meaning of what they did say that is controlling.[91]

A BRIEF SUMMARY OF SOME POPULAR AND HELPFUL RULES OF CONSTRUCTION FOR INTERPRETATING LAND DESCRIPTIONS

General Rules of Construction[92]

Every deed, otherwise valid, will be considered to have intended to convey an estate of some nature.[93] Therefore, every attempt should be made to uphold the deed whenever possible.

---

[87] *Henry v. White*, 60 So.2d 149, 257 Ala. 549 (1952).

[88] *Corn v. Branche*, 240 P.2d 537, 74 Ariz. 356 (1952); *Aller v. Berkeley Hall School Foundation*, 103 P.2d 1052, 40 Cal. App.2d 31 (1940); *Pierce v. Freitas*, 280 P.2d 67, 131 Cal. App.2d 65 (1955); *Mason v. Peabody Coal Co.*, 51 N.E.2d 285, 320 Ill. App. 350 (1943); *Bruen v. Thaxton*, 28 S.E.2d 59, 126 W.Va. 330 (1943).

Even a clearly stated intention is ineffectual, if the instrument does not meet the legal requirements as to the language used. *Long v. Holden*, 112 So. 444, 216 Ala. 81 (1927), see 52 ALR 536.

The courts are compelled to identify the intent of grantor with his plain language. *Gaston v. Mitchell*, 4 So.2d 892, 192 Miss. 452 (1941).

Where parties to a deed are deceased, the court must be guided by the language contained in the deed in determining the intention of the parties. *Hoppes v. American Nat. Red Cross*, Com. Pl., 128 N.E.2d 851(Ohio 1955).

The nature of the subject matter may be an important factor in construing a deed when the language is such as to require interpretation. *Brooke v. Dellinger*, 17 S.E.2d 178, 193 Ga. 66 (1941).

[89] Ibid.

[90] *Sullivan v. Rhode Island Hospital Trust Co.*, 185 A. 148, 56 R.I. 253 (1936).

[91] *Urban v. Urban*, 18 Conn. Sup. 83 (1952).

[92] See Wilson, *Interpreting Land Records*, for an extensive explanation and discussion of rules of construction.

[93] *Penienskice v. Short*, 194 A. 409, 38 Del. 526 (1937).

Where descriptions set forth in deeds are not ambiguous, they must be followed.[94] When, and only when, the meaning of a deed is not clear, or is ambiguous or uncertain, will a court of law or equity resort to established rules of construction to aid in the ascertainment of the grantor's intention by artificial means where such intention cannot otherwise be ascertained. Unlike a settled rule of property, which has become a rule of law,[95] rules of construction are subordinate and always yield to the intention of the parties, particularly the intention of the grantor, where such intention can be ascertained. Since all rules of construction are in essence only methods of reasoning which experience has taught are best calculated to lead to the intention of the parties, generally no rule will be adopted that tends to defeat that intention.[96]

The modern tendency is to disregard technicalities and to treat all uncertainties in a conveyance as ambiguities to be clarified by resort to the intention of the parties as gathered from the instrument itself, the circumstances attending and leading up to its execution, and the subject matter and the situation of the parties as of that time. Substance rather than form controls. Hence, in the construction of deeds, surrounding circumstances are accorded due weight. In the consideration of these various factors, the court will place itself as nearly as possible in the position of the parties when the instrument was executed. Where the language of a deed is ambiguous, the intention of the parties may be ascertained by a consideration of the surrounding circumstances existing at the time of the execution of the deed. In this connection, it has been said that in interpreting a deed, the choice of words used to grant the land may show the aptitude of the scrivener and is a pertinent factor.[97]

---

[94] Mich. 1958. *Gawrylak v. Cowie*, 86 N.W.2d 809, 350 Mich. 679.

Ala. Civ. App. 1986. Rules of construction should not be used in construing a deed if intention of parties to the deed, especially the grantor, can be ascertained from the document itself. *Alabama Medicaid Agency v. Wade*, 494 So.2d 654.

Vt. 1939. Rules for construing deeds are adopted for the sole purpose of removing doubts and obscurities so as to get at the meaning intended by the parties, and where there is no doubt or obscurity, there is no room for construction. *Kennedy v. Rutter*, 6 A.2d 17, 110 Vt. 332.

[95] Unlike a rule of construction, a settled rule of law is one which fastens a specific import and meaning upon particular language employed in a deed and states arbitrarily the legal effect which such language will have, attaching thereto a specific and unimpeachable intention, even though the parties employing the language may have had and may have evinced quite a different intention. Such rules therefore ingraft certain meaning upon language employed in a deed and determine what effect is given to such language in law. In other words, a rule of property is to be applied automatically as a resultant of the language used, and the court will not refuse to apply such rule merely on the surmise that the grantor did not intend that his phraseology operate in the way which the rule makes it operate. 23 Am Jur 2d, Deeds, § 224, Settled Rules of Property.

[96] 23 Am Jur 2d, Deeds, § 221.

Cal. App. 1962. A deed must be construed from its four corners. *Palos Verdes Corp. v. Housing Authority of Los Angeles County*, 21 Cal. Rptr. 225, 202 C.A.2d 827.

A deed must be read as a whole and every part thereof given effect if possible in order to arrive at the true meaning of the parties, and until such rule has been exhausted resort should not be had to artificial and arbitrary rules of construction. *Burchfield v. Hodges*, 29 Tenn. App. 488 (1946).

[97] Ibid.

Rules of construction are particularly apposite where all clauses of a deed cannot be harmonized and one of several possible meanings must be selected as expressing the true intent.[98]

The construction of deeds presents a question of law for the court to decide. It is the court's duty to interpret a deed in the light of the law in existence at the time of its execution and delivery, which must be read into and become an enforceable part thereof.

The rules for the construction of deeds are essentially those applicable to other written instruments and to contracts generally. But while, as a general consideration, the same broad rules govern the construction of both deeds and wills, in some jurisdictions deeds are more strictly construed than wills.[99]

When the intention of the parties is uncertain, resort must be made to well-settled but subordinate rules of construction, to be treated as such and not as rules of positive law.[100] All rules of construction are but aids in arriving at the grantor's intention, and rules of construction may be applied only where the application of the rule with respect to the intent of the parties does not banish all doubt concerning the conclusions to be drawn from the language of the conveyance and the circumstances attending its formulation. In construing a deed, it is not permissible to interpret that which has no need of interpretation.

Each instrument must be construed in the light of its own language and peculiar facts, and it has been said that, since the language employed in deeds varies so materially and so much, precedents are rarely controlling in a concrete case, except as they may furnish general aiding rules. Nevertheless, it has been said that the construction of a deed must be governed by the strict rules of the common law, and where a deed expresses two conflicting intentions, it must be construed according to the rules of construction.[101]

When a deed is ambiguous, the construction most favorable to the validity will be adopted.[102]

> *In construing a deed, it is not permissible to interpret that which has no need of interpretation.*

Some of the more frequently applied rules of construction are:

- *Intent* of the parties is the controlling consideration.[103]

---

[98] Ibid.

[99] 23 Am Jur 2d, Deeds, § 223.

[100] 26 C.J.S., Deeds, § 82.

Vt. 1912. Rules for construing deeds must be treated as subordinate, and not as rules of positive law. *Johnson v. Barden*, 83 A. 721, 86 Vt. 19, Ann Cas 1915A, 1243; see also *deNeergaard v. Dillingham*, 187 A.2d 494, 123 Vt. 327 (1963).

[101] Ibid.

[102] *Earle v. International Paper Co.*, 429 So.2d 989 (Ala. 1983).

[103] The intention of the grantor must be interpreted from the deed itself. *Lancaster & J. Elec. Light Co. v. Jones*, 75 N.H. 172 (1909).

Generally, secret intentions of grantor not communicated to grantee are not binding on the grantee. *Cawthon v. Cochell*, 121 S.W.2d 414 (Tex. Civ. App., 1938).

- In ascertaining the intention of the parties, separate deeds or instruments executed at the same time and in relation to the same subject matter, between the same parties, may be taken together and construed as one instrument.[104]

- A deed must be construed as a whole, and a meaning given to every part thereof.[105]

- Words in a deed are presumed to have a purpose.[106]

- Documents are to be viewed in light of the surrounding circumstances.[107]

- Descriptions are to be construed according to their plain terms, and words given ordinary meaning.[108]

- A document should be construed in light of the law in existence at the time it was executed.[109]

- A deed should be construed with reference to the actual state of the land at the time of its execution.[110]

- Technical words should be construed according to their technical.

- meaning.

- Words that have definite legal significance must be given their legal effect.

- A specific, or particular, description controls a general one.[111]

---

[104] Several instruments or deeds of the same date between the same parties may be construed together. *Jackson v. Dunsbagh*, 1 Johns. Cas. 91 (N.Y. 1799).

[105] In ascertaining grantor's intention, all the provisions of the deed are to be considered without undue preference to any of its parts so that the grantor's intention may be carried out. *Keller v. Keller*, 123 S.W.2d 113 (Mo. 1938).

[106] In construing deed, effect must be given to each word, clause, or term employed by parties, rejecting none as meaningless or surplusage. *Woods v. Seymour*, 183 N.E. 458, 350 Ill. 493 (1932).

Words in deeds are presumed to have been used advisedly, and every word must be given effect where not inconsistent with other language used. *Edwards v. Edwards*, 52 S.W.2d 657 (Tex. 1932).

While effect should be given to every word of written instrument whenever possible, it is never permissible to insert meaning to which parties did not agree. *Tubb v. Rolling Ridge, Inc.*, 214 N.Y.S.2d 607, 28 Misc.2d 532.

[107] When instrument does not, by its terms, clearly and plainly describe land affected, or where it is phrased in language susceptible of more than one construction, intention of parties is to be ascertained, not solely from words of instrument, but from its language when read in light of circumstances surrounding transaction. *Thomas v. Texas Osage Co-op Royalty Pool*, 248 S.W.2d 201 (Tex. 1952).

[108] Words in a deed must be given their ordinary and popular meaning, unless used in a technical sense, or the context shows that they are used in a different sense. *Wood v. Mandrilla*, 140 P. 279, 167 C. 607 (1914).

[109] Deed should be construed in light of law existing when deed was executed. *Stuart v. Fox*, 152 A. 413, 129 Me. 407 (1931).

[110] A deed of land conveys the property described in its existing state; and is to be construed in all its parts with reference to the actual, rightful state of the property conveyed, at the time of the conveyance, unless some other time is expressly referred to. *Dunklee v. Wilton R. Co.*, 24 NH 489 (1852).

[111] Particular description in deed governs general, and where courses, distances, monuments, lot and range designations or surrounding lands enable boundaries to be readily determined, such definitely bounded parcel is the lot conveyed, although perhaps inconsistent with some general words of description or statements of quantity. *Spooner v. Menard*, 124 Vt. 61 (1963).

- References are part of the description.[112]
- Descriptions are to be construed against the grantor and in favor of the grantee.[113]
- A false or erroneous description, or statement, may be disregarded.[114]
- A *meaning and intending clause* will not limit or enlarge the grant.[115]
- *Punctuation* is ordinarily given slight consideration in construing a deed.[116]
- Words will control punctuation marks rather than the other way around. However, punctuation will be resorted to in order to settle the meaning of an instrument, after all other means fail.[117]
- Courts generally take judicial notice of commonly used initials and abbreviations in descriptions.[118]
- Written words in a deed control printed words.[119]

---

[112] A description in a deed of the land conveyed may refer to another deed or map and the deed or map to which reference is thus made is considered as incorporated in the deed itself. *McCollough v. Olds*, 41 P. 420, 108 C. 529 (1895).

Map or plat referred to in deed becomes part of deed and need not be registered. *North Carolina State Highway Commission v. Wortman*, 167 S.E.2d 462 (N.C. App., 1969).

[113] Language of deed being that of grantor, any doubt as to its construction should be resolved against him. *James v. Dalhart Consol. Independent School Dist.*, 254 S.W.2d 826 (Tex. Civ. App., 1953). Rule that only in case of ambiguity in deed is it to be construed most strongly against grantor is rule of last resort to be applied only when all other rules for construing an ambiguous deed fail to lead to satisfactory clarification of instrument and is particularly subservient to paramount rule that intention of parties must be given effect, insofar as it may be ascertained, and to rule that every part should be harmonized and reconciled so that all may stand together and none be rejected. *Gibson v. Pickett*, 512 S.W.2d 532 (Ark. 1974).

[114] Any particular of a description may be rejected, if it is manifestly erroneous, and enough remains to identify the land intended to be conveyed. *Arambula v. Sullivan*, 16 S.W. 436, 80 Tex. 615 (1891); *Cartwright v. Trueblood*, 39 S.W. 930, 90 Tex. 535 (1897); *Standefer v. Miller*, 182 S.W. 1149 (Tex. 1916); *Hunt v. Evans*, 233 S.W. 854 (Tex. 1921).

[115] Words in a deed after description by metes and bounds, "being the same premise whereon * * * now actually reside," held not a second description, but mere identification. *Marcone v. Dowell*, 173 P. 465, 178 C. 396 (1918).

A clause in a deed, at the end of a particular description of the premises by metes and bounds, "meaning and intending to convey the same premises conveyed to me," is merely a help to trace the title, and does not enlarge the grant. *Brown v. Heard*, 27 A. 182, 85 Me. 294 (1893).

[116] Punctuation is ordinarily given slight consideration in construing deed. *Berry v. Hiawatha Oil & Gas Co.*, 108 S.W.2d 497 (Ky. 1946).

[117] In construing deeds, words control punctuation marks, and not punctuation marks words. *Teachers' Retirement Fund Ass'n of School Dist. No. 1 v. Pirie*, 34 P.2d 660, 147 Or. 629 (1934).

[118] Doctrine of "judicial notice" operates to admit into evidence, without formal proof, those facts which are matter of common and general knowledge and which are established and known within limits of jurisdiction of court. *Palmer v. Mitchell*, 206 N.E.2d 776, 57 Ill. App.2d 160 (1965).

[119] If there be any conflict in the deed between the printed part and the part written in, the latter will control in construing it. *Miller v. Mowers*, 81 N.E. 420, 227 Ill. 392 (1907); *Wilson v. Harrold*, 123 N.E. 563, 288 Ill. 388 (1919).

- Prior deeds in chain of title may be looked to.[120]
- Later words may be resorted to.[121]

When ambiguities appear in the description,[122] and when discrepancies arise between adjoining descriptions or between the description and the physical evidence of the boundaries as it exists on the ground, rules of construction are applied to determine intention. The rules, which are based on reason, experience, and observation and pertain to the weight of evidence, state the order of preference and relative importance of calls in a grant. This order of priority is:

1. Lines actually surveyed and marked prior to the original conveyance control over calls for monuments.[123]

[120]If a deed ambiguously or uncertainly describes land, prior deeds in the chain of title may be looked to to secure the true description. *Davis v. Seybold*, 195 F. 402 115 C.C.A. 304 (W.Va., 1912).

All instruments in a chain of title when referred to in a deed will be read into it. Scheller v. Groesbeck, 231 S.W. 1092 (Tex.Com.App., 1921); reversing 215 S.W. 353; *Easley v. Brookline Trust Co.*, 256 S.W.2d 983 (Tex.Civ.App., 1953).

[121]Words which are added to the latter part of a deed, for the sake of great certainty, may be resorted to to explain preceding parts which are not entirely clear. *Wallace v. Crow*, 1 S.W. 372 (Tex. 1886).

[122]The term *ambiguity* is interpreted as connoting any doubt, uncertainty, double meaning, or vagueness that is inherent in the descriptive words themselves, or that may arise in the application of the description to its subject, the surface of the earth. 68 ALR 4, *Admissibility of parol evidence to explain ambiguity in description of land in deed or mortgage.*

When, and only when, the meaning of a deed is not clear, or is ambiguous or uncertain, will a court of law or equity resort to established rules of construction to aid in the ascertainment of the grantor's intention by artificial means where such intention cannot otherwise be ascertained. Unlike a settled rule of property which has become a rule of law, rules of construction are subordinate and always yield to the intention of the parties, particularly the intention of the grantor, where such intention can be ascertained.

Since all rules of construction are in essence only methods of reasoning which experience has taught are best calculated to lead to the intention of the parties, generally no rule will be adopted that tends to defeat that intention. *23 Am Jur2d § 224, 221*. However, the modern tendency is to disregard technicalities and to treat all uncertainties in a conveyance as ambiguities to be clarified by resort to the intention of the parties as gathered from the instrument itself, the circumstances attending and leading up to its execution, and the subject matter and the situation of the parties as of that time. Substance rather than form controls. Hence, in the construction of deeds, surrounding circumstances are accorded due weight. In the consideration of these various factors, the court will place itself as nearly as possible in the position of the parties when the instrument was executed, and where the language of a deed is ambiguous, the intention of the parties may be ascertained by a consideration of the surrounding circumstances existing at the time of the execution of the deed. In this connection, it has been said that in interpreting a deed, the choice of words used to grant the land may show the aptitude of the scrivener and is a pertinent factor. *Beduhn v. Kolar*, 202 N. W. 2d 272, 56 Wis.2d 471(1972).

[123]Requirements for the Control of Lines Marked and Surveyed:

(a) Lines must be marked prior to or at the time of the conveyance. *Woodbury v. Venia*, 114 Mich. 251, 72 N. W. 189 (1897).

(b) Lines marked must be adopted by the grantor, either directly, or indirectly through reference to a plat or map showing them, or by incorporating them into deeds. *Missouri, K. & T. Ry Co. of Texas v. Anderson*, 36 Tex. Civ. App. 121, 8] S. W. 781 (1904).

2. Calls for fixed monuments control over calls for adjoinders.[124]
3. Calls for adjoiners control over calls for course and distance.
4. Calls for course and distance control over calls for quantity.[125]

RELATIVE IMPORTANCE OF CONFLICTING ELEMENTS[126]

While sometimes believed to be a hard and fast set of rules to be adhered to in all situations and under all circumstances, this order may be changed, and any element may control depending on the circumstances and the intent of the parties.

In determining boundaries of a tract of land, it is not permissible to disregard any call if it can be applied and harmonized in any reasonable manner.[127] In so doing, conflicts may arise within a description such that it is not possible to harmonize all of the calls. However, the courts have agreed on a classification of and gradation of calls in a grant, survey, or entry of land, and has even been applied to field notes,[128] by which their relative importance and weight are to be determined. These rules are not artificial, built on mere theory, but are the results of human experience. They are not conclusive, imperative, or universal, but, like other *rules of construction*, they are adaptable to circumstances, or only rules of evidence, or merely helpful in determining which of the conflicting calls should be given controlling effect. A

(c) Must be identifiable, otherwise they lack the essential qualifications of monuments. *City of Eldora v.Edgington*, 130 Iowa 151, 106 N. W. 503 (1906).

(d) Where they do not agree with course and distance, evidence of their actual location must be clear and convincing. *Albert* v. *City of Salem et al.*, 39 Ore. 466, 65 Pac. 1068 (1901).

[124]Where there are no monuments contradicting the measurements on a parcel of land, and no substantial reason to establish their inaccuracy, course and distance control.

Considering the problem for a different point of view, mention of adjoiners could be dropped from the rules of construction. A call for an adjoiner unequivocally exhibits an intention that the property described extends only to the line of abutting property called for as an adjoiner. Hence, no need to apply title rules to construct intention. The problem then, is resolved to the question: Where is the boundary line of the adjoining property? The best proof of its location lies in its monumentation and delineation upon the ground. The rules of construction, absent adjoiners, should then be applied to the description of the adjoining property. Transfer of concentration from the property primarily concerned, to the property called for as the adjoiner seems [to this author] to be the key to the solution. *Griffin*, "Retracement and Apportionment as Surveying Methods for Re-establishing Property Corners.".

[125]In *Fortenbury et al v. Cruse, et al, 199 S.W. 523 (1917)*, where the description was incomplete, a call for quantity played a large part in determining the location of boundaries. *Rioux v. Cormier*, 75 Wis. 566, 44 N. W. 654 (1890): Where it is clear that intention is to convey a certain quantity of land, that intention is decisive and controlling.

[126]See Wilson, *Interpreting Land Records* for an extensive treatment of this topic.

Actual location on ground of original lot lines will control, if ascertainable. *Neill v. Ward*, 103 Vt. 117 (1930).

[127]*Boardman v. Reed*, 6 Pet. (U.S.) 328, 8 L.Ed 415 (1832), *Bostick v. Pernot*, 265 S.W. 356, 165 Ark. 581 (1924).

[128]*U.S. v. Champion Papers, Inc.*, 361 F.Supp. 481 (Tex. 1973).

call that would defeat the intention of the parties will be rejected regardless of the comparative dignity of the conflicting calls, and, where calls of a higher order are made by mistake, the calls of a lower order may control, as indicating the intention of the grant.[129]

The rules for the order of conflicting description elements are founded on the principle that those elements are to control in which error is least likely to occur.

> Lines actually run on the ground control; when not so run and marked they are to be located by the calls of the description properly construed, including references to other surveys and the lines of adjoining parcels.
>
> —C.J.S. Boundaries, § 14

Corners and Marked Lines[130]

    A. Natural Monuments[131]

    B. Artificial Monuments[132]
        Maps, Plats,[133] and Field Notes[134]
        Adjoiners[135]
        Metes and Bounds

    C. Courses
        Angles

    D. Distances[136]

    E. Area

---

[129] *Miller v. Southland Life Ins. Co.* Civ. App. 68 S.W.2d 558 (Tex. 1934).

[130] In ascertaining boundaries of land conveyed the line actually run controls everything if it can be traced. *Broomershine v. Stocklager*, 1 Dayton 38 (Ohio 1867).

[131] Markers not mentioned in the description have not the authentication qualifying them as monuments, in the true sense, and so have no force to dispute courses and distances set out in the deed. The fact that the boundary line claimed followed a natural landmark was of little consequence where it was not named in the description of deed although it easily could have been. *Haklits v. Oldenburg*, 201 A.2d 690, 124 Vt. 199 (1964).

[132] Where monuments mentioned in a deed are identified, they control both courses and distances given, whether they were seen by the parties to the deed or not. *Anderson v. Richardson*, 92 Cal. 623 (1892).

[133] A plat is a representation of land on paper, appealing to the eye by means of lines and memoranda rather than by words. *Thompson v. Hill*, 73 S.E. 640 (Ga. 1912).

[134] A "map" is a picture of a survey, "field notes" constitute a description thereof, and the "survey" is the substance and consists of the actual acts of the surveyor, and, if existing established monuments are on the ground evidencing such acts, such monuments control because they are the best evidence of what surveyor actually did in making the survey and are part at least of what surveyor did. *Outlaw v. Gulf Oil Corp.*, 137 S.W.2d 787 (Tex. Civ. App., 1940).

[135] Where land is described in a deed by reference to abutting land, and abutting lines can be accurately determined, the line of the adjacent tract becomes a monument. *Groth v. Johnson's Dairy Farm, Inc.*, 124 NH 286 (1983); *Kennett Corp. v. Pondwood, Inc.*, 108 NH 30 (1967).

[136] Distance being regarded as more unreliable than course, it follows that, the greater the distance, the greater probability of error, and vice versa. *Matador Land & Cattle Co. v. Cassidy-Southwestern Commission Co.*, 207 S.W. 430 (Tex. Civ. App., 1918).

Reversing Course

Courses may be run in reverse order where by doing so a difficulty can be overcome and the known calls harmonized. Such practice should be used only when calls cannot be satisfied or the description fails to close, when following them in the order stated. As the Texas court stated:

> The reversal of calls is but another rule of construction. It may be used to correct the defect, but it does not destroy or eliminate it. It is of no greater dignity or controlling effect than the rule that a general description must be called to aid a doubtful or defective particular description. Both rules serve the same purpose, to discover the real intention.[137]

The court also stated: "The only reason for reversing calls in field notes is to follow better the surveyor's footsteps, and mere running of lines according to course and distance does not locate the surveyor's footsteps in absence of any marked lines or established corners."[138] The court went on to say in *Ayers v. Watson*: "If an insurmountable difficulty is met with in running the lines in one direction, and is entirely obviated by running them in the reverse direction, and all the known calls of the survey are harmonized by the latter course, it is only a dictate of common sense to follow it."[139]

In *Ralston v. Dwiggins*, the Kansas court stated: "The footsteps of the surveyor may be traced backwards as well as forwards and any ascertained monument in the survey may be adopted as a starting point where difficulty exists in ascertaining the lines actually run."[140]

Case #20

NONCOMPATIBLE SURVEYS

Richmond Cedar Works v. West

Va. 533 (1929)

Where surveys were not made by the same parties, were not contemporaneous nor approximately so, but were thirty-four years apart, one survey cannot be used as evidence to fix a line in the other survey.

---

[137] *Gulf Production Co. v. Spear*, 84 S.W.2d 452, 125 Tex. 530 (1935), reversing Civ. App. 76 S.W.2d 558 (1934).

[138] *Howell v. Ellis*, Tex. Civ. App., 201 S.W. 1022 (1918).

[139] 11 S.Ct. 201, 137 U.S. 584, 34 L.Ed. 803 (Tex. 1890)

[140] 225 P. 343, 115 Kan. 842 (1924).

Case #21

RACINE V. EMERSON

85 Wis. 80

55 N.W. 177 (1893)

> *Great care must be used in reference to resurveys since surveys*
> *made by different surveyors seldom wholly agree.*

This argument was between a recent survey of a street line and the original survey of the street line from which improvements were based on. In analyzing the problem, the court stated that the ruling question in the case was "Where is the east line of the street in front of the lot in question, according to the original plat? 'It is not, "where is such line according to any subsequent survey or plat?" It continued:

> All resurveys or subsequent surveys are of no effect except to determine that question. A resurvey that changes lines and distances and purports to correct inaccuracies or mistakes in the old plat is not competent evidence in the case. There are only two questions [in this case]: (1) where is the true line fixed by the original plat? (2) is the fence in question on that line? A resurvey that changes or corrects the old survey and plat can never determine the first question. A resurvey must agree with the old survey and plat to be of any use in determining it.

> A subsequent survey was ordered by the city in this case to "correct the old plat, to straighten the streets, and make a better plat than the old one." "Resurveys for the lawful purpose of determining the lines of an old survey and plat are generally very unreliable as evidence of the true lines. The fact, generally known and quite apparent in the records of courts, is that two consecutive surveys by different surveyors seldom, if ever, agree; and the greater number of surveys, the greater number of differences and disagreements will occur. When two surveys disagree, the correct one cannot be determined by still another survey. It follows that resurveys are of very little use in a case as this, except to confuse it.

In a previous similar case, *Miner v. Brader*, there were two surveys, and they disagreed. The court had to resort to the evidence of a practical location of the lines by monuments.[141] Monuments set by the original survey in the ground, and

---

[141] 65 Wis. 537 (1886).

named or referred to in the plat, are the highest and best evidence. If there are none such, then stakes set by the surveyor to indicate corners of lots or blocks or the lines of streets, at the time or soon thereafter, are the next best evidence. The building of a fence or building according to such stakes, while they were present, become monuments after such stakes have been removed or disappeared, and the next best evidence of the true line."

> It is a matter of common knowledge that surveys made by different surveyors seldom, if ever, completely agree and that, more than likely, the greater the number of surveys the greater the number of differences.

> *—Erickson v. Turnquist*
> 77 N.W.2d 740 (Minn.)

PUBLIC LAND SURVEY SYSTEM (PLSS)

Not only has there been a standard manual of instructions for the survey of public lands, but there have numerous sets of special instructions. For successful retracement, the investigator must understand under which set of instructions the survey was done under. From one set of instructions to another, the unit of distance used, the length of chain, the basis of direction, the type of marker set and whether it had witnesses or not, as well as other rules, may be different.

Knowing the instructions in place at the time of the survey, and presuming that the surveyor followed those instructions,[142] the investigator could expect to follow in the previous footsteps by following the same instructions. In other words, you could expect to locate the previous survey based on those instructions.

Particular attention should be paid to special surveys, such as mineral surveys, homestead entry surveys, donation land claims, town sites, and ranchos. Each has its own idiosyncrasies and methods for retracement.

The mind-set: The surveyor followed the instructions and did what he
was supposed to do. Believe it, and the success will be higher than if
you believe otherwise.

---

[142]Until it be otherwise shown, it is presumed the surveyor did his duty in making a survey. *Phillips v. Ayres*, 45 Tex. 601 (1876).

Presumption existed that original survey actually surveyed all the lines, ran the course and distances and marked the boundaries as called for in the field notes. *U.S. v. Champion Papers, Inc.*, 361 F.Supp. 481 (1973).

Case #22

SURVEY DONE CONTRARY TO THE ACCEPTED RULE

Glenn v. Whitney

P.2d 257 (Utah 1949)

In this case, the respondent contended that the survey, as made, was legally insufficient for several reasons. One reason was that under the rule as announced in *Henrie v. Hyer*,[143] the surveyor was required to run his survey of all sections from the east boundary of the township westward and was not at liberty to run it eastward from the west boundary as to sections 19 and 20 as he did. The rule to be found in that case upon which respondent relies follows.

> "Resort should be had, first, to the monuments placed at the various corners when the original government survey of the land was made, provided they are still in existence and can be identified, or can be relocated by the aid of any attainable data. But if this cannot be done and a survey becomes necessary, this must be made from the east, and not the west, boundary of the township." Citing *Mason v. Braught*, 33 S.D. 559, 146 N.W. 687.

> It would appear that reason for the rule requiring a resurvey to be run from the east boundary westward is to establish uniformity in accordance with an established method of survey having been adopted by government surveyors. Such uniformity is desirable in order to preserve, as nearly as possible, the amount of land included within each government patent based upon a government survey. This being true, it is clear in this case, that the rights of the original purchasers of land in sections 19 and 20 were based upon a survey that commenced from the west boundary of the township. In addition, there was at least one monument on the west township line which could be identified. Under circumstances such as these, to give effect to the purpose for which the rule contended for was devised, we hold that the survey of sections 19 and 20 made from the west boundary of the township to the east was proper.

> While it is true, as urged by the respondent, that because of the curvature of the earth, boundary lines of townships in the northern hemisphere converge slightly to the north and the shortage, if any, in a given township is to be taken up in the westernmost tier of sections, according to established practice, it appears that both government surveyors, in running their surveys of the township here involved took up their shortages where their surveys from the east and west met. Although the rights of the parties, through their predecessors in interest, became established according to an unusual method of surveying, the only way in which to preserve these rights is to follow the survey as originally made.

[143] 92 Utah 530, 70 P.2d 154 (1937).

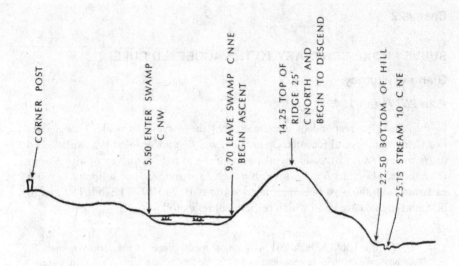

Figure 10.3 Section line profile reconstructed from field notes. From Lappala, *Retracement and Evidence of Public Land Surveys.*

Using Topographic Calls

Topographic calls in field notes may be useful to a retracement surveyor. At the least, whether they are used to relocate lost or obliterated corners, they serve two functions:

1. They are a check on the work done by the surveyor and how good his or her measurements were across the terrain.
2. They provide an index of any error to assist in searching for corners.

> *Following footsteps means just that, following the footsteps of the original surveyor, not some intermediate surveyor. Nor does it mean piecing together information from surrounding surveys.*

SUMMARY

Intention of the grantor as recited in the deed and as interpreted with the aid of the rules of construction is the controlling determination to be made in locating boundaries by resurvey. The highest and best proof of this intention, ordinarily, lies not in the words of expression in the deed, but rather, in the work upon the ground itself, where the survey was made prior to the conveyance. The surveying method of retracement is applied to locate these points upon the ground. In conducting the retracement, care should be exercised to retrace the lines as they were originally run, and not where subsequently and erroneously re-established by some intermediate owner or holder. The location of a boundary line must be determined as of the time of its creation. The retracement should proceed from some known point recognized and accepted by the original survey to hypothecate the unknown location of the property corner. Retracement should apply rules of construction to contradictory evidences of intention.

—*Robert J. Griffin, "Retracement and Apportionment as Surveying Methods*
for Re-establishing Property Corners,"
Marquette Law Review 43 (1960): 484

CHAPTER 11

CORNERS

> Though a corner in a description be not marked by any
> visible object, it is sufficient where it is susceptible of
> precise location by aid of the compass.
>
> *—Hartshorn v. Wright*
> Fed. Cas. No. 6,169 (Pet. C.C. 64) U.S., 1813

> In an old settled country, the principal work of the surveyor
> is to retrace old boundary lines, find old corners, and
> relocate them when lost.

> A corner is not lost so long as its position can be determined
> by evidence of any kind without resorting to surveys from
> distant corners of the same or other surveys. Often after
> making a survey from a distant corner, the surveyor will
> come upon some traces or evidence which will enable him
> to determine the true position of the corner he is seeking. It
> is an uncertain way at the best to locate corners by running
> lines and measuring from distant corners, and should only
> be resorted to in absence of better proof of the original
> location of the corner sought.
>
> *—Francis Hodgman*
> A Manual of Land Surveying

When considering a corner, there are three possibilities:

1. There never was a marker set.
2. There is a marker or remnants thereof.

3. There was a marker and it is now gone—but is it lost or obliterated?

> Where no corner was ever made, and no lines appear
> running from the other corners towards the one desired, the
> place where the courses and distances will intersect is the
> corner. IF, however, a marked line can be found, it shall be
> pursued as far as may be done; but, if it does not extend to
> the intersection, then the course of the patent shall be taken
> until the intersection is made.
>
> —*Wishart v. Cosby*
> 8 Ky. (1 A.K. Marsh.) 832 (1818)

> It is only in the absence of all monuments and marks upon
> the ground and in the total failure of evidence to supply
> them that recourse can be had to calls for courses and
> distances as authoritative.
>
> —12 Am. Jur 2d, § 73 and cases cited therein

> In ascertaining boundary, the rule is to find the lines and
> corners, if they ever were made, and, if not, to take as data
> such as have been made; and if there are no monuments to
> govern, to take the course and distance called for.
>
> —*M'Nairy v. Hightour*
> 2 Overton 302 (Tenn. 1814)

The goal is to find the corner, or its location. Anything less may be classed as a failure, and one must rely on second best.

There is no rule which will rigidly and inflexibly apply to all cases for restoring lost corners and boundary lines except this—that the aim of the surveyor should always be to find the exact spot where the original corner or line was located. The thing to find out is not where the corner or line ought to have been, but where it actually was.

> Do not give up a corner as lost while any means of finding
> its exact location are left untried.
>
> —*Francis Hodgman*
> A Manual of Land Surveying

A *corner* has been defined as the point or place where two converging lines, sides, or edges meet; an angle. Also, it is a piece designed to form, occupy, mark, protect, or adorn a corner of anything.[144] Thus, many equate a *marker* to a *corner*, when often it is not. The boundary corner is the point of intersection of two boundary lines; a marker is something placed by someone to call attention to the corner. A marker may or may not occupy the position of the corner. In fact, some markers have no relationship to any corners, but occasionally they are used based on the assumption that they do mark the corners. Horseshoe pins, mailbox posts, and burial markers for pets all fall into this category. In addition, an accessory may have been placed never intending to be at the location of the corner, but is a witness to a corner and indicates where the corner is located. Other property corners also act as witnesses to a corner in addition to true "witness corners."

A *corner* has been defined by the court system as the intersection of two converging lines or surfaces; as an angle, whether internal or external; as the "corner" of a building, the four "corners" of a square, the "corner" of two streets. A mere variation in a line does not constitute a "corner."[145]

> The terms "corner" and "monument" are often used largely in the same sense, though a distinction should be noted to clarify the difference. The term corner denotes a point determined by the survey process, whereas a monument is the physical structure erected for the purpose of marking the corner point upon the earth's surface.

> —Bureau of Land Management Manual of Instructions, 1947, § 349.

In the case of *Arneson v. Spawn,*[146] the court summarized the rule: If the stakes or monuments placed by the government in making the survey to indicate the section corners and quarter posts can be found, or the place where they originally were placed can be identified, they are to control in all cases.

In *United States v. Champion Papers*, the court stated:

> Time has destroyed the original witness trees marked by surveyor J.S. Collard. However the calls contained within the field notes for two pine witness trees at the south corner of the survey can still be identified with reasonable certainty. Two stump holes have been identified by credible evidence as being at the intersection of old and well-marked boundary lines. Construing the field notes for the patent of the Riggs survey in light of all the surrounding circumstances, these calls are the most certain and positive factors available in locating the footsteps of the original surveyor[147]

In *Champion,* the original marks had disappeard, but the places where they were originally placed could be identified.

---

[144] *Webster's New Collegiate Dictionary.*

[145] *Christian v. Gernt. et al.*, Tenn. Ch., 64 S.W. 399 (1900).

[146] 49 N.W. 1066 (S.D. 1891).

[147] *U.S. v. Champion Papers, Inc.*, 361 F.Supp. 481 (Tex. 1973).

CLASSIFICATIONS OF CORNERS

Corners can be defined in a number of ways.

Corner. The intersection of two converging lines or surfaces; an angle, whether internal or external; as the "corner" of a building, the four "corners" of a square, the "corner" of two streets. A mere variation in a line does not constitute a "corner." *Christian v. Gern, et al.*, 64 S.W. 399 (Tenn. Ch., 1900).

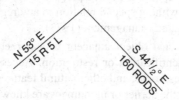

Figure 11.1   A Property Corner.

Corner, Closing. A corner at the intersection of a surveyed boundary with a previously established boundary line. In the survey of the public lands of the United States, when a line connecting the last section corner ad the objective corner on an established township boundary departs from the astronomic meridian by more than an allowable deviation, the line being surveyed is projected on cardinal to an intersection with the township boundary, where a closing corner is established and connection made to the previously established corner.

Corner, Existent. One whose position is identifiable by evidence of monument, its accessories, or description in field notes, or can be located by acceptable supplemental survey record, some physical evidence, or testimony. *Reid v. Dunn*, 20 Cal. Rptr. 273 (1962).

Corner, Lost. A point of a survey whose position cannot be determined, beyond reasonable doubt, either from traces of the original marks or from acceptable evidence or testimony that bears upon the original position, and whose location can be restored only by reference to one or more interdependent corners. *Manual of Instructions,* Bureau of Land Management (1973).

Corner, Meander. A corner marking the intersection of a township or section boundary and the mean high-water line of a body of water. Also a corner on a meander line.

Corner, Nonexistent. A corner which has never existed cannot be said to be lost or obliterated and established under the rules relating to the establishment of lost or obliterated corners, but should be established at the place where the original surveyor should have put it. *Lugon v. Crosier,* 240 P. 462, 78 Colo. 141.

Corner, Obliterated. One at whose point there are no remaining traces of the monument or its accessories, but whose location has been perpetuated, or the point for which may be recovered beyond reasonable doubt by the acts and testimony of the interested landowners, competent surveyors, other qualified local authorities, or witnesses, or by some acceptable record evidence. *Manual of Instructions,* Bureau of Land Management (1973).

Corner, Witness. By conventional usage, a monumented point usually on a line of the survey and near a corner. It is employed in situations where it is impracticable (or impossible) to occupy the site of a corner. A *witness point* is a monumented station on a line of a survey and is used to perpetuate an important location more or less remote from and without special relation to any regular corner. *Manual of Instructions,* Bureau of Land Management (1973).

Corner Accessory. A physical object adjacent to a corner, to which the corner is referred for future identification or restoration. Accessories include bearing trees, mounds, pits, ledges, rocks, and other natural features to which distances or directions, or both, from the corner or monument are known. Accessories are part of the monument and, in the absence of the monument, carry the same weight.

Corners, Double. Usually the two sets of corners along a standard parallel; the standard township, section, and quarter-section corners placed at regular intervals of measurement; additionally, the closing corners established on the line at the points of intersection of the guide meridians, range and section lines of the sureys brought in from the south. In other cases, not fully in conformity with the rectangular plan, two corners, each common to two townships only, instead of one corner of the four townships. Similarly, two corners, each common to two sections; and two quarter-sections corners, each referring to one section only. The term is sometimes used to denote two lines established on the ground although the field-note record indicates only the one line.

Monument. Some tangible landmark established to indicate a boundary.[148] Other characterizations have been "a permanent landmark established for the purpose of indicating a boundary, or boundaries,"[149] "a fixed place on the earth,"[150] and "the visible mark or indication left on natural or other objects indicating a line or boundary of a survey."[151] To be classed as monuments, objects have been required to have certain physical properties, such as visibility, permanence, and stability, and definite location, independent of measurements.[152] Monuments are of two types, natural and artificial.

---

[148] *Cornelious v. State*, 113 So. 475, 22 Ala. App. 150; *Parran v. Wilson*, 154 A. 449, 160 Md. 604 (1931); *Whitcomb v. Milwaukie*, 121 P. 432, 61 Or. 292 (1912).

[149] *Thompson v. Hill*, 73 S.E. 640, 137 Ga. 308 (1912).

[150] *Coombs v. West*, 99 A. 445, 115 Me. 489 (1916), 2 ALR 1424.

[151] *Grier v. Pennsylvania Coal Co.*, 18 A. 480, 128 Pa. 79 (1889).

[152] *Parran v. Wilson*, 154 A. 449, 160 Md. 604 (1931); *Koch v. Gordon*, 133 S.W. 609, 231 Mo. 645 (1910).

Natural Monuments. An object permanent in character that is found on the land as it was placed by nature.[153] Natural monuments include lakes, ponds, rivers, streams, rocks (where not placed by humans), springs, trees, and landforms.

Artificial Monuments. A landmark or sign placed by the hand of humans.[154] Such monuments include streets, roads, stakes, pins, pipes, posts, concrete bounds, stone bounds, drill holes, and like items.

Reference Monuments. An accessory, used in situation where the site of a corner is such that a permanent monument cannot be set or where the monument would be liable to destruction, and bearing objects are not available.[155]

A monument does not have to be marked, and can be merely a point. For instance, a boundary line between two marked points may not be marked but can be a monument, especially when named, as in "line of Smith." A point mathematically computable has also been held a monument.[156]

Extreme care must be exercised not to make assumptions or draw conclusions too quickly by equating some marker or monument to a corner. Some markers occupy the location of a corner as they may have been set there originally or have been placed correctly. Others are mere witnesses to the corner or have been placed erroneously due to incorrect interpretation of evidence or error in the mathematics or computations used to arrive at the position of the corner.

An undocumented stone without any proof may be just a plain, ordinary stone.

Occasionally, in rocky areas, where there is underlying ledge or the true corner position is in an inconvenient location, such as the middle of a road, in a driveway or a stream, a marker is placed *near* a corner but not noted as such. In reality, such would be a witness to the corner, but may be close enough to the corner so as to be misleading, perhaps even within the realm of reasonable error.

Distinguishing between Lost and Obliterated Corners

When a corner is unmarked because it never was marked even though it was established, or it was marked and the marker or its remains or other evidence has disappeared, it needs to be marked. The rules are clear: With private surveys, courses and distances may be used in the absence of monuments, but since they are among the most unreliable calls,[157] they may be used only when there is total and complete failure of finding the established corner.

[153] *Parran v. Wilson*, 154 A. 449, 160 Md. 604 (1931); *Timme v. Squires*, 225 N.W. 825, 199 Wis. 178 (1929).

[154] Ibid.

[155] *Manual of Instructions*, Bureau of Land Management (1973).

[156] *Matthews v. Parker*, 299 P. 354, 163 Wash. 10 (1931). The center of a section is not a physical government monument, but it is a point capable of mathematical ascertainment, "thus constituting it, in a legal sense, a monument call of the description."

[157] *Stafford v. King*, 30 Tex. 257 (1867); *Riley, administratrix, & c v. Griffin, et al.*, 16 Ga. 141 (1854).

Figures 11.2 and 11.3 These two photographs are of stones appearing to be town bounds but were found to be about 500 feet out of position. They are marked in typical fashion and dated 1955, but a search of the records revealed that no municipal boundary work was reported for that year. They are likely fraudulent, or at the very least placed without authority. Besides, they are obviously incorrect. They may mark something, but they do not mark the municipal boundary as they might lead a person to believe without undertaking further investigation.

Corners are established by law, sometimes based on the acts of
humans. Corner markers, or monuments, are set by people, or by
nature and then adopted by people.

The Montana court, in the case of *Myrick v. Peet,* stated: "Before courses and distances can determine the boundary, all means for ascertaining the location of the lost monuments must first be exhausted."[158] All means all, not just some. "Monuments are facts; the field notes and plats indicating courses, distances, and quantities are but descriptions which serve to assist in ascertaining those facts."[159] "Marks on the ground constitute the survey; courses and distances are only evidence of the survey."[160]

In the Public Land Survey System, with government corners established by Congress, there are two choices, depending on whether the corner is deemed lost or obliterated. With a lost corner, proportioning is appropriate. With an obliterated corner, proportionate measurement is not. Far too often a corner is deemed lost, when it should be termed obliterated, and therefore the rule of apportionment is inappropriate. Besides that, it leads others, including surveyors, to believe the line has been re-created, when in fact a new, totally incorrect, line has been established that is not the boundary or the property line.

The court stated in *U.S. v. Doyle*:

> For corners to be lost, they must be so completely lost that they cannot be replaced by reference to any existing data or other sources of information, and before courses and distances can determine boundary, all means for ascertaining location of the lost monuments must first be exhausted. Means to be used to locate lost monuments or corners include collateral evidence such as fences that have been maintained, which should not be disregarded by surveyor, and artificial monuments such as roads, poles, and improvements may not be ignored; surveyor should also consider information from owners and former residents of property in the area.[161]

The court in *U.S. v. Doyle* also referenced the case of *Mason v. Braught*, which stated that for corners to be lost, "they must be so completely lost that they cannot be replaced by reference to any existing data or other sources of information."[162] Along with that case, the court said: "see advisory comments of supplemental manual,[163]

---

[158] 180 P. 574 (Mont., 1919).

[159] *Martin v. Carlin*, 19 Wis. 477, 88 Am. Dec. 696 (1865).

[160] *Hunt v. Barker*, 27 Cal. App. 776, 151 Pac. 165 (1915); *Woods v. Johnson*, 264 Mo. 289, 174 S.W. 375 (1915).

[161] 468 F.2d 633 (Colo. 1972).

[162] 146 N.W. 687 (S.D. 1914).

[163] Restoration of Lost or Obliterated Corners and Subdivision of Sections. A supplement to the Manual of Surveying Instructions, containing a discussion of practices followed by the Bureau of Land Management, prepared especially for the information and guidance of county and local surveyors. 1952.

page 10" [which states,] "if there is some acceptable evidence of the original location of the corner, that position will be employed. A line should not be regarded as doubtful if the retracement affords recovery of acceptable evidence."

Effect of Not Following Original Footsteps

Two decisions are particularly noteworthy bearing on conclusions that either did not follow the rules or attempted to shortcut the process. The first, *Hagerman v. Thompson*, stated that "the purpose of a resurvey is to ascertain lines of the original survey and original boundaries and monuments as established and laid out by survey under which parties take title to land, and they cannot be bound by an resurvey not based on survey as originally made and monuments erected."[164] In this case there were three surveys presented to the court, which, after evaluation, said: "The three surveys in question here were resurveys, binding on no one, unless one of these perchance should ultimately in a proper proceeding be found to be correct. Which one of these resurveys is correct is a question of fact."

The second decision is that of *Williams v. Barnett,* which had to do with an inappropriate, unenforceable boundary agreement. The court stated in this case that "resurveys in no way affect titles taken under a prior survey."[165]

PRINCIPLES, BASED ON COURT DECISIONS

> Where footsteps of the surveyor are not found, it is the
> court's duty to ascertain the surveyor's intention from his
> field notes and the circumstances and conditions
> surrounding the survey.
>
> —*Howell v. Ellis*
> 201 S.W. 1022 (1918)

> The purpose of a resurvey is merely to ascertain lines of
> original survey, and original boundaries and monuments,
> and parties cannot be bound by any survey not based on that
> originally made and monuments erected thereunder.
>
> —*Day v. Stenger*
> 274 P. 112 (Idaho, 1929)

> The purpose of a resurvey is to trace footsteps of original
> surveyor, and when marks of his footsteps are found, they

[164]235 P.2d 750 (Wyo. 1951).
[165]287 P.2d 789 (Cal. App., 1955).

control, but when they cannot be found, old use and
occupancy and old recognition must suffice.

—*Ballard v. Stanolind Oil & Gas Co.*
80 F.2d 588 (Texas, 1935)

A survey establishing a line between adjacent landowners
will not revive the right of an original owner against an
established boundary, since all the survey does is establish
the line and not the title.

—*Grell v. Ganser*
155 Wis. 381 (1949)

A precisely accurate resurvey cannot defeat ownership
rights flowing from original grant and boundaries originally
marked off.

—*U.S. v. Doyle*
468 F.2d 633 (1972)

Consequences from an Error of Location

In addition to not affecting valid titles taken under the original survey, and the obvious result of misleading parties who rely on such work, an error of location may present serious consequences for surveyors. Not only has the retracement surveyor produced an incorrect product that is likely contrary to the contract for services, but another professional who relies on this incorrect or inadequate product may be equally at fault. The Wisconsin decision of *Ivalis v. Curtis v. Harding* had to do with a section line incorrectly located by a county surveyor.[166] The line was originally surveyed and marked (established) between 1859 and 1863, and was erroneously located in 1915. The title documents for both parties to this action were drawn based on the 1915 survey, which parties believed to be the dividing line between government lots 8 and 9. The error was perpetuated by a surveyor in 1971. This surveyor was later found negligent for erroneously locating the correct line, despite the fact that he pointed out that other surveyors commonly relied on the monuments set in the 1915 survey, including the opposing surveyor in this case on other occasions. The court suggested that those surveyors may also be negligent in their activities, but such was irrelevant in this case.

Today, a large problem exists in the survey profession in finding that such lines, believed to be correct and relied on by other surveyors, are in fact not correct, and have resulted in problems with property improvements (on the wrong property, over the boundary, or in violation of setback requirements), or subsequent survey work done in reliance on the incorrect survey. Like a row of dominoes, when one falls, they

[166]496 N.W.2d 690, 173 Wis.2d 751 (1993).

all fall. *Things are only as good as their basis, and if the basic survey is incorrect, all subsequent surveys relying on it will also be incorrect.*

> Any procedure in scientific testing is only as good as the scientist's ability to perform such a test. A standard procedure will not guarantee the reliability of the testing results. It is not unprecedented to learn that incorrect or fraudulent results were reported by forensic scientists.
>
> —*Dr. Henry C. Lee*
> Cracking Cases

A retracement surveyor is one more in a long list of forensic scientists, looking for clues and evidence with which to reach conclusions.

A strong recommendation is to not rush to apply rules of apportionment, that is, treat a corner as lost, until all (at least all *reasonable*) efforts for recovery have been exhausted. Using methods of apportionment means determining a corner position mathematically, based on probabilities. Everyone should know that is not the way the original surveyor established the corner, or if mathematics for some reason were employed, the measurements were based on them, and all measurements, contain error. Measurements are not perfect, and they are not based on the laws of probability, even though they are often interpreted as if they were.

As difficult and as time-consuming as the retracement process may seem, failure to find the corner, or its original position, means the troubles have only begun. Reliance on courses and distances means dealing not only with their inherent errors but with the errors introduced by the persons who made the measurements, along with current errors in reproducing them. And there are two types of errors, cumulative and compensating. Identifying, separating, and dealing with these errors demands a great deal of analysis, patience, and perhaps expertise. (Treatment of errors is part of Chapter 15.) In addition, adopting apportionment as a "quick and easy" alternative may mean compromising the process, ultimately resulting in a failure to properly identify the title lines and the correct corner locations. At best, the result is based on probability and total faith on the reliability of the measurements used.

Double Corners

In many surveys, the field notes and plats indicate two sets of corners along township boundaries, and frequently they are along section lines where parts of the township were subdivided at different dates. In such cases, there are usually corners of two sections at regular intervals and closing section corners established later on the same line, at the points of intersection of a closing lines. The quarter-section corners on such lines usually are controlling for one side only.

In more recent surveys, where the record calls for two sets of corners, those that are the corners of the two sections first established, and the quarter-section corners relating to the same sections, will be employed for the retracement and will govern both the alignment and the proportional measurements along that line. The closing section corners, set at the intersections, will be employed in the usual way, that is, to govern the direction of the closing lines.

Case #23

Double Corners outside the PLSS

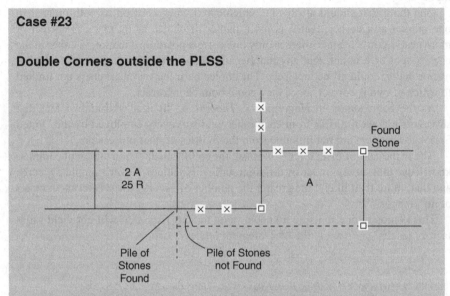

Figure 11.3 Found and not found evidence by previous surveyor.

Researching the subject tracts of an ownership and their abutting parcels resulted in a sketch of the evidence recited in the descriptions. A recorded survey was found and examined during the field retracement. There were two piles of stones found near to one another, one marking the corner of one of the tracts, the other marking a corner of a parcel that a previous owner had a life estate in. Upon her death, the title reverted, and the corner markers for that parcel became meaningless. Unfortunately, the surveyor apparently was not aware that there were two piles of stones, as he selected the incorrect one for his property corner.

Multiple Corners

Often called pincushions and porcupines, another type of double corner, or even triple or more markers in the same area, which is perhaps justified in the minds of some, is, at best, confusing. Where there are two legitimate corners near to one another, as in the preceding example, or other instances in the Public Land Survey System where double corners can legitimately exist, what is not acceptable is multiple markers at or near a property corner. In the extreme, there have been known to be more than a dozen markers in a very small area, all purporting to mark a corner. All the surveyors came up with their own solution as to the corner location, depending on what they relied on for evidence, how good their measurements were, how much error their work had, how they dealt with the error, and perhaps a host of other considerations. There is only one corner position, regardless of how many markers there might be at the location.

One thing that should always be remembered when confronted with more than one marker at a corner: that it is not a matter of choice as to which is correct or more nearly correct. Since there is only one correct position, a marker is either in the correct spot or it is not, and any number of markers at or near a position, whether one or many, could all be incorrect. The choice between two markers is not limited to which of two is correct, since they could both be incorrect.

As the court stated in *Hagerman v. Thompson*, "It is a well-known fact that surveyors are apt to differ from each other, and surveyors employed by the United States government are not immune from the frailties of their profession."[167]

And in the case of *Erickson v. Turnquist*, the court stated: "It is a matter of common knowledge that surveys made by different surveyors seldom, if ever, completely agree and that, more than likely, the greater the number of surveys the greater the number of differences"[168]

This is mostly due to using a known point to start from that will not yield satisfactory results, or errors in the measurement process.

---

[167]235 P.2d 750 (Wyo. 1951), citing *United States v. State Investment Co.*, 264 U.S. 206.
[168]77 N.W.2d 740 (Minn.).

CHAPTER 12

DIRECTIONS AND DISTANCES

Course and distance are the most unreliable calls; distance is less reliable than course, because of the mistakes of the officers, over which the locator has no control; but of natural and artificial objects the locator can take note on the ground; hence the general rule, that course and distance yield to natural objects; while, under certain circumstances, course and distance may control, yet generally they are but guides to the other calls.

—*Stafford v. King*
30 Tex. 257 (1867)

It has ever been held that the marks on the ground constitute the survey; that the courses and distances are only evidence of the survey.

Where a surveyor when making an actual survey marks division lines on the ground by monuments, such lines control calls and distances indicated on his map, and in deeds conveying land according to it.

—*Andrews v. Wheeler*
103 P. 144, 10 Cal. App. 614 (1909)

It is only in the absence of all monuments and marks upon the ground and in the total failure of evidence to supply them that recourse can be had to calls for courses and distances as authoritative.[169]

—12 Am. Jur. 2d Boundaries, § 73

[169] *M'Iver v. Walker*, 17 US (4 Wheat) 444 (Tenn. 1819), 4 L.Ed 611; *M'Iver v. Walker*, 13 US (9 Cranch) 173, (Tenn. 1815), 3 L.Ed 694; *Bryan v. Beckley*, 16 Ky. (Litt. Sel. Case) 91 (1809); *Budd v. Brooke*, 3 Gill (Md.) 198 (1845); *Collins v. Clough*, 222 Pa. 472, 71 A. 1077 (1909).

Relying on measurements is a last-resort effort, and many courts have said that it is only in the absence of the corner that resort may be made to courses and distances. Since it is a known fact that measurements contain error, and the more measurements the more likelihood of error and the error being cumulative, measurements are likely to be unreliable.

Failure to find the location of a corner means the troubles have only begun. Reproducing a corner position by relying on measurements means dealing with the errors of the original surveyor, the errors of the retracement surveyor, and if there are any intermediate surveys, their errors as well. Finding the position of the corner means the exercise is over.

The next decisions deal with the use of measurements, and when it is appropriate.

> Before courses and distances can determine a boundary, all means for ascertaining the location of lost monuments must be first exhausted; and, where there is a conflict between monuments and courses and distances, the latter must yield to the former.
>
> ——*Myrick v. Peet*
> 180 P. 574 (Mont., 1919)

The authorities recognize that for corners to be lost:

> [t]hey must be so completely lost that they cannot be replaced by reference to any existing data or other sources of information.
>
> —*Mason v. Braught*
> *supra* 146 N.W. at 689, 690

> Before courses and distances can determine the boundary, all means for ascertaining the location of the lost monuments must first be exhausted.
>
> —*Buckley v. Laird*
> 493 P.2d 1070, 1075 (Mont.); Clark, Surveying and
> Boundaries § 335, at 365 (Grimes ed. 1959)

> A "map" is a picture of a survey, "field notes" constitute a description thereof, and the "survey" is the substance and consists of the actual acts of the surveyor, and, if existing established monuments are on the ground evidencing such acts, such monuments control because they are the best evidence of what surveyor actually did in making the survey and are part at least of what surveyor did.
>
> —*Outlaw et al. v. Gulf Oil Corporation et al.*
> 137 S.W.2d 787 (Tex. 1940)

> Courses and distances, depending for their correctness on a great variety of circumstances, are constantly liable to be incorrect; difference in the instrument used, and in the care of surveyors and their assistants, lead to different results.

Courses and distances are pointers and guides, rather to ascertain the natural objects of boundaries.

—Riley, administratrix, & c v. Griffin, et al.
16 Ga. 141 (1854)

Definitions

Direction. An instruction about how to do something, such as reach a destination.[170] In surveying, it is a guide for moving from one point to another.

Courses. The direction of a line run with a compass or transit and with reference to a meridian.[171]

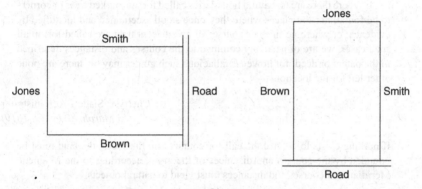

Figures 12.1 and 12.2   Same parcel, described two different ways in the chain of title. In reality, it is probably oriented northwest-southeast, but one person saw it as oriented east-west, while another saw it as oriented north-south.

While the terms "courses" and "directions" are often used interchangeably, "direction" is more encompassing, and may include general as well as specific instructions. A direction may take the form of a bearing or an azimuth, with degrees of angle included, or may be general, such as "northerly" or "northwardly." When the former is used, the basis of the reference meridian must be determined; and in the latter case, the direction is based on the observer's perception or recollection of the general direction. What someone says the direction is and what it actually is may not agree, and whatever system that the person is operating with, even if it is not correct by other standards, must be taken into consideration.

Direction also is generally expressed by stating that a line is parallel with another line or perpendicular to the last described line. A direction may also be defined

[170] *Standard College Dictionary.*
[171] *M'Iver v. Walker*, 9 Cranch. (US) 173, 3 L Ed 694 (1815).

through the use of an *angle,* either expressed in degrees or in general terms, such as "thence at a right angle."

Distance. The space between two objects; measure of separation in place; hence, length.[172] "Distance" and "length" are often used interchangeably, and the use of the word "distance," in land measurement, generally implies "length."

> The rules which have been established by the decisions of the courts, for settling questions relative to the boundary of lands, have grown out of the peculiar situation and circumstances of the country; and have been moulded to meet the exigencies of men, and the demands of justice.—
>
> These rules are,
>
> 4. Where there are no natural boundaries called for, no marked trees or corners to be found, nor the places where they once stood ascertained and identified by evidence; or where no lines or courses of an adjácent tract are called for; in all such cases, we are of necessity confined to the courses and distances described in the patent or deed: for however fallacious such guides may be, there are none other left for the location.
>
> > —Cherry v. Slade's Administrator
> > *3 Murph. 82 I N.C., 1819)*

> If nothing exists to control the call for courses and distances, the land must be bounded by the courses and distances of the grant, according to the Magnetic Meridian: but courses and distances must yield to natural objects.   .
>
> When there are no natural boundaries called for, no marked trees or courses to be found, nor the places where they once stood, ascertained and identified by the evidence; or where no lines or courses of an adjacent tract are called for, in all such cases, Counts are of necessity confined to the courses and distances described in the grant or deed.
>
> Courses and distances occupy the lowest, instead of the highest grade, in the scale of evidence, as to the identification of land.
>
> Courses and distances, depending for their correctness on a great variety of circumstances, are constantly liable to be incorrect; difference in the instrument used, and in the care of surveyors and their assistants, lead to different results.
>
> Courses and distances are pointers and guides, rather to ascertain the natural objects of boundaries.
>
> > —*Riley, administratrix, & c v. Griffin, et al.*
> > 16 Ga. 141 (1854)

---

[172] *Webster's New Collegiate Dictionary.*

Courses Favored over Distances

Since courses tend to be more reliable than distances, as discussed next, they generally take priority when there is a conflict between the two. However, depending on circumstances, either may prevail given a certain set of conditions.

In *Strafford v. King,* the Texas court explained it this way:

> Of all these indicia of the locality of the true line, as run by the surveyor, course and distance are regarded as the most unreliable, and generally distance more than course, for the reason that chain-carriers may miscount and report distances inaccurately, by mistake or design. At any rate, they are more liable to err than the compass. The surveyor may fall into an error in making out the field-notes, both as to course and distance, (the former no more than the latter,) and the commissioner of the general land office may fall into a like error by omitting lines and calls, and mistaking and inserting south for north, east for west. And this is the work of the officers themselves, over whom the locator has no control. But when the surveyor points out to the owner rivers, lakes, creeks, marked trees, and lines on the land, for the lines and corners of his land, he has the right to rely upon them as the best evidence of his true boundaries, for they are not liable to change and the fluctuations of time, to accident or mistake, like calls for course and distance; and hence the rule, that when course and distance, or either of them, conflict with natural or artificial objects called for, they must yield to such object, as being more certain and reliable.
>
> There is an intrinsic justice and propriety in this rule, for the reason, that the applicant for land, however unlearned he may be, needs no scientific education to identify and settle upon his land, when the surveyor, who is the agent of the government, authoritatively announces to him that certain well-known rivers, lakes, creeks, springs, marked corners, and lines constitute the boundaries of his land. But it would require some scientific knowledge and skill to know that the courses and distances called for are true and correct, and with the aid of the best scientific skill mistakes and errors are often committed in respect to the calls for course and distance in the patent. The unskilled are unable to detect them, and the learned surveyor often much confused.
>
> Although course and distance, under certain circumstances, may become more important than even natural objects—as when, from the face of the patent, the natural calls are inserted by mistake or may be referred to by conjecture and without regard to precision, as in the case of descriptive calls—still they are looked upon and generally regarded as mere pointers or guides, that will lead to the true lines and corners of the tract, as, in fact, surveyed at first.
>
> *The identification of the actual survey, as made by the surveyor is the desideratum of all these rules. The footsteps of the surveyor must be followed, and the above rules are found to afford the best and most unerring guides to enable one to do so.*[173] [Emphasis added.]

[173]30 Tex. 257 (1867).

Case # 24

Johnson v. M'Millan,
Strob. Law, 143 (S.C., 1846)

> *Distances may be increased, and sometimes courses departed from,*
> *in order to preserve the boundary; but the rule authorizes no other*
> *departure from the former, than such as is necessary to preserve*
> *the latter.*

One of the questions in this case was whether a certain field was within the grant described, known as the M'Nair grant. The court stated:

> In locating land, the object is to ascertain what was intended to be included within the description given in the grant. It is a question of intention; and in arriving at the intention, certain rules have been suggested by experience, which have grown into canons, and are usually called the rules of location. These rules are well laid down in *Bradford v. Pitts,* 1 Mill. Con. Rep. 115,[174] and have been repeated in a great number of cases since. The great principle which runs through them all, is, that where you cannot give effect to every part of the description, that which is more fixed and certain shall prevail over that which is less so: thus natural marks, such as rivers, creeks, rocks, and the trees, marked by the surveyor, or the lines of adjacent lands, when ascertained, are more conclusive evidence of what was intended, than distances or even courses; and therefore one of the rules of location universally recognized is, that "course and distance must yield to actual marks, whether natural or artificial,—but in the absence of these, course and distance must determine the location." *Coats v. Matthews*, 2 N. & M'C., 99. In *Colclough v. Richardson*, 1 M'C., 167, it is said, "a correct location consists in the application of any one or all of these rules to the particular case; and when they lead to contrary results, that must be adopted which is most consistent with the intention apparent on the face of the grant."

The court went on to say that when the plat in this case was examined:

> The apparent intention is to include within it all the lands lying within certain lines, running certain courses and distances from the black jack corner on Kenneth Morrison's land on the one side, to Alexander M'Intosh's land on the other. Morrison's land is found as represented; but when the lines are run along the courses called for, the distances will not carry you to M'Intosh's land. In such case, as the intention is clear that M'Intosh's land is to be

[174]S.C., 1818.

the boundary on that side, the distances must be increased until the boundary is reached. In that way, courses, distances, the form of the plat, and every thing indicating the intention of the grantor, is preserved, except that in conformity with the rule, the length of the lines is increased in order to preserve the boundary.

This case is noteworthy in that it considers and discusses construing a grant against the grantor. It distinguishes between cases of that nature and this particular set of circumstances, and the fact that if the call for the abutting tract was absent, the outcome would have been different for a different reason, and based on a different rule.

CHAPTER 13

DEALING WITH DIRECTIONS

> Allowances should be made for the variation of the
> magnetic needle from the true meridian as well as for the
> greater precision of modern surveying.
>
> —*State v. West Virginia Pulp & Paper Co.*
> 152 S.E. 197 (W.Va., 1930)

> In running line established in 1885, allowance must be
> made for variation of needle.
>
> —*McCourry et al. v. McCourry*
> 105 S.E. 166 (N.C., 1920)

> When a course is resorted to for want of a better guide to
> find the terminus or boundary of a tract of land, it is the
> course as it existed at the time to which the description of
> the tract of land refers. If it appears that because of the
> magnetic variation, that course is not the same with that
> which the needle now points out, it is the duty of the jury to
> make allowance for such variation, in order to ascertain the
> true original line. However the needle may vary, the
> boundaries of the land remain unchanged.
>
> —*Norcom v. Leary, 3 Iredell*
> (25 N.C.) 49, 1842

Noah Barker, former Land Agent for Maine, spent some serious time analyzing
magnetic observations and engaged in much correspondence with others on the topic.

The case of *Wells v. Jackson Iron Manufacturing Company*[175] was the subject of great analysis and correspondence by and between well-known scholars of the time.

Professor Quimby was concerned about the use of the magnetic needle in surveying to the point where he wrote extensively about it.[176] He stated in his introduction:

> It may seem of little importance to reproduce what has been so long known, when nothing specially new can be added; but an examination of the records of surveys made within the last fifty years will show that there is need either of more general knowledge on this subject, or of a better use of what is known. It is quite unusual to find in any of these records the slightest reference to magnetic declination; and there is reason to believe that surveyors sometimes rely too implicitly upon the needle in retracing old lines by their former magnetic bearings. It will appear by the behavior of the needle that, while it is a valuable aid, it can never be depended on for such purposes, and should, in all cases, be used with caution, and only when extreme accuracy is not required.

Quimby went on to say that "as an instrument for the determination of the true bearings of lines, it is evident that the magnetic needle can be of little value except as we are able to determine accurately its declination, or the angle it makes with the true meridian." He listed these rules:

- The declination is not the same in all places.
- For a given place it is subject to a secular change of unknown period, but requiring at least several hundred years for its completion.
- It has a diurnal change, with a maximum and minimum for each day.
- It has also an annual maximum and minimum, changing with the seasons of the year.
- It is subject to irregular disturbances, being more or less affected by every meteorological change.

Quimby continued by saying:

> The most difficult problem ever presented to the surveyor is that which asks him to retrace a lost line, with but one point known, and the bearing from some old deed. To add to his perplexity, the parties in interest are usually too much excited by the apprehension of being robbed of a square rod of rocky pasture, or of swamp rich in mud and brakes, to be able to give correctly such facts as might be serviceable in the solution of the problem. In such case, if the parties cannot be induced to agree upon a second bound and thus determine the line, there is no way but to "*run by the needle,*" after making due allowance for change in declination since the previous survey. Running in this way may lead to the

---

[175] citation Wells v. *Jackson Iron Mfg., Co.*, 47 N.H. 235, 90 Am. Dec. 575 (1866).

[176] E.T. Quimby, A *Paper on Terrestrial Magnetism*, 1874.

discovery of some old landmark, nearly obliterated, and thus settle the dispute; but if not, though the error in the bearing is likely to be 15′ to 30′, it is *better than a lawsuit*; and if, in such case, the parties in their ignorance believe that to be "true as the needle to the pole" is to be true enough it is certainly an occasion where "'tis folly to be wise."

How to Tell Whether Magnetic or True

While the true test is matching the calls to the monuments on the ground, sometimes clues lead us to believe one way or the other. First, there is a presumption, particular with older surveys, that the bearings were observed by using a magnetic compass and taken from the magnetic meridian.[177]

Second, the wording used in descriptions, returns of survey, on maps, and the like may provide a clue. It should be relatively easy to determine, since (1) if bearings are true, they will not change and should lead directly to the corners, bearing ordinary measurement errors in mind, and (2) if the bearings are magnetic, the simple conversion followed by verification should result in finding corners. Some usual wording used by early surveyors follows.

> **Due.** Means "exact"; "without error."[178] As used in the description in a deed, requiring a line to be run due north, means "exactly," and adds nothing to the description. The point of a compass, if due north, is exactly north, and so is simply north.[179]

Noah Barker wrote in his essay, "I understand 'due north,' to mean '*true*, or exact north,' coinciding with the true meridian; and 'due west' means '*true* west.' The magnetic courses are continually changing, and the terms 'due north, &c.' cannot, with propriety, be applied to 'magnetic north, &c.' " The essay also contained a letter from Alexander Wadsworth, Esq. Civil Engineer, &c. of Boston, "who has had a long practice in surveying, a part of which has been in New Hampshire." He said in the letter, "the magnetic course is a line constantly and ever varying, and it would therefore be an absurdity to call it the due north line; for the '*due*,' or 'true' north and south line never varies."

- **With the Compass.** Noah Barker addressed this phrase in his essay, saying: "that the lines in common land-surveying may be run, and generally are run in this country, with an instrument called "the compass;" and by making the

---

[177] *Taylor v. Fomby*, 22 So. 910, 116 Ala. 621 (1897); *Bryan v. Beckley*, 16 Ky. (Litt. Sel. Cas.) 91 (1809).

[178] Noah Barker, An Essay on the Cardinal Points, taken from Johnson's *Dictionary of the English Language*.

[179] *Wells v. Jackson Iron Mfg., Co.*, 47 N.H. 235, 90 Am. Dec. 575 (1866).

proper allowance for declination, the true bearings of lines are taken with it. But when we say that the true, or fixed lines, such as due North, or true meridian lines are run "with the compass," we must not be understood as saying that such lines are run "by compass," or "by the magnetic needle," for these expressions are as unlike in their meaning as would be the two expressions,—"a man was slain with his guide;" and, "a man was slain by his guide."

- **By the needle**
- **According to the needle**
- **As observed**
- **Presenting the declination on the map, either in words, or two arrows**

This does not, by itself, indicate which was used, but the fact that surveyors indicated the declination demonstrates that they were aware of it and knew how to work with it. The only way they could know would be to determine a true line and compare their compass observation to it.

- **According to the magnetic meridian**
- **According to the true meridian.**

With these last two it is obvious how surveyors referenced their directions. What might need to be further examined should things not fit properly or the search is unsuccessful, is whether the surveyors made any allowances or computations correctly. Occasionally a surveyor is found to have used the wrong declination or set off the wrong angle on the instrument. Substituting east for west, or vice versa, is a common error.

Figure 13.1   North arrow on an 1806 plan, showing the magnetic variation.

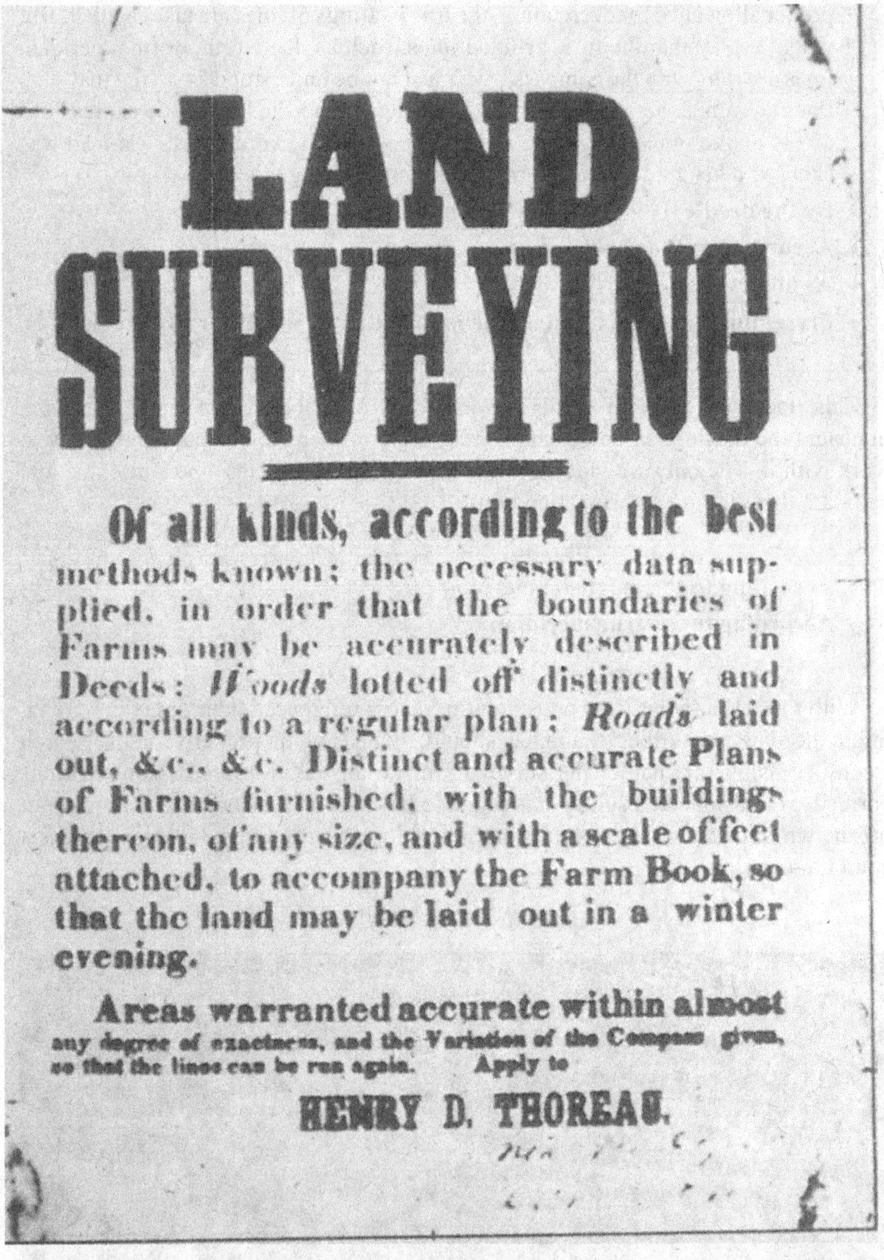

Figure 13.2 Henry David Thoreau's advertisement for surveying services. Note his statement about magnetic variation: "the Variation of the Compass given, so that the lines can be run again."

Required by Statute. It is important to review the statutory requirements, or other survey requirements or instructions, in place at the time an original survey was made. These instructions often detail the equipment and methodology to be used at a point in time.

> Sed quære—Vide act of Virginia, Chan. Rev. p. 23, ch. 12—I Litt: E.L.K. p. 389—Finnie v. Clay, vol. 2, p. 352.
>
> By the ftatute of Virginia of 1772, it is enacted, that after the firft of January 1773, every furveyor fhall, under the penalty of five pounds, "*return* all his or their original or new furveys, and *protract* and *lay down their plats by the true*, and not by the artificial or magnetic meridian" —and moreover exprefs in their return of each furvey the true quantity or degree of the variation aforefaid, and whether it be eaft or weft.
>
> According to the *true meridian*, the expreffiion *Eaft* is intelligible to furveyors and mathematicians; it is always the fame, whether ufed in 1780 or 1813—in January or December; but according to the magnetic needle, the courfe East is *changeable*, varying in different years, and progreffing in its variation from the month of January to the month of December. To afertain what was the precife variation in any given year or month, is now very difficult, if not impoffible; but it is always practicable to afcertain the variation at the time being, and to execute a furvey by the true meridian. By making entries relate to the *true* meridian, all are referred to one uniform ftandard, and whether *entered* and *furveyed* in the fame year, or in 1779 and in 1813, the refult will be the fame.
>
> *—Vance v. Marshall*
> 6 Ky. (3 Bibb) 148 (1813)

There are two distinct reasons for dealing with magnetic bearings:

1. To recover a line described and the corners
2. To mark a corner, or re-set a monument[180]

Recovering a Described Line

In order to recover a line designated with a bearing, certain considerations must be made. Although the direction is presented in a description, there is no assurance that the observation was made on that date, but likely it was made close to that date. Without additional information, it is usually safe to use that date for declination conversion, as any differences are likely to be so small as to be inconsequential.

---

[180]In some cases this corner would be marked for the first time, where no monument was actually set the first time, such as where a description recites direction and distance and does not call for a monument, or calls for a "point."

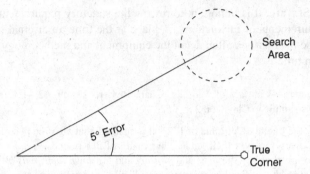

Figure 13.3 Searching in the wrong area due to angular error caused by failure to account for the change in declination.

The North Carolina court recognized this fact and addressed it in the case of *Greer v. Hayes*.[181] It said:

> Nothing else appearing, the calls in a deed must be followed as of the date thereof. Where it clearly appears upon the face of the deed or where the evidence shows, that a line as established on a prior date was adopted and was copied in the deed according to the courses and distances thereof, it is necessary to take into consideration the variations of the magnetic needle in locating the same.

The court went on to elaborate: "If it appears that the call of the deed relates to the line as it existed on the date the line was originally established or surveyed, the inquiry is as to the true location of the line as it was on that date."

In recovering such an existing line or looking for an established corner, it is crucial to refine the correction for magnetic variation as much as possible. This will reduce the amount of error, in theory reducing the search area. By doing this, much greater success will result. There is no second best; the goal is to locate the corner. It is not sufficient to "just come close."

> A court, in adjudicating upon surveys, is bound to notice judicially the magnetic variation from the true meridian. *Bryan v. Beckley*, 16 Ky. 91 (1809), and noted a constant change in variation between magnetic North and true North. *Parker v. T.O. Sutton & Sons*, 384 S.W.2d 433 (1964).

The error introduced by not correcting for declination is equivalent to 92 feet horizontal, per degree of arc, over a distance of one mile. Simple mental calculation demonstrates that significant error can result if there is inherent error of a degree or

[181]216 N.C. 396 (1939).

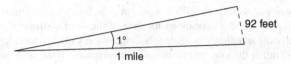

Figure 13.4 Diagram of the error in a mile of distance per degree of arc.

two. In open country, most of the time it may not be extremely critical, but in very brushy country, or in swamps and wetlands, a few feet can sometimes be critical and spell the difference between success and failure.

An error of 50 feet is often critical. In areas thick with vegetation or covered by swamp, an error of 50 feet might result in a failure to locate a marked corner, even if the marker still exists. Rotted-off stakes, covered-over stones, and the like are difficult to find when the conditions are perfect; often they are nearly impossible with gross error. At the least, a difficult corner is apt to take a lot of time to recover.

SIGNIFICANT COURT DECISIONS RELATIVE TO MAGNETIC VARIATION

> To restore lost lines and corners.
>
> Course and distance not to be departed from, but in cases of necessity.
>
> Distances must first yield.
>
> Allowances to be made for variation of the needle.
>
> Court bound to take notice that there is a magnetic variation from the true meridian.
>
> Surveyors general took their courses from the magnetic meridian.
>
> The variation of the magnet since the original survey, is the allowance to be made.
>
> The variation of the magnetic meridian from the true meridian is recognized by the statutes and by the former opinion of this court. [S]uch variation . . . is one of those principles acknowledged by scientific men, which this court are bound to notice as relative to surveys, as much as they would be bound to notice the laws of gravitation, the descent of the waters, the diurnal revolution of the earth, or the change of seasons, in cases where they would apply.
>
> *—Bryan &c. v. Beckley*
> 16 Ky (Litt. Sel. Case 91) (1809)

> If nothing exists to control the call for courses and distances, the land must be bounded by the courses and distances of the grant, according to the Magnetic Meridian: but courses and distances must yield to natural objects.
>
> *—Riley, administratrix, &c. v. Griffin, et al.*
> 16 Ga. 141 (1854)

> Nothing else appearing, the calls in a deed must be followed as of the date thereof, and it is only when it appears on the face of the deed or from other evidence that such calls and distances relate to a former survey made with reference to the magnetic rather than the true meridian, that variations in the magnetic pole will be computed as of the date of the former survey.

—Greer v. Hayes
216 N.C. 396 (1939)

Judicial Notice of Declination

The court will not take judicial notice that there has been a change in declination. In the case of *Bryan v. Beckley*, the court stated that "a court, in adjudicating upon surveys, is bound to notice judicially the magnetic variation from the true meridian."[182] That court stated later, in *Vance v. Marshall*, that "the court cannot judicially take notice that any variation of the magnetic meridian has occurred between the date of the entry and the date of the survey—the variation between given periods is a fact to be proved." [183]

Marking a Corner that Has No Marker

This is even more critical than the former, since there is nothing for comparison. The corner is a point, and the rule is to place the marker at the point or replace the marker if there was one set in the past. And it must be placed or replaced at the *exact* location where it was established. To do this, a number of factors regarding the original surveyor's directions must be considered, such as:

- Adjustment of instrument
- Local attraction
- Inherent errors
- Carelessness

There is only one true corner, and one position for that corner, at any given point. Reproducing that point or place a marker at its position requires patience and skill.

The marker's position can be checked through comparison with known corners, either on the subject parcel or from abutting tracts.

[182] 16 Ky. (Litt. Sel. Cas.) 91, 12 Am. Dec. 276 (1809)
[183] 6 Ky. (3 Bibb) 148 (1813)

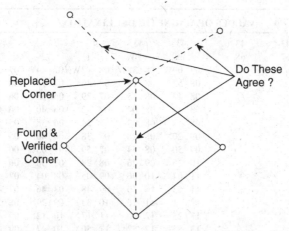

Figure 13.5 Verifying a determined corner position with known corner positions.

Correcting for Change in Declination

By taking the parcel back in the records as far as possible, or to the originating description, the investigator can determine the approximate year of the observations. The date of the document is not necessarily the same as the date of the survey, but usually it is close enough that any error will be negligible. Using the table covering the dates, the earliest and the recent, the differences in declination can be observed, then applied to the directions of interest. The latitude and longitude can be scaled from a topographic quadrangle map or derived from any one of several Internet and map sites. Until the creation of computer software to do the calculations, the method was to graphically depict two coordinate axes, true and magnetic, with the declination angle, apply the early magnetic bearing, and compute the true bearing. Then, taking a second set of coordinate axes using the present declination, the investigator would apply the true bearing, which does not change over time, and compute the present magnetic bearing.

Tables can be ordered from this source:

U.S. Department of Commerce
National Geophysical and Solar-Terrestrial
Data Center (D62)
Boulder, Colorado 80302

or

www.ngdc.noaa.gov

VALUES OF MAGNETIC DECLINATION

LAT	43	43	43	43	43	43	43	43	43
LONG	66	67	68	69	70	71	72	73	74
1750				09 09W	08 38W	08 12W	07 44W	07 23W	07 02W
1760				08 52	08 18	07 46	07 14	08 48	06 22
1770				08 44	08 05	07 29	06 53	06 22	05 50
1780				08 47	08 04	07 22	06 40	06 03	05 25
1790				08 57	08 09	07 24	06 38	05 55	05 13
1800				09 20	08 28	07 38	06 46	05 59	05 11
1810				09 50	08 54	07 59	07 03	06 11	05 19
1820				10 25	09 26	08 28	07 30	06 34	05 38
1830				11 07	10 06	09 05	08 04	07 06	06 07
1840				11 53	10 50	09 48	08 46	07 46	06 45
1850				12 41	11 36	10 33	09 29	08 28	07 26
1860				13 25	12 20	11 17	10 12	09 10	08 07
1870				13 55	12 52	11 50	10 47	09 47	08 46
1880				14 21	13 21	12 23	11 24	10 27	09 29
1890				14 40	13 43	12 47	11 49	10 54	09 57
1900				15 05	14 11	13 17	12 22	11 27	10 32
1905				15 32	14 38	13 44	12 49	11 53	10 57
1910				16 01	15 07	14 13	13 17	12 21	11 24
1915				16 28	15 35	14 41	13 45	12 48	11 50
1920				16 48	15 55	15 01	14 05	13 08	12 09
1925				17 11	16 19	15 25	14 29	13 32	12 34
1930				17 29	16 37	15 43	14 48	13 51	12 53
1935				17 43	16 50	15 56	15 01	14 04	13 06
1940				17 48	16 54	16 00	15 04	14 06	13 07
1945				17 53	16 59	16 03	15 06	14 07	13 07
1950				17 48	16 53	15 57	15 00	14 00	13 00
1955				17 47	16 52	15 57	14 59	14 00	13 00
1960				17 46	16 52	15 58	15 01	14 02	13 03
1965				17 40	16 48	15 55	15 00	14 04	13 06
1970				17 35	16 45	15 54	15 01	14 06	13 10
1975				17 31	16 45	15 57	15 06	14 14	13 21
1980				17 33W	16 50W	16 05W	15 18W	14 29W	13 39W
1985	19 05W	18 34W	18 01W	17 26W	16 47W	16 07W	15 24W	14 39W	13 53W
1990	19 05W	18 37W	18 07W	17 34W	16 56W	16 21W	15 40W	14 58W	14 14W

Table 13.1 Table of declination values covering the latitude and longitude of the subject parcel.

Use of the Internet Site

If a computer is available along with Internet access, this site can be consulted:

www.ngdc.noaa.gov/seg/geomag/geomag.shtml

If you do not know the precise location, you can derive it by inputting the address or the ZIP code into this site (or a similar one):

http://zip4.usps.com/zip4/welcome.jsp

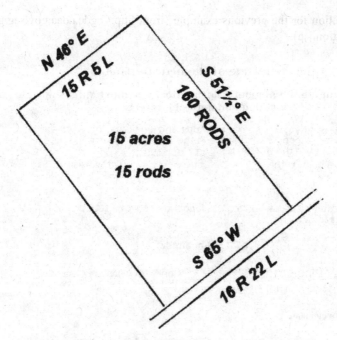

Figure 13.6 1832 description from a creating deed.

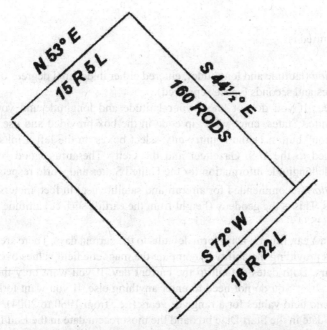

Figure 13.7 Bearings corrected to 1982 using Table 13.1.

Computation for the previous example (from http://ngdc.noaa.gov/seg/geomag/jsp/Declination.jsp):

Estimated Value of Magnetic Declination

To compute the magnetic declination, you must enter the location and date of interest.

If you are unsure about your city's latitude and longitude, look it up online! In the USA try entering your zip code in the box below or visit the U.S. Gazetteer. Outside the USA try the Getty Thesaurus.

Search for a place in the USA by Zip Code: [   ]   Get Location

---

Enter Location: (latitude 90S to 90N, longitude 180W to 180E). See Instructions for details.

Latitude: 43   ⊙ N ○ S **Longitude:** 71   ○ E ⊙ W

Enter Date (1900-2010): Year: 1982   Month (1-12): 1   Day (1-31): 1

Compute Declination

Declination = 15° 53' W changing by 0° 2' W/year

Input required is:

1. *Location* (latitude and longitude), entered either in decimal degrees or degrees minutes and seconds (space separated).

   Note: If you do not know your latitude and longitude and you live in the United States, enter your zip code in the box provided and use the "Get Location" button or the country-city select boxes on the left. Links are also provided to the U.S. Gazetteer and the Getty Thesaurus, good sources of latitude/longitude information for the United States and world respectively.

2. *Elevation* (recommended for aircraft and satellite use) in feet, meters, or kilometers. This is the geodetic (height from the earth's surface) altitude based on the WGS84.

3. *Date* in Year, Month, Day (form defaults to the current day). There are two date entries providing the ability to compute the magnetic field values over a range of years. Both dates default to the current day. If you want only the current field values, you do not need to enter anything else. If you want to know the magnetic field values for a range of years (i.e., from 1960 to 2004), enter the oldest date in the Start Date box and the most recent date in the End Date box.

4. *Date Step Size* (used only for a range of years) is the number of years between calculations. For example, if you want to know the magnetic field values from

1960 through 2004 for every two years, enter 1960 for the Start Year, 2004 for the End Year, and 2 for the Step Size.

5. To compute your field values, hit the *Compute!* button.

Results include the seven field parameters and the current rates of change for the final year:

Declination (D) positive east, in degrees and minutes

Annual change (dD) positive east, in minutes per year

(from http://ngdc.noaa.gov/seg/geomag/magfield.shtml)

The computation will automatically be made for dates after 1900. However, if the original observation is prior to 1900, the graphic computation is still necessary. To make this more efficient, you can derive the table. Taking the previous example:

Location		Degree	Minute		Date
Zip Code:	Northern Latitude:	43	00	Start Date:	1832
Get Location	Western Longitude:	71	00	End Date:	2006

Compute Historical Declination

Results:

Year	Declination	Year	Declination
1832	9° 15′ W	1940	16° 1′ W
1840	9° 49′ W	1945	16° 4′ W
1850	10° 34′ W	1950	15° 58′ W
1860	11° 17′ W	1955	15° 57′ W
1870	11° 51′ W	1960	15° 58′ W
1880	12° 24′ W	1965	15° 56′ W
1890	12° 47′ W	1970	15° 55′ W
1900	13° 18′ W	1975	15° 57′ W
1905	13° 45′ W	1980	16° 6′ W
1910	14° 14′ W	1985	16° 0′ W
1915	14° 41′ W	1990	16° 6′ W
1920	15° 1′ W	1995	16° 12′ W
1925	15° 26′ W	2000	16° 2′ W
1930	15° 44′ W	2005	15° 43′ W
1935	15° 57′ W	2006	15° 38′ W

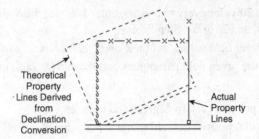

Figure 13.8 Comparison of theoretical position of property with actual position.

˙While this process will work very effectively, investigators commonly find that the derived bearings do not exactly fit the physical evidence at the site. This is generally because the original surveyor's observations were affected by one or more additional influences that caused a rotation in the resultant figure. Consequently, in many cases, an additional adjustment, or correction, is required.

This only means that the original surveyor's compass was affected by or "pulled off" by things besides the magnetic pole. Local attraction that might be inherent at the site, such as minerals in the soil or items on the person, such as a chain, chaining pins, a revolver or ax, can introduce further attraction and throw a compass needle off. In more recent times, overhead wires have become a problem.

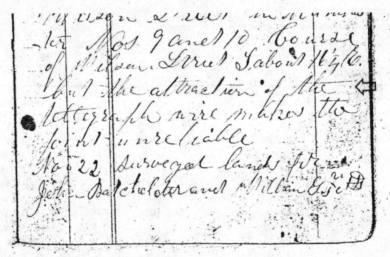

Figure 13.9 Section of surveyor's notes where he reports that because of the overhead telegraph wires, his point is unreliable.

Getting onto the Previous Surveyor's System

This is where profiling can help: knowing the previous surveyors, what kind of equipment they used, from their notes and from other surveys they performed. For example, in one instance, in the 1950s a large survey firm bought a new, sophisticated transit. Just before they used it for the first time they set of the present declination, which, after that, they never changed.

Another time a surveyor was selling his antique brass transit at a yard sale. A surveyor bought it at an outrageously low price, whereupon the seller informed him that it was manufactured with a defect—E and W had been printed backward on the face of the compass. He told the surveyor to make sure he converted his bearings because of that.

At least, there is some insight as to what one might expect in trying to follow what he did on the ground.

Returns of Survey, Allowances

Early surveyors were instructed to "give good measure" so they often put in an allowance for "uneveness of ground," wet areas, poor quality land and "swag of chain." In order to know how the original survey was performed, it is necessary to either get the original surveyor's field notes to compare with today's survey, or figure out his methodology through comparison with a known survey. Occasionally one will come across a return of survey, where the surveyor reported how he made his measurements.

May ye 21st 1738. Then Finished the Surveying and Laying out of a Township of ye Contents of Six miles Square. To Satisfie a Grant of ye Great and General Court of ye Province of ye Massachusets Bay made ye 16th Day of January 1737 on the Petition of Samuel Haywood and others and their Assotiates; Lying on the Easterly Sid of a Great Hill Called Manadnock Hill between Said Hill and a Township Laidout to ye Inhabitants of Salam and others who Servid in ye Expedition to Canada anno 1690 and Lyeth on the Southerly branch of Contokock River near the hed there of said Branch Runing throughit. It began att a Black Burch tree ye South East Corner and from thence it Ran West Six Miles and Sixty Eight Rods by a line of marked trees to a Spruse tree marked for ye South west Corner, from thence it Ran

north by a line of markd trees six Miles and Sixty Rods to a Stake and Piller of Stons ye northwest Corner and from thence it Ran East by a line of Markd trees Six Miles and Sixty Eight Rods to a Stake and heep of Stons the Northeast Corner and from thence Straight to where it began Six Miles and Sixty Rods. the Lines above said Contains ye Contents of six miles Square with ye alowance of one Chane in thirty for Sagg of Chane and fifty acres for apond

ℙ Joseph Wilder Jun Surveyr

Figure 13.10 A return of a survey made in 1738, in which the surveyor allowed "one chain in 30 for sag of chain." Given a six mile–long line, that would translate to 6 miles × 80 chains per mile, or 480 chains, divided by 30, or 16 chains total, × 4 would equal 64 rods, or × 16.5, which would equal 1,056 feet of allowance. Not factoring such procedure into today's measurements could introduce considerable error, resulting in some wildly distorted distances when trying to apply excess, deficiency, or seniority of title.

Compass Variation

Since differences exist among compasses, which compass should be used as the guide in laying out or retracing any particular grant of land? The truth is, the best compass ever made only *approximates* toward the true bearings to be taken with it. Investigators must therefore experiment with each compass and ascertain its *variation*—not from another compass, which may be equally or even more defective—but from the *true meridian* accurately found by observation, of a celestial object, the range of monuments already established on a true meridian, or some other *true* course.[184]

Determining a True Meridian

To determine a true meridian, observe Polaris, the sun, tie into geodetic marker or into a known line, such as from a railroad, or a highway survey.

Using the site www.geocaching.com, it is possible to acquire information on a government control position in the area.

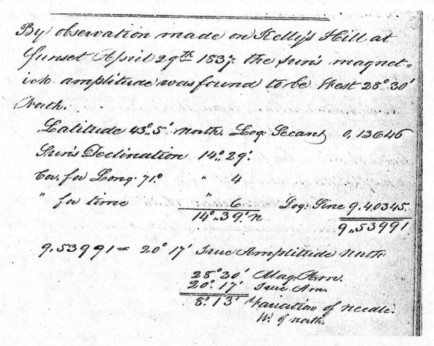

Figure 13.11   A surveyor's observation and calculation of compass variation at a given location and a point in time.

[184]Noah Barker, *An Essay on the Cardinal Points*. With this particular discussion, he pointed out that had he a difference of 15 minutes between the bearings indicated on two compasses on the same line, it would make a difference, on the west line of a survey he performed of a large original grant, of 4 chains 20 links, or nearly 17 rods.

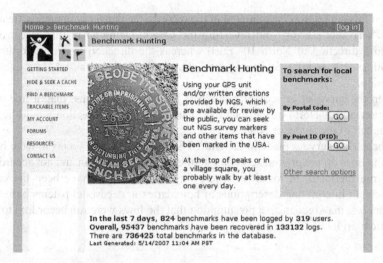

Figure 13.12   Example from www.geocaching.com. Screenshot courtesy of Groundspeak, Inc. The Groundspeak Geocaching Logo is a trademark of Groundspeak, Inc. Used with permission.

Local Attraction

In addition to inherent error in using a compass, introduced error can be even more troublesome. Attracting metals on the person or part of the landscape can cause serious error in compass readings. In some instances, differences have been found of more than five degrees within a distance of 40 rods. Such an error at the end of the line is nearly 60 feet.

Early Expression of Bearing

In early surveys, it was not uncommon when a line was close to the east-west meridian to express its direction in reverse. For example:

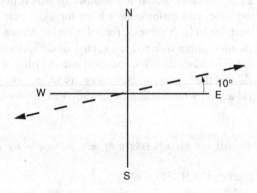

Figure 13.13   Bearing recited as East 10° North.

When this situation is encountered in the records, the danger is in merely treating it as a "scrivener's error" and reversing the letters. Doing so in this case would result in a direction of North 10° East, whereas the direction is really North 80° East (90° − 10° = 80°). Swapping the letters has introduced a 70° error into the description.

When the surveyor or scrivener stated E 10° N, he meant just that, a line with a direction of 10° North of the East meridian. Today such direction would be the complement of the angle, or N 80° E.

In the earlier surveys made in the United States, the practice seems to have been to run the lines on the ground according to the magnetic meridian and not according to the true meridian. In localities where this is assumed to have been the case, the courts, in interpreting descriptions of boundaries in deeds and patents based on such surveys, may recognize a presumption that the lines were run according to the magnetic meridian.[185]

Presumption of Magnetic Bearings

Courses in a deed (grant) are to be run according to the magnetic meridian, unless something appears to show that a different mode is intended in the instrument.[186] The referenced case was an action to determine the title to the summit of Mt. Washington, New Hampshire, which both parties claimed under different grants. Rejecting the contention that use of the term "due north" meant north by the true meridian, the court declared that as a matter of history and of common knowledge, in that jurisdiction private and even town boundaries had almost uniformly been run out according to the magnetic meridian and held it to be part of the common law of that state that the courses in deeds of private lands are to be run according to the magnetic meridian when no other is specially designated. Evidence as to the actual intention of the parties or of the surveyor would not be permissible, stated the court, nor would the present method of surveying the public lands of the United States bear on the question raised by the present case, since that method was adopted at a comparatively recent date and long after the system of surveying private boundaries in the state had been established by long-continued usage. Neither would the manner in which public boundaries, such as those between states and nations, have been surveyed bear on the question in issue, stated the court, for in such surveys, regard is had to accuracy, permanency, and certainty of verification rather than to the expense or difficulty of the method or its practical convenience or adaptation for common use in ordinary surveying. The court accordingly held that the courses in question were to be construed as referring to the magnetic meridian, notwithstanding the addition of the word "due" to their description.

[185] *Taylor v. Fomby*, 22 So. 910, 116 Ala. 621 (1897); *Bryan v. Beckley*, 16 Ky. (Litt. Sel. Cas.) 91 (1809).

[186] *Wells v. Jackson Iron Mfg. Co.*, 47 N.H. 235 (1866).

Court bound to take notice that there is a magnetic variation from the true meridian.

Surveyors generally took their courses from the magnetic meridian

—Bryan & c. v. Beckley,
16 Ky. (Litt. Sel.· Case 91) (1809)

In *Goodwin v. Greene,* the North Carolina court stated that in determining the location of a disputed line between the lands of the parties involved, proper magnetic variations since 1898 should be allowed.[187] In *McKinney v. McKinney,* the Ohio court declared that although the public lands in the state had been originally surveyed in part according to the true meridian and in part according to the magnetic meridian, it was the customary practice in that jurisdiction for contracts and private surveys to speak with reference to the magnetic meridian.[188]

Relating to the True Meridian

Some courts have held or recognized that in the absence of statute or any internal indication to the contrary, the description in a deed is presumed to related to the true meridian.

Case #25

REFERENCE TO GOVERNMENT SURVEY

In *Reed v. Tacoma Bldg. & Sav. Ass'n.,*[189] an ejectment action wherein the main question was as to the actual location on the face of the earth of a certain boundary line as described in a deed to the plaintiff and whether such deed should be construed to speak with relation to the true meridian or according to government survey. The boundary in question was described in a deed to the plaintiff as "Commencing at a point 60 rods west of the northeast corner of section 8, in township 20 north, of range 3 east; running ..."; and it appeared that the north line of said section 8 as established by government survey did not run due west, as indicated by the true meridian, but diverged from the true west line to the north. The questions in this case were (1) the actual location on the face of the earth of the north line of the northeast quarter of section 8, as the same runs west from the northeast corner of said section, and (2) whether the deed should be construed to

---

[187]237 N.C. 244, 74 S.E.2d 630 (1953).
[188]8 Ohio St. 423 (1858).
[189]2 Wash. 198, 26 P. 252 (1891).

mean west according to the true meridian or west according to the government survey. The instructions to the jury were that the line should be run according to the government survey, to which the defendant objected. The judge indicated to the jury that the presumption was conclusive, that west in said deed meant west according to the government survey.

The appeals court stated:

> We do not think that the mere reference to the northeast corner of section 8, as it referred to in the deed, is sufficient to raise the presumption that the parties intended to be governed by the United States surveys, but that it was referred to simply as a known point, the same as any monument or specific permanent object might be referred to. If the language of the deed had been, '60 rods west of the northeast corner of section 8 on said section line,' then the word 'west' would have been construed in connection with the section line, and the presumption would have been that the conveyance was made with reference to the established government survey; but the language is quite different. If the words of a deed are ambiguous, or susceptible of two constructions, testimony will be allowed to prove the meaning of the deed, although this manner of ascertaining the intention of the parties must not be invoked when the description can be ascertained from the deed itself. Title to land must not rest upon the fallible memory of witnesses when it can be avoided.

Appellee argued that there is nothing doubtful in the language of this deed; that the word "west" means "west"; that it is in no sense ambiguous and will not admit of any construction explain its meaning. The court stated:

> However, considering the importance of this decision to the public, and recognizing the probability that the city of Tacoma was uniformly platted and located according to one or the other of the theories urged here, and that structures and improvements amounting to many millions of dollars have been made throughout the city which would probably be affected by this decision, its results reaching far beyond the parties to this action, we are of the opinion that public policy demands that the custom of surveyors, in locating town plats in the state of Washington, and especially in the city of Tacoma, may be submitted to the jury to aid them in construing the intention of the parties to the deed. The jury should be instructed that, under the language of the deed in question, the presumption is that the north line of Cavender's first addition commences at a point 60 rods west, according to the true meridian, of the northeast corner of section 8, in township 20 north, of range 3 east, and running thence west, according to the true meridian, etc.; but that such presumption may be rebutted by extraneous testimony.

Apart from statute or special instruction, the question whether a description in a title document relates to the magnetic or to the true meridian is to some extent a question of geography. In those states along the eastern seaboard that comprise the area included in the original Thirteen Colonies, it was the early practice to survey according to the magnetic meridian. In some of these states, in the absence of any statutory requirement to the contrary, and where there is no indication in the instrument of a contrary intention, there is a presumption that courses relate to the magnetic meridian.

Special Instructions

By Act of Congress, approved May 8, 1796,[190] providing for the sale of public lands of the United States in the territory northwest of the Ohio River and above the mouth of the Kentucky River, what is known today as the Public Land Survey System, or the rectangular system of surveys, was adopted. This system requires that the public lands shall be divided by north and south lines run according to the true meridian and by others crossing them at right angles. In all of the western states with the exception of Texas[191] and small portions of western lands originally included within Mexican grants, in Florida, and in Alaska (prior grants), the lands have been surveyed and are generally described in accordance with the rectangular system established by Congress on the true meridian.[192] However, metes and bounds surveys and descriptions are often superimposed on the PLSS, in the form of ranchos, homestead entry surveys, mineral surveys, town sites, donation land claims, and other unique titles. Small portions of Kentucky and Tennessee were divided in rectangular fashion under special instructions. Special instructions may be found in a variety of locations, so investigators must be aware of how surveys were performed in a given area in order to conduct an effective retracement. Ohio is a collection of several land survey systems, including the beginnings of the U.S. Rectangular System. Therefore, investigators must take particular care to be aware of which system of surveying they are retracing.

Rectangular surveys are found in many, if not most, states, and in the original Thirteen Colonies the surveys are orderly, but not according to a specific set of instructions. They vary from town to town or township to township, and it is necessary to understand how a particular town was laid out. Original maps are scarce and field notes almost nonexistent, which forces investigators to undertake a sometimes arduous task of reconstructing large areas in order to know where to begin to

---

[190]43 USCS § 751.

[191]Several states had special instructions at various times; consult the appropriate statute or case law.

[192]*Manual of Instructions*, Bureau of Land Management, 1973, § 2-17, 2-18. The direction of each line of the public land surveys is determined with reference to the true meridian as defined by the axis of the earth's rotation. Bearings are stated in terms of angular measure to the true north or south. The Manual of 1890 prohibited the use of the magnetic needle except in subdividing and meandering, and then only in localities free from local attraction. The Manual of 1894 required that all classes of lines be surveyed with reference to the true meridian independent of the magnetic needle.

search for original evidence. Some past surveys did not go to the necessary extent in their retracement work, resulting in a failure to find original evidence and setting monuments inappropriately at locations other than original corners. This situation gives rise to conflicts in titles, and today's retracement surveyors must not only be on guard regarding erroneous survey information, but also know what course of action to pursue when such conflicts are uncovered.

Several courts, for example, have stressed that bad surveys do not affect good titles. In *U.S. v. Doyle,* the court stated: "Precisely accurate resurvey cannot defeat ownership rights flowing from original grant and boundaries originally marked off." [193] In *Hagerman v. Thompson*, the Wyoming court said: "The purpose of a resurvey is to ascertain lines original survey and original boundaries and monuments as established and laid out by survey under which parties take title to land, and they cannot be bound by any resurvey not based on survey as originally made and monuments erected."[194] And the California court stated in *Williams v. Barnett*: "Resurveys in no way affect titles taken under a prior survey."[195]

One rule is that, if nothing else appears, the calls in a description must followed as of the date thereof, yet where it clearly appears upon the face of description, or where the evidence demonstrates, that a line as established at a prior date was adopted and was copied in the description according to the courses and distances thereof, it is necessary to take into consideration the variations of a magnetic needle in locating the same.[196]

In a number of cases involving the boundaries of lands, the courts have held or recognized that, in the absence of any indication within the instrument of a contrary intention, the description in a deed is to be read as relating to the magnetic meridian.[197]

"Early surveyors expressed in their plats and certificates of survey courses designated by the needle; and if nothing exists to control the call for course and distance, the land must be bounded by the courses and distances of the patent, according to the magnetic meridian," declared Justice Marshall in an action for ejectment turning upon the location on the ground of a patent for 5,000 acres granted by the state of North Carolina to the plaintiff's predecessor in title.[198]

If nothing exists to control the call for course and distance, the land must be bounded by the courses and distances of the grant, according to the magnetic

[193] *U.S. v. Doyle*, 468 F.2d 633 (1972).

[194] *Hagerman v. Thompson*, 235 P.2d 750 (Wyo. 1951).

[195] *Williams v. Barnett*, 287 P.2d 789 (Cal. App., 1955).

[196] *Greer v. Hayes*, 216 N.C. 396, 5 S.E.2d 169 (1939), and numerous others.

[197] 43 USC § 751; *Riley v. Griffin*, 16 Ga. 141 (1854); Doe ex dem. *Taylor v. Roe*, 11 N.C. (4 Hawks) 116 (1825), *M'Iver v. Walker*, 9 Cranch (U.S.) 173, 3 L. Ed. 694 (1815), *Wells v. Jackson Iron Mfg. Co.*, 47 N.H. 235 (1866), *Reed v. Tacoma Bldg. & Sav. Assoc.*, 26 P. 252, 2 Wash. 198 (1891).

It is part of the common law in many states that courses in deeds of private lands are to be run according to the magnetic meridian when no other is specifically designated. *Wells v. Jackson Mfg. Co.*, 47 N.H. 235 (1866).

[198] *M'Iver's Lessee v. Walker*, 13 U.S. 173, 3 L. Ed. 694 (1815).

meridian.[199] The court added in this case that it was the practice of surveyors to express in the plots and certificates of survey the courses that are designated by the needle. However, as stated in the case of *Myricks v. Peet*,[200] "before courses and distances can determine a boundary, all[201] means for ascertaining the location of lost monuments must be first exhausted; and, where there is a conflict between monuments and courses and distances, the latter must yield to the former."

Which Meridian Intended May Be a Question of Fact

Some courts have taken the view that, with respect to the location of boundaries of land, whether the description of a deed relates to the magnetic meridian or to the true meridian is a question of fact, there being no presumption that either the one or the other is intended.

Parol evidence of a general local custom to run boundary lines by the magnetic meridian was held properly admitted in the California case of *Jenny Lind Co. v. Bower & Co.*[202] This was an action to recover damages resulting from an alleged trespass, wherein the adjacent landowners had entered into a written agreement as to the dividing line between their respective mining claims. The line in question was described as beginning at a designated point and running "thence north 23° and 15' west, 643 feet to a pine stake, and thence north 45° west to Devil's Cann..." The contention was that such evidence should not be received since it tended to vary the terms of a written agreement, whereupon the court stated that such evidence was in fact properly received, not to contradict or vary the meaning of the term "north," but to ascertain the sense in which it had been used by the parties. Noting that the term has two meanings, one common and the other technical, and that nonprofessional people generally mean, in stating courses, the lines indicated by the compass without making any allowance for any variation in the needle, the court observed that even professional surveyors, as appeared from the evidence taken in the case, would not consider the true meridian as intended unless specially so informed.[203]

The court in *Milliken v. Buswell* declared that in weighing evidence of recent surveys of ancient lines, consideration must be given to the variance of the needle

[199] *Riley v. Griffin*, 16 Ga. 141 (1854).

[200] 180 P. 574 (Mont. 1919).

[201] Emphasis on the word "all" should not be necessary, but for the sake of clarification, "all" means "the whole of; when referring to amount, quantity, extent, duration, quality, or degree." *Webster's Dictionary*.

[202] 11 Cal. 194 (1858).

[203] In the case of ambiguity, where the statement of word(s) or phrase, is subject to more than one meaning, extrinsic evidence may be relied on to explain, so long as it does not add to, subtract from, contradict or vary the words used. *Peacher v. Strauss*, 47 Miss. 353 (1872).

in determining a magnetic course.[204] In this case the plaintiff had sought to establish his title to a certain parcel of land, which was described as

> a certain lot or parcel of land situated in Oxford, in the County of Oxford and State of Maine, and located on cement road leading from South Paris to Portland [presently Route #26] and commencing on Northerly corner of lot formerly owned by Amos Smith and running South 17 rods; thence West 19 rods; thence North 17 rods; thence East 19 rods to the point of beginning.

A registered land surveyor introduced as a witness by plaintiff testified that in an attempt to prove where the plaintiff's land would lie on the face of the earth, he endeavored to locate the northerly corner, which was the starting point in plaintiff's description, and that to do so, he resorted to an earlier deed to a predecessor in title to plaintiff, which conveyed a larger tract from which plaintiff's parcel was subsequently carved out, which earlier deed made reference to a stone corner situated on what was [presently known at Route 26]. According to the calls in the earlier deed, a line was to run from this stone corner at the road on a course of south 45° west to another road. The witness testified in detail as to the examination of other deeds of conveyance involving other lots in the general area, his tracing of lines on the ground, and his discovery of markers on the ground that led him to conclude that the stone corner mentioned in the earlier deed represented the same northerly corner referred to in the description of the deed to plaintiff, and of his further conclusion that a line which he plotted from the stone corner to the second road, which line ran south 55° 50′ west, was the same line that in the earlier deed was described as running south 48° west. The witness explained the discrepancy in the two bearings by attributing it to the declination of magnetic north over the years. Since the description in the plaintiff's deed, as well as the original description in an earlier deed, indicated that a parallelogrammatic (*sic*) lot was intended by the parties, the witness plotted the southern and western lines respectively parallel to the northern and eastern lines in laying out plaintiff's lot on the ground. The contention of the defendants was that the deed required the courses of plaintiff's lot to be determined by the "true" meridian, that is, by the sidereal or astronomical meridian, and not by the magnetic meridian. The court below concluded that the plaintiff had borne the burden of proof of establishing the location of his lot, and on appeal, the defendant below contended that the trial court had erred as a matter of law in so holding. Pointing out that in identifying descriptions in deeds, the use of the words "north" or "northerly," "east" or "easterly," "south" or "southerly," "west" or "westerly" does not always indicate a direction that is due north, east, south, or west, the court quoted with approval a statement from an earlier case decided by the Supreme Court of the United States, to the effect that it is the practice of surveyors to express in their

[204] 111 A.2d 111 (Me., 1973)

plats and certificates of survey courses that are designated by the needle and that if nothing exists to indicate a contrary intention, the land must be bounded by the courses according to the magnetic meridian. The court held that in the present case, there was ample evidence to support the finding of the court below that the compass bearings pertinent to the issues represented the magnetic meridian as opposed to true north. This conclusion was supported by the testimony of two surveyors that it was unlikely that "true" north was used as the basis of the ancient survey to which the original deed referred. Further, noted the court, the use of "true" north would necessitate a change in the location of Route 26, which otherwise would run through the parcels in question, contrary to the position of the parties who both agreed that their respective lots were bounded on the east by this road rather than being traversed by it.

Whether a description referenced the magnetic meridian or the true meridian was a question of fact as to which extraneous (extrinsic) evidence might be received was the subject of *McKinney v. McKinney*.[205] The court noted that in that jurisdiction the public lands had originally been surveyed according to both meridians. The fact that an original survey was made by the magnetic meridian would raise a strong, perhaps a conclusive, presumption that all deeds and contracts referring as well to the original as to subdivision lines had reference to the magnetic meridian, stated the court, which further declared that even where the original survey was made by the true meridian, calls of courses of new subdivision lines, unless so made in reference to the original courses as to manifest an intention to be controlled by original courses, would in general be run by the magnetic meridian, since that was the general and customary usage in running lines. Each case, stated the court, must depend so much on the calls, the connection of the subdivision lines with the original true meridian lines, and other facts and circumstances that no definite rule of construction could be prescribed.

In *Martin v. Tucker,* the court refused to presume that either a magnetic or a true course was intended by a deed description of the line in question, which ran "due south."[206] The court declared that in each case the trier of fact must determine, from the evidence presented, the course to be followed. The court held that the evidence warranted the conclusion of the court below that in 1906, when the deed was executed, it was the needle of the magnetic compass which controlled and set out the disputed boundary line, and refused to disturb the trial court's finding that "due south" was equivalent to "magnetic south" in the description in question. The expert testimony apparently found persuasive by the trial court was that of a retired civil engineer who had surveyed and platted lands in the area for some 50 years, and in whose opinion the 1906 deed had been prepared by a layman and the "due south" referred to therein was the magnetic meridian. He further testified that, around the turn of the twentieth century, most deeds described the direction in which the surveyor's needle

[205] 8 Ohio St. 423 (1858)
[206] 111 R.I. 192, 300 A.2d 480, 70 ALR3d 1215 (1973).

pointed; that recourse to true bearings, with the additional work of determining the magnetic declination and of computing the true bearing, was only undertaken for larger surveying projects; that of all the jobs he had performed during his career, only three or four had employed the use of the true meridian; and while agreeing that in a given instance a reference in a deed describing "due south" could refer to the true south, he believed such was not the case with the line in question in this proceeding.

Use of "North," "Due North," or Similar Designation

In some cases involving the boundaries of land as affected by the description in a deed, it has been urged that a course described as running to one of the cardinal points of the compass, especially when qualified the term "due"—that is, a course of "north," "due north," or the like—imports reference to the true meridian. This contention has usually been rejected.

Reference to the case of *McCourry v. McCourry*[207] is helpful in that in a decree in an 1885 partition described a boundary between two tracts in this way: "From a poplar standing on east bank of said creek, thence west 190 poles, passing near the spring to a stake on the top of the William Griffith ridge." When a surveyor attempted to run the line in question according to "west 190 poles from a poplar," the other party would not permit the survey or to make allowance for the variation of $1°$ for every 20 years, which, according to the expert testimony in the case, was the proper and customary allowance for the variation in the compass, the contention being that the line should run due west without allowance for variation. Adverting to testimony of the surveyors in the case that it had been the custom of all surveyors in that locality since 1805 to allow $1°$ variation of the needle for every 20 years, and that a line which was laid off in 1885 to run due west would run N. $88\frac{1}{4}°$ west at the time of the suit, owing to such variation in the needle, the court discussed at some length scientific knowledge respecting the variation of the magnetic pole and held that, owing to scientific facts, the line in dispute, which was properly laid out and "due west 190 poles" in 1885 could at the present be identified only by setting the compass "N. $88\frac{1}{4}°$ west," and found no error in the judgment of the court below establishing the line according to the latter bearing.

Referring back to the Ohio case of *McKinney v. McKinney*, the court stated that no fixed judicial construction can be adopted for interpreting the words "west" or "due west," and their meaning must frequently depend on and be controlled by extraneous facts. Part of the description read "beginning at the center of the east line of the land owned by the said Thomas McKinney, thence running due west to the Great Miami River, containing, or supposed to contain, one hundred and fifty-nine acres, be the same more or less." The court noted that the original survey was a part of the

[207] 180 N.C. 508, 105 S.E. 166 (1920).

lands of the state and had been made with reference to the magnetic meridian, that other lands in the state had originally been surveyed by the true meridian, but that in contracts and private surveys the magnetic meridian had been generally adopted. The court declared that where an original survey had been made by the true meridian, and contracts made and deeds executed for parts of such surveys calling for and adopting the survey in their descriptions, clearly such calls of the original courses referred to the true meridian, and that it was equally clear that if the contracts or deeds thus made called for courses originally surveyed by the magnetic meridian, then such calls referred to the magnetic meridian. The court further stated that in the subdivision lines, and in contracts of sale and deeds for parts of sections originally surveyed by the true meridian, subdivision lines having no reference to the original lines would, in general, be surveyed by the magnetic meridian, as such was the usual mode of surveying lands in all parts of the state. In this case, stated the court, the jury might properly inquire and determine whether the calls of the courses of the original survey were for the true or the magnetic meridian and whether the original purchase from the government, using the words "due west" were or were not doing so with a view to a subdivision of the section between them with reference to the original courses. The lower court was mistaken, held the court, in not receiving evidence showing the surrounding circumstances under which the contract for conveyance was made and under which the words were used by the parties. The court continued that whether the lands in question were originally surveyed by the true meridian or not was a fact, perhaps slight, which the parties were entitled to show in giving construction to the words "due west," which words did not admit of a uniform judicial construction.

Also going back to *Wells v. Jackson Iron Mfg. Co.*, the New Hampshire court stated that the use of the word "due" did not indicate that the course thus described is to be run by the true meridian. The word "due" means merely "exactly" and in fact adds nothing to the description of the point of compass, for "due north" is exactly north, and so is simple "north," stated the court, observing that as the designation of the points of compass is conventional, the word "due" applies with equal propriety to those points as referred to either meridian. The court added that it found nothing in the strict meaning of the term, in its popular accep6tation, or in its legal or scientific use, or in the usages or history of the particular jurisdiction to indicate that the term "due" especially relates to the sidereal meridian.

In was held, however, in the case of *Richfield Oil Corp. v. Crawford* that unless other terms of a deed or admissible extrinsic evidence show that a different method was intended by the parties, a "due north" call should be surveyed on an astronomical basis.[208] As the Rhode Island court pointed out in *Martin v. Tucker*, the rule of *Richfield* follows a line of cases from that jurisdiction which can be referred to a local statute declaring that where a course is described as "north," "south," "east," or "west," a true course is intended.

From a practical standpoint, sometimes a test is needed to determine which meridian the surveyor made reference to. Even after corrections have been made properly,

---

[208] 39 Cal.2d 729, 249 P.2d 600 (1952).

there may still be a difference between the result and what the surveyor either reported or set. This means getting into, or onto, the particular surveyor's system.

> Where a line is found marked and is admitted or proved to be a line of the grant as originally surveyed, which line exhibits a variation from what the needle pointed when the survey was made, to ascertain the other lines as originally surveyed and called for in the grant, but which are not marked, they should be run with the same variation.
>
> If, in running the lines of a grant, one line be found marked, which is admitted or proved to be a line of the grant, and which will run with a variation from the calls of the grant, if no other marked lines be found, the other calls should be run with the same variation as that found on the marked line, to ascertain the land granted.
>
> *—Sevier v. Wilson*
> 7 Tenn. (Peck) 146 (1823)

Robert Griffin stated to go from the known to the unknown. One method is to survey some known parcel by this surveyor and make a comparision, or compare with something already known.

Case #26

MAKING THE CORRECTION FOR VARIATION ACCORDING TO THE CORRECT DATE

Greer v. Hayes

216 N.C. 396 (1939)

> Nothing else appearing, the calls in a deed must be followed as of the date thereof, and it is only when it appears on the face of the deed or from other evidence that such calls and distances relate to a former survey made with reference to the magnetic rather than the true meridian, that variations in the magnetic pole will be computed as of the date of the former survey.
>
> Plaintiff offered in evidence a deed dated 1862 containing in the description the following courses: "to the mouth of Tanyard Branch, Hayes' corner; thence north 17 degrees west with Harper's line 22 poles to a stake in the bend of said branch, said Harper's corner; thence north $3^3/_4$ degrees west with said Harper's line 8 poles to a stake." Also offered in evidence another 1862 deed between the same two parties as before, in which the calls in part were as follows: "Beginning on a persimmon corner, the northeast corner of said

Shell's tract in Harper's line; thence North $3^3/_4$ degrees west with said line 101 poles and 17 links to a stake in said line or rock." The plaintiff contended that the recited calls in these two deeds, when fitted together, constitute the same calls and approximately the same distance as the courses and distances of the disputed line and that the disputed line should be fixed and established by the calls of those deeds, making allowance for variation of the magnetic needle since 1862.

Where it clearly appears on the face of the deed or where the evidence shows that a line as established on a prior date was adopted and was copied in the deed according to the courses and distances thereof, it is necessary to take into consideration the variations of the magnetic needle in locating the same. *McCourry v. McCourry,* 180 N.C. 508; 105 S.E. 166. If it appears that the call of the deed relates to the line as it existed on the date the line was originally established or surveyed, the inquiry is as to the true location of the line as it was run on that date. *Geddie v. Williams,* 189 N.C. 333, and cases there cited; 4 R.C.L., 112; *Rodman v. Gaylord,* 52 N.C. 262.

The surveyor in this case testified that in running the line, he allowed "the proper variations according to the [two aforementioned] deeds." He further testified:

The only way I used the [aforementioned] deed was to determine the age. When I ran by her deed I came 50 feet below where she contends. I did not make any investigation as to where R.J. Ervin's corner was. In surveying I left out one call in the deed. If I had run that call it would have taken me about 200 feet away from the creek. I did not run the lines of the deed. This little dotted line beginning at the letter A (the corner as contended by plaintiff) represents a fence and a ditch on one side of it. It has been cultivated up to here."

The court continued that this record indicates that the charge to the jury with explanation was harmful to the defendant. There was nothing to warrant an inference that the calls of the line were copied from or had reference to the 1862 deed, or to justify the conclusion that the line was to be ascertained by such calls as of the date of that deed. And yet the jury fixed the line of the deed run by the surveyor, by making the proper allowances for the variations of the magnetic needle since the date thereof, as the line controversy, so that the line as established runs north 13 degrees 30 minutes west 22 poles; thence north no degrees 15 minutes west 109 poles and 17 links, rather than north 17 degrees west 22 poles; thence north $3^3/_4$ degrees west 109 poles and 17 lines as called for in plaintiffs deed—a variation of $3^1/_2$ degrees. The line as thus established deprives the defendants of their street frontage and gives to the plaintiff lands defendants had heretofore cultivated, the yards to the houses they had built on the land and a part of one of the houses. Whereas, if the line was surveyed from the beginning corner contended for by the plaintiff according to the calls of her deed as established in the deed of 1901, it will extend, approximately at least, along a fence and ditch up to which the defendants have heretofore cultivated.

Case #27

FAILURE TO CONSIDER MAGNETIC VARIATION

Henrie v. Hyer, et al.

92 Utah 530 (1937)

> A survey was not incorrect for failure of surveyors to consider variation of magnetic needle, where survey was not a magnetic needle survey, and bearing of magnet needle was relied on only for checking purposes by surveyors, who had information and ability necessary to make survey from monuments in place or from a solar or polaris observation.

While many decisions have stated that magnetic declination must be considered, there are instances where that is unnecessary, such as the case here. Knowing the instructions a surveyor operated under or knowing how a surveyor did the work will dictate the procedure and subsequently the procedure to follow in the retracement.

In this case, the court determined that there was no evidence pertaining to the position of the original corner, which therefore fell within the definition of a lost corner. It stated that whether the corner were an obliterated corner or a lost corner the result must be the same. The court agreed that the procedure followed by one of the surveyors was correct and the line and corner he established resulted in the correct position of the corner in question.

However, the plaintiff challenged one part of the procedure. In the argument and in a memorandum opinion of the court, it was charged that there was a failure of the surveyors to take into consideration the variation of the magnetic needle. The court responded:

> The surveys indicated this was not necessary. The surveyors had all the information and ability necessary to make an angle resurvey from the monuments in place or from a solar or polaris observation, independent of the magnetic needle. The survey, as indicated, was not a magnetic needle survey and the magnetic bearing of North 18° East was a matter of information and in no sense controlling upon the resurvey for the re-establishment of either an obliterated or a lost corner. The bearing of the magnetic needle from true north and south is a matter to be ascertained after and from the determination of true north by other methods. True north is not ascertained or determined from the bearing of the magnetic needle. A compass is merely a graduated circle with a magnetized needle swinging upon a pivot and properly balanced so as to permit it to swing and respond to the magnetic attraction. It is an instrument helpful for finding general directions but not reliable when accuracy is required, nor is it used except for checking purposes in surveys.

Case #28

"DUE NORTH" WAS ASTRONOMIC

Richfield Oil Corp. v. Crawford

39 C.2d 729; 249 P.2d 600 (1952)

One of the issues in this case was a boundary dispute. In 1948 a quit claim deed contained a description that led to the basic question:

> Beginning at a point...marked "Cuyama Rancho C-No. 31"...thence N. 65°10′24″ West along said SW line...a distance of 2,877.60 feet; thence due North 13,295.04 feet to true point of beginning....

This "due north" description was used two more times in the conveyancing. Richfield surveyed the legal description of the property by running the line 2877.60 feet along the boundary of the ranch established by Fitzgerald, from whose survey

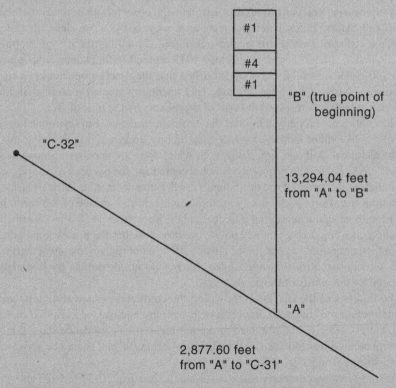

Figure 13.14   Diagram accompanying *Richfield Oil Corp. v. Crawford*.

the description was compiled (the line C-31 to C-32) to point A and then turning an angle of 65°10'24" on the transit at that point and surveying the line from A to B, the true point of beginning for parcels one and four. This method, know as a basis of bearings survey, assumed for purposes of the survey that the southwest line of the ranch had a bearing of 65°10'24" at point A. Defendants' survey used the same method as Richfield to find the location of A. At that point, however, they surveyed the call "thence due North 13,295.04 feet to true point of beginning" by running the line "true north" in the same manner as the township boundary would have been surveyed under the description in the original sublease, that is, based on an observation of the North Star. (See 5 Thompson on Real Property, § 2824.) Since the southwest line of the ranch did not actually have a bearing of 65°10'24" at point A, the difference of the two surveys at point B was approximately 11 feet, east and west.

The trial court permitted Richfield and defendants to present considerable extrinsic evidence regarding the proper method of surveying the A to B line. Substantial evidence supported the trial court's conclusion that defendants' method was correct. Dr. Thomas, a professor at the California Institute of Technology and president of the American Society of Civil Engineers in 1950, testified that good engineering practice required an astronomical observation to survey the "due north" call. Alfred Jones, chief engineer and surveyor of Los Angeles County for 10 years, stated that ordinarily a "due north" call should be run as "true north." R.V. Pearsall, a licensed surveyor since 1913, agreed with Thomas and Jones. Richfield called several experts who testified that the description should be surveyed by a basis of bearings method. This testimony created a conflict in the evidence, which was resolved in favor of defendants by the trier of fact.

The trial court properly admitted the extrinsic evidence. Surveyors and civil engineers, like other experts, may give testimony on questions involving matters of technical skill and experience with which they are peculiarly acquainted. *several cases cited.* The testimony is not accepted for the purpose of varying or contradicting the terms of the deed, but to aid the trial court in its difficult task of translating the words of the deed into monuments on the surface of the earth, in accord with accepted surveying practices. Thus, as early as 1858, this court held that surveyors could be examined on the question whether the professional practice in the community was to run a "north" call "true north" or "magnetic north," since without the extrinsic evidence the court could not determine the boundary indicated by the words of the deed.

The findings of the trial court, based on the conflicting expert testimony and its interpretation of the deed, are consistent with the boundary description in the deed. Unless other terms of a deed or admissible extrinsic evidence show that a different method was intended by the parties, a 'due north' call should be surveyed on an astronomical basis. *several cases cited.*

The base line was properly surveyed along the line from C-31 to C-32 and not by use of the 65°10'24" bearing described in the deed, since the call to the known boundary line of the ranch prevailed over the call to its astronomical bearing. There is no reason why the 65°10′24″ bearing, rejected in locating point A, must as a matter of law be used in surveying the remaining calls in the description. Richfield points out that in many cases a "due north" or "north" call has been surveyed on other than an astronomical basis, *several cases cited*, but in each of the cited cases other parts of the deed or the findings of the court based on admissible extrinsic evidence required a rejection of the astronomical method of surveying the call. In the present case, substantial evidence supports the finding of the trial court that defendants' method of survey was correct. The decision of the lower court was affirmed.

REFERENCES

70 ALR3d 1220.   *Boundaries: description in deed as relating to magnetic or true meridian..*

CHAPTER 14

DEALING WITH DISTANCES

> The science of geometry and mathematics is exact. The infinite depths of stellar space are measured with such exact nicety that the position of stars and planets can be calculated to the fraction of a second of time.... How can it be that in the ascertainment of one line of so small an area, bounded by four lines only, a difference of from 8 to 24 feet arises? It is evident that the methods pursued, and not a defective science, have brought about the different results, different maps.
>
> *—Warren v. Boggs*
> 111 S.E. 331 (W.Va., 1922)

> It is a matter of common knowledge that surveys made by different surveyors seldom, if ever, completely agree and that, more than likely, the greater number of surveys the greater number of differences.
>
> *—Erickson v. Turnquist*
> 77 N.W.2d 740, 247 Minn. 529 (1956)

Unless controlled by other calls, calls for distance must be strictly observed. Some have been misled by this statement, in that in the absence of monumentation and the inability to locate a corner, the distance was relied on to locate the corner. It is the distance as used by the creating surveyor, under the conditions and circumstances at the time, that is controlling. Some have made incorrect assumptions, such the rod being 16 HF feet in length, when in fact it is not, and making assumptions with certain other dimensions, when the original surveyor made allowances, had inherent error, and the like. The Kentucky court in *Violet v. Bowman* alluded to that consideration by stating: "Where the distances on the lines called for are not expressly given, or are vague and indefinite, they must be ascertained by construction, giving full effect to the apparent intention of the locator and to the descriptive words of the grant or conveyance."[209] Whereas courses are frequently corrected for errors, so must distances be given the same consideration.

HOW WERE THE ORIGINAL MEASUREMENTS MADE?

In laying out roads and lines of townships, it was usual for surveyors to make large measure, of which, however, there is no certain standard. Therefore, a "local correction" may be necessary for a given survey. The only way to known for certain is to study surveyors' field notes or returns of survey.

Noah Barker, in his essay, explained that

> some allow one in thirty, for the swagging of the chain.
>
> The length of a man's arm to every half chain, has been allowed for inequality of surface. The half chain is most convenient in thick woods; but some have very absurdly used a line; and if any allowance is made for its contraction by moisture, it must be arbitrary. Surveyors are often sworn to go according to their best skill and judgment; this they may do with great sincerity, and yet, for want of better skill, may commit egregious mistakes.

Another explanation was made this way:

> The longer lines exceeded the record by 20, 30, and even 40 rods. It has been explained by old surveyors in the following manner: the chain bearers added to the length of every chain; when the foremost man had drawn his chain straight from the hand of the rear man at the last pin, he took the end of the chain in one hand and a pin in the other hand and stepping as far as he could in advance he reached forward with the pin and dropped it. This method would increase each chain length about the measure of a man's stature and the excess for a mile line would be about 30 rods.[210]

there's allowed about one Rod in thirty for uneven land and Swag of Chain, also there is allowed one hundred acres for a farm all ready Granted to Coll. Josiah Willard with five hundred acres for ponds—

᛭ Josiah Willard Surveyor

Figure 14.1 Part of a return of a 1736 survey by Josiah Willard stating how he did his survey.

---

[209] 1 A.K. Marsh. 282.

[210] *Manchester Historic Assn Collections*, G. Waldo Brown (1911), Volume VI, Page XVII.

Without that information, retracement surveyors must find known points and derive the "local correction." Without doing this, measurements will be unreliable and will either (1) cause a failure to find an existing original marker or (2) cause a new marker to be set in the wrong location. Most retracement surveyors have personally witnessed this by finding a newly set marker not far away from an original monument.

CONSEQUENCES OF IGNORING CORRECTIONS

Looking at previous a return of survey, modern surveyors who encounter excess or deficiency when retracing a lot subdivided from the original tract might easily be misled. This would be critical enough in searching for original evidence, but it is exceedingly critical in setting a marker for the first time. Gross error can easily result if less than half of the information is taken into account. For example, there is an allowance of 1 rod in 30 in this survey. Assuming the distance to be as stated, when it actually is longer, will result not only in excess at the outset but extreme conflicts later as tracts are further subdivided. When gross errors are found in either distance or area, further checking is in order to uncover the error. Without doing so, it is highly unlikely that the survey being performed will be correct, unless completely by accident. There are just too many variables to take statements and measurements on faith without some comparison or verification.

No doubt a bigger problem is today's surveyor retracing not only an original survey, but encountering past retracements that either made the wrong assumptions or did not take into account the allowances made by the first surveyor. This problem is widespread, resulting from incomplete retracements in an attempt to save time and money. A common problem found today is where, in a subdivision, past surveyors did their work on a lot-by-lot basis and did not take into account or perhaps did not even consider other lots in the block or the subdivision, thereby leaving any excess, deficiency, or mistakes for the next unwary surveyor. Without a clue or knowledge of how previous surveyors failed to do everything they should have, it is common to grossly underestimate the amount of work necessary to do it correctly or to unravel the bad work of others.

The only way to approach projects such as these is to work chronologically, determining the original survey work and creation of title(s), then following the resurveys and retracements one at a time, in order, to conclude what is reliable and what is not. Review the court decisions: Incorrect surveys are not binding on parties,[211] and boundary agreements can be executed only under the right conditions and when not

---

[211] Precisely accurate resurvey cannot defeat ownership rights flowing from original grant and boundaries originally marked off. *U.S. v. Doyle*, 468 F.2d 633 (1972).

Resurveys in no way affect titles taken under a prior survey. *Williams v. Barnett*, 287 P.2d 789 (Cal. App., 1955).

contrary to the statute of frauds. Those that do not follow legal requirements are likely to be void.[212]

Case #29

OVERRUN OF DISTANCE ON ORIGINAL SURVEY

Owen v. Bartholomew

26 Mass. 520 (1830)

> A usual practice of surveyors of land laid out by the proprietors of a town, to overrun the exact measures, is admissible to show that the boundaries of an ancient grant by the commonwealth in an adjacent town, and described by courses and distances, exceeded the distances given.
>
> For the same purpose, evidence is admissible that at the time of an ancient grant by the commonwealth, it was the uniform practice, in surveying such grants, to give large measure.

In this case, the Commonwealth of Massachusetts owned all the land between Williams's grant on the east and the line of Mount Washington on the west, on May 17, 1755, when it granted to Austin and Shears, from whom the tenant derived his title,

> one hundred and fifty-four acres of said land lying west of Sheffield, and is the grantees' improvement, bounded easterly on Williams's grant and the northeast corner of Drake's land, thence running west nine degrees north on land of said Drake 285 poles, thence running north nine degrees east 87 poles, thence east nine degrees south 280 poles, until it comes to 20 rods northerly of said Williams's grant, then on the west side of said grant to the bounds first mentioned.

The demandant claimed a parcel of land marked on a plan, on the ground that enough was left to the eastward of it and west of Williams's grant to satisfy the conveyance to Austin and Shears, according to an exact measurement of the length of the lines given in that grant. The tenant claimed to a line about 30 rods farther to

---

[212]In the absence of a real dispute, an agreement, purporting to establish the boundary between lands of adjacent proprietors at a line known by both to be incorrect, the result of which, if given effect, must be to transfer to one lands which both know do not belong to him, is without consideration, and within the statute of frauds and void. *Myrick v. Peet*, 180 P. 574 (Mont., 1919).

Doctrine that coterminous owners may by agreement, implied from acquiescence, establish and fix their mutual boundary lines, is applicable only where the true boundary line is otherwise unknown or uncertain and cannot be used where true boundary line is known or means of knowledge is within reach. *Williams v. Barnett*, 287 P.2d 789 (Cal. App., 1955).

the west than the demandant located the grant, and he disclaimed the land between that line and the line of Mount Washington.

The committee, on the same day, made grants to Drake and to Joseph and John Owen southward of the grant to Austin and Shears. On November 27, 1771, the commonwealth granted to Petite a tract of land to the northward of the grant to Austin and Shears, which came to within a few rods of the Lanesborough Spring, a known boundary now existing.

In 1823 the tenant petitioned the legislature to grant him the land described in his petition, between the line of Mount Washington on the west and the land granted to Austin and Shears on the east. The legislature authorized the sale of that land, and the demandant bought one-third of it. The demandant contended that the land he purchased went 30 rods to the eastward of the line to which the tenant claimed. The question the jury was to settle was: Where were the westerly bounds of the land granted to Austin and Shears? The west line of the grant to Williams, as laid down in the plan, was considered at the trial as substantially correct. There was evidence tending to prove that those who claimed under Austin and Shears had occupied and claimed under the original grant, the premises demanded, and that a walnut tree stood at the southwest corner of the premises demanded, which was mentioned in the deed of Fellows to Benton of June 7, 1777. The tenant offered evidence tending to show that those who claimed under the grants of the commonwealth to Drake and Joseph and John Owen had occupied and claimed, under their grants of the same day and of the same direction of the compass (north nine degrees east) up to the same exhibited westerly line to which the tenant claimed. The tenant offered a deed from Nathaniel to William Owen, dated in 1788, referring to a hemlock tree, which it was said (and there was parol evidence to prove) stood on the line produced, to which the tenants' claim extended. The demandant objected to all this testimony as irrelevant, contending that the monuments referred to were erected long after the grant to Austin and Shears, and had no tendency to prove the westerly line of the grant of the Commonwealth to them.

The tenant introduced the evidence of a survey of the town of New Marlborough, to prove that the land laid out by the proprietors of that town generally overran the length of the lines given. To this evidence the demandant objected, as it did not relate to grants from the commonwealth; but the judge overruled the objection.

The tenant also produced evidence to prove that the ancient grants in Mount Washington uniformly overran the exact measures. The demandant objected to this evidence, contending that in the absence of monuments, the lines should be ascertained by exact measurement. But the jury was instructed that they were to consider the overmeasurements of the ancient grants in settling this boundary and that they were now to locate the grant to Austin and Shears as they believed upon the evidence it would have been located in 1755, when it was made. It was proved by skillful surveyors that the land included in that grant was rough and mountainous. If the line were measured up hill from Williams's grant to Mount Washington, it would go about four rods to the westward of the line to which the demandant claimed; if measured downhill, it would make four rods less; so that exact mathematical certainty was not to be obtained. The jury were then to take into consideration the uniform usage of giving large measure at that time, and wee to say, upon the whole matter,how far to the westward the grant to Austin and Shears should extend. The walnut tree and the hemlock tree, and the occupation of the several grantees of the Commonwealth, in a continuous line from north to south, under the grants of 1755, were to be taken into consideration in ascertaining the place of the dividing line between the lands then granted and the lands since granted to the demandant and his associates.

The question was whether the grant to Austin and Shears included the whole or a part of the land in controversy, and if a part, what part. This was left to the jury, and the demandant moved to set aside the verdict, because evidence, which was objected to at the trial, was improperly admitted.

The court do not mean to intimate, that where the grant of the Commonwealth is definite in its terms and application, evidence *aliunde* is admissible to control it; but where it is uncertain what land is intended, extrinsic evidence must be admitted. We think this is a case proper for the admission of such evidence. The tenant claims under an ancient grant. The deed refers to no precise monuments. It refers to the grant to Williams, the west line of which was uncertain at that time, and to the grantees' improvements, and to the north line of the grant to Drake. The situation and extent of the land could not be ascertained from this description, except by resorting to other evidence than the grant itself. We think all the deeds were rightly admitted. Those to Drake and Joseph and John Owen, made on the same day with the grant to Austin and Shears, were of the same length of line, east and west, and any evidence which would show the west line of those grants would prove that of the land in question. The Court are all of opinion, that where it is proper to go into evidence *aliunde*, it is proper to consider contiguous grants. Evidence of a location at the time would show the land intended to pass, and in the case of *Makepeace* v. *Bancroft*, 12 Mass. R. 469, monuments erected soon after the grant, were admitted to be proved for the purpose of showing the extent and limits of the grant. Whether the monuments were erected simultaneously or subsequently, would affect the weight, but not the nature of the evidence. The deed of Fellows to Benton, in 1777, was competent evidence, as it tended to settle the west line of the land in controversy. So the deed from Nathaniel to William Owen, of 1788, though it would have less weight, being later, is yet of some force, as referring to a monument previously established.

Evidence also was introduced, to show that in adjacent townships there was a tendency to overrun in the location of grants. We think this evidence was rightly admitted, for where lines are referred to, mathematical lines are not intended, but such as are used by people conversant with the business of surveying. The evidence of this description had reference to the town where this land is situated, or to a neighboring town, and where probably the same surveyors were employed.

Figure 14.2   A line expressed as N 64° W, 6 miles, 8 Tallies & 8 Chains, or 2,096 rods.

WHAT UNIT OF MEASUREMENT WAS USED?

There are multiple definitions for almost any unit, and it depends on the area and use of a unit of measurement as to what its definition is.[213] There are numerous definitions of the Spanish *vara,* several of the French *arpent,* several of the English rod, several of the foot, several of the acre, and a number of the chain and corresponding link.

Commonly there is more than one definition of the rod (also known as pole and perch), depending on where and how it was used.

Fortunately, the description in Figure 14.2 provides a way to make a conversion to determine the units this surveyor was using: 6 miles × 5280 feet + 48 chains [8 tallies + 8 chains] × 66 feet = 31680 + 3168 = 34848 divided by 2096 rods = 16.6 ≈ 16.5 feet per rod. Otherwise, the conversion would have to be derived, using expressed distances from the record compared with known distances.

WHAT TYPE OF CHAIN?

When a measurement is expressed in chains, or chains and links, the type of chain must be identified. Usually the type can be determined through comparison on a known line. At certain times, some locations designated, either through survey instruction or by statute, what type of chain was to be used. Even so, chains break, are repaired, and there is no guarantee that everyone followed all the instructions implicitly. A wide variety of chains have been found, and preserved. A few are shown in Figure 14.3.

HOW DID THE SURVEYOR MEASURE?

Were there any problems with the measuring device, whatever it was? What type of terrain and vegetation was encountered? Abrupt changes in elevation, erratic topography, and the presence of wetlands or bodies of water all affect the results that might be expected. Break a long-surveyed line down into parts and treat each section, with its particular characteristics, separately.

[213] Chapter 2 of *Interpreting Land Records* examines the issue of unit of measurement used in great detail.

Figure 14.3   A variety of chains, with links ranging from 7.92 inches (standard two-pole Gunter's chain in front) to one with links that are more than two feet in length (the rear). Included here is a vara chain, a metric chain, an engineer's chain, and a Mexican railroad chain. From the collection of Francois "Bud" Uzes.

Straight Line or Other Method

It is generally agreed that distance with reference to provisions in a deed is to be determined by a straight line, in the absence of any apparent contrary intent of the parties to the instrument. In fact, many courts have ruled that a presumption exists that a line described as running from one point to another, unless a different line is described in the instrument or marked on the ground, is to be a straight line, so that by ascertaining the points at the angles of a parcel of land, the boundary lines can at once be determined.[214] The rule of surveying, as well as of law, is to reach

[214] *Halstead v. Aliff*, 78 W. Va. 480, 89 S.E. 721 (1916).

the point of destination by the line of shortest distance,[215] and lines should never be deflected, except in order to conform to the intention of the parties.[216] However, the legal presumption may be overcome where a boundary line has actually been marked on the ground.[217]

When an ascertained object is called for in a description, it must be reached by a straight line without regard to an erroneous description by course and distance.[218]

A line described as running toward one of the cardinal points of the compass should be presumed to run directly in that course unless its direction is controlled by some object.[219]

"Thence to a corner" means a direct line from the former point to the latter.[220]

A line "to range with" another, from the end of which it begins, must follow the path of the other line when extended and be a continuation of it.[221] And ordinarily, a boundary line marked for only part of the distance should be continued in the same direction for the full distance.[222]

Constitutional or Statutory Provision

Decisions are not in agreement as to the proper method of determining distance in the construction of constitutional provisions or statutes. In some jurisdictions, a straight line on the horizontal plane has been considered the proper method of measurement.

Not a Straight Line

In some jurisdictions, the courts have recognized other means than the straight line as a method for measuring distances. None of these means have to do with property lines; rather they relate to jurisdictional lines.[223]

Unevenness of ground to be allowed for.

> Horizontal measure to be attained, being the basis of the art of surveying.
>
> —*Bryan &c. v. Beckley*
> 16 Ky (Litt. Sel. Case 91) (1809)

---

[215] *Bartlett Land, etc., Co. v. Saunders*, 103 U.S. 316, 26 L.Ed. 546 (N.H., 1880); *Leonard v. Smith*, 111 La. 1008, 36 So. 101 (1904).

[216] *Platt v. Jones*, 43 Cal. 219.

[217] *Willoughby v. Willoughby*, 20 Ky.L. 1061, 48 S.W. 427 (1898); *Seneca Nation v. Hugaboom*, 132 N.Y. 492, 30 N.E. 983 (1892); *Hough v. Horne*, 20 N.C. 369 (1839).

[218] *Campbell v. Branch*, 49 N.C. (4 Jones L.) 313 (1857).

[219] *E.E. McCalla Co. v. Sleeper*, 105 Cal. App. 562, 288 P. 14 (1930); *Green v. Palmer*, 68 Cal. App. 393, 229 P. 876 (1924); *Vermont Marble Co. v. Eastman*, 91 Vt. 425, 101 A. 151 (1917); *Hagan v. Campbell*, 8 Port. 9, 33 Am. Dec. 267 (Ala. 1838).

[220] *Bryant v. Vinson*, 3 N.C. (2 Hayw.) 3 (1797).

[221] *Lilly v. Marcum*, 214 Ky. 514, 283 S.W. 1059 (1926).

[222] *Banks v. Talley*, Del. Super., 194 A. 362 (1937).

[223] 54 ALR 781, IIb.

Slope or Horizontal Measurement

A number of cases have dealt specifically with the question whether the horizontal measure or surface measure is the proper method of determining the extent of the boundaries of a tract of land. As pointed out in some of the cases, the selection of one method or the other in surveying large tracts of mountain land could make a material difference in the amount of land actually encompassed within the boundaries of such tracts. Likely for this reason, most court cases have involved large tracts of mountain land, and in the majority, horizontal measurement has been held to be the proper method. However, other cases have held or at least recognized that surface measurement was the proper method where such was the custom of the locality or was dictated by the circumstances.

Effect of Local Custom or Practice

In the case of *Justice v. McCoy*,

> where the proper method of measuring the depth of a city lot which fronted on one street and extended part of the way up a hillside toward a parallel street was involved, and the evidence tended strongly to support the theory that the depth of the lot should be measured along the surface of the hillside, the court, in holding that such method of measurement was proper, said that it found authority for the proposition that a surface measurement was proper where it was a custom of the locality, or where it was dictated by the circumstances of the case.[224]

That surface measurement may properly be used in making surveys where it is the custom of the locality or is dictated by the circumstances is supported by the case of *Boyton v. Urian*.[225] In this case the boundaries of two tracts of land had been surveyed in 1794, and there was some question as to whether the boundary between them had ever been marked on the ground. The court merely held that that the owner of one of the tracts was only entitled to have one of boundary lines measured in the same way as it had been measured by the surveyor who made the original survey, and not by horizontal measurement.

In the North Carolina case of *Duncan v. Hall,* the court recognized that in making surveys of mountainous regions, the early surveyors, at least, had universally adopted surface measurement.[226] Since this has been found to be true in several states, it may often be considered whenever mountainous terrain is a consideration.

In *Stack v. Pepper,* the same court a year later stated that it was generally known that in all the early surveys of entries in the state, and in most of the later ones,

[224]332 S.W.2d 846 (Ky., 1960), see 80 ALR2d 1206.
[225]55 Pa 142 (1867).
[226]117 N.C. 443, 23 S.E. 362 (1895).

surface measurement had been adopted.[227] The court further stated that it would be presumed that lands embraced in grants and deeds were originally surveyed by surface measurement, but that such presumption would prevail only where it appeared feasible and reasonable to pursue such a course. The court approved a survey that had omitted from the length of a boundary line the distance up the face of a sheer precipice over which the lines ran, as it appeared the surveyor who had made the original survey had also omitted this measurement.

The gist of the North Carolina decisions seems to be that while there may be a presumption that surface measure was used in the early surveys of mountain lands, nevertheless, when the rights of the parties have not become fixed by a survey actually made, or by the passage of time or otherwise, and there is a current controversy necessitating an adjudication of the rights of the parties and requiring a survey of a tract of mountain land, the proper mode of conducting such survey is by using horizontal measurement, which is the method approved by the authorities as correct.

Horizontal Measurement

In *McEwen v. Den*,[228] involving a patent for 5,000 acres of mountain land in Tennessee, the grant provided that it should "begin on the south bank of Coal creek, four poles below Rowling's mill; thence running south with the foot of Walden's ridge, 894 poles, to a stake at letter H, in Henderson & Co.'s Clinch river survey; then west, crossing Walden's ridge, 894 poles to a stake; then a direct line to the beginning." It appeared that only the first line called for had been surveyed and marked and that it fell short of 894 poles, being only about 800 poles long, and the surveyor who had made the survey testified that the first line was the only line he had run and that the other three lines had only been platted, that the proper method of making surveys was by horizontal measurement, but that he had not been in the habit of making them in that way and in surveying the line in question had used surface measurement, and that the custom of the country was to adopt surface measurement, and that he had made the survey in accordance with such custom. The court said that the grantee was bound to abide by the first line as surveyed and marked, but that the other lines would be governed by a legal rule, which local custom could not change. If such custom should be recognized as law governing surveys, it would prevail in private surveys in cases of sales of land, when the purchaser who bought a certain number of acres might, by surface measurement across a mountain, lose a large portion of the land he had paid for; and that such would be the case as to the present grantee if he should be restricted to surface measure, whereas, by the terms of the patent, the government had granted to the extent of lines approximating horizontal measurement. The court said

[227] 119 N.C. 434, 25 S.E. 961 (1896).

[228] 24 How.(U.S.) 242, 16 L. Ed. 672 (Tenn., 1861).

that in measuring the second line, it should be run from the terminus of the first line at right angles thereto a distance of 894 poles, that the method of measuring would be to level the chain, as was usual with the chain carriers when measuring up and down mountainsides, or over other steep acclivities or depressions, so as to approximate, to a reasonable extent, horizontal measurement, this being the general practice in surveying wild lands in Tennessee, and the reasonable certainty of distance and approximation to a horizontal line being a matter of feet for the jury to determine; that the third line should then be run from the terminus, as thus ascertained, of the second line and parallel to the first line a distance of 894 poles, the chain being leveled as above indicated; and that the fourth line should then be run from the terminus of the third to the point of beginning near Bowling's mill.

In the North Carolina case of *Gilmer v. Young*, the boundaries of two large tracts of land were not in dispute.[229] The question presented was whether, in surveying these tracts to determine their acreage, surface measure or horizontal measure should be used, since payment was to be made by the acre, and the contract for purchase provided that an accurate survey be made to determine the number of acres in each. The court stated that all the standard authorities were against the method of surface measurement, that, this being admitted, there seemed to be but little for it to consider, and that when the state granted its western domain in large bodies there was no evidence whatsoever that it adopted surface measurement, and there was no ground for presuming that it did so in spite of the fact that the authorities agreed that the horizontal method of measurement was the correct one.

Also, the Tennessee court in the case of *Bleidorn v. Mountain Coal and Min. Co.*, which involved several conflicting and overlapping entries and grants of mountain land, held that horizontal rather than surface measure was proper in conducting a survey and that the law construed a call for distance as a call for a point ascertained by horizontal survey.[230]

Even though, it appears that the question of whether surface or horizontal measure should be used in making a survey was not an issue in *Bryan v. Beckley*, which was an action involving the location of lost corners and liens of a survey made in 1774, the court recognized that allowance should be made for the unevenness of the ground over which the lines ran and said that reasonable accuracy could never be attained in completing the business without good instruments, careful chain carriers, and such allowance, or other safeguard, for unevenness of surface as would be equivalent to horizontal admeasurement, upon which the art and rules of surveying were founded.[231]

In the previous case, *Beckley v. Bryan* (an earlier appeal of *Bryan v. Beckley*), the court also recognized that in running the lines of the survey, a proper allowance should be made for the unevenness of the ground over which the lines ran.[232]

[229] 122 N.C. 806, 29 S.E. 830 (1889).

[230] 89 Tenn. 166, 15 S.W. 737 (1890).

[231] 16 Ky (Litt. Sel. Cas.) 91, 12 Am. Dec. 276 (1809).

[232] 2 Ky. (Sneed) 91 (1801)

In the Connecticut case of *Lew v. Bray*,[233] the court did not determine whether horizontal or surface measurement was the correct method of determining the extent of a boundary line, but it did affirm the action of the lower court in finding, on the evidence, that horizontal measure had been used in measuring the western boundary of the plaintiff's land. The case had to do with the plaintiff's purchase from another party of a parcel bounded on the south and east by streets and on the west by a line running north 168 feet from the southern boundary. The part of the property over which this western boundary ran was rough and uneven, so that the surface at a point about halfway between the end points of the line was about 12 feet below the level of the ground at the ends; and before the deed had been delivered, the plaintiffs and the grantor's agent had measured the western boundary and had placed a stake in the ground at its northern terminus. According to the plaintiff, the boundary had been measured in a horizontal line, and they and their grantors erected a fence from this stake along the northern boundary of the property to the street on the east. The defendant then purchased from the plaintiffs' grantor the property lying north of their property. The defendant, contending that plaintiff's western boundary had been measured along the surface of the ground, removed the fence and erected a stone retaining wall, which was found to encroach upon plaintiff's property. The area of encroachment was in the shape of a triangle being 1 foot $9\frac{1}{2}$ inches at its base on plaintiffs' western boundary and extending 110 feet to a point at the street on the east.

Surface Measure

A few cases have either held or at least recognized that surface measurement was used and was appropriate where it is the custom of the locality, or if the circumstances of the case dictate.

In the case of *Duncan v. Hall*,[234] the court recognized that in making surveys in undulating or mountainous sections, the early surveyors, at least, universally adopted surface measurement. In running long lines from the top of one high and precipitous mountain to that of another, the area or acreage sold by the state to its citizens would have appeared much less than it actually was if level measurement had been adopted in laying off large grants.

Variation or Combination of Methods

Even though generally approving surface measurement as the correct method of making a survey, the courts in a few instances have recognized that a departure may

---

[233] 81 Conn 213, 70 A. 628 (1908).
[234] 117 N.C. 443, 23 S.E. 362 (1895).

be made from strict surface measurement under some circumstances, as where the line crosses a deep depression or a sheer precipice. The Pennsylvania case of *Boynton v. Urian* and the North Carolina case of *Stack v. Pepper*, both discussed previously, took this into account. In the latter case, the court stated that it was a fact generally known and acknowledged that in all of the early surveys of entries made in the state, and in most of the later ones, the surface and not the level or horizontal method of measurement was shown to have been adopted. This was the general rule, and the courts took notice of the fact and presumed that lands embraced in grants and deeds were originally measured in this way, both because it was a matter of general knowledge that such had been the custom and because the judicial annals of the state were corroborative of that fact. But the court said that while the presumption was generally that a survey of the surface was contemplated and adopted by the parties to a deed, that presumption prevailed only where it appeared feasible and reasonable to pursue that course, and it was held that in surveying a tract of land, the boundary of which ran up the face of a perpendicular cliff, a surveyor had adopted a proper method of surveying in running the line to the base of the cliff and then continuing the survey from the top of the cliff, after walking around to it, without including in the length of the boundary line the distance up the face of the cliff.

Following the footsteps or reproducing what the original surveyor did will dictate the course of action to be followed. As the court stated in *Hart v. Greis*,[235] in a suit involving a boundary question, search must be made for the footsteps of the original surveyor. When found, the case is solved.

[235] 155 S.W.2d 997 (Tex. 1941).

REFERENCES

54 ALR 781   *Distance as determined by straight line or other method.*

80 ALR2d 1208   *Boundaries: measurement in horizontal line or along surface or contour*

CHAPTER 15

DEALING WITH MATHEMATICS

Torture numbers and they'll confess to anything.

—*Gregg Easterbrook*
author/lecturer

Perfect numbers like perfect men are very rare.

—*René Descartes*
(1596–1650)

ERRORS

In reviewing original surveys or surveys subsequent to the original, it is sometimes necessary to analyze the mathematics of the work. Some surveys were adjusted to achieve perfect closure while others were not. One test that people sometimes make is to find whether the description in a land record actually closes mathematically. Whether it does or not has no reflection on the error, or types of errors, that are inherent in the description. /Any figure can be made to close mathematically by adjusting directions and distances to get a perfect fit. Any method can be employed in order to do this, even though there are several standard methods. The only way to gain insight into what previous surveyors did mathematically is to compare their surveys with a current one. And even then, there is no guarantee that both surveyors did not make the same errors in the same place or that either, or both, will not have compensating errors.

Brinker and Wolf nicely summarized what needs to be kept in mind about errors:

- No measurement is exact
- Every measurement contains errors
- The true value of a measurement is never known
- The exact error present is always unknown.[236]

[236] Brinker and Wolf, *Elementary Surveying*, 7th ed. (1984).

204

Even so, the goal is locate the corner being sought. If it is marked, the goal is to find the marker. If it is not, the goal is locate the point where there would be a marker if one existed.

Types of Errors that Can Be Expected

A number of inherent errors in previous surveys will affect going from one corner to another. These include errors in distance, errors due to poor closure, alignment errors, error due to correction or failure to correct for slope, errors due to temperature and sag of chain, especially in longer lines, and carelessness in measuring, among others. Since usually we do not know what previous surveyor did, especially if we are dealing with an older survey, we must consider all or any of these factors.

Definitions

Accuracy. Denotes absolute nearness of measured quantities to their true values.
Precision. The degree of refinement or consistency of a group of measurements. The measurement of how close observations or results come to the average value(s).

Understanding of the difference between these two terms by shooting at a target or by doing a survey.

a. Shooting at a Target

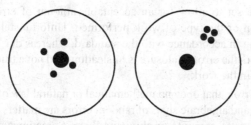

Figures 15.1 and 15.2   Comparing accuracy with precision. The shot pattern on the left is accurate but not precise. The shot pattern on the right is precise but not accurate.

b. Doing a Survey

Suppose measurements were made of a parcel of land so carefully that the result was a very small amount of error, or a very small error of closure. However, as good as the work was, the surveyor was on the wrong parcel of land. This survey is precise but not accurate.

To increase accuracy, greater precision must be used in the instruments, methods, and observations. Precision therefore can be defined as the degree of perfection used in the instruments, methods, and observations. Accuracy is the degree of perfection obtained. Since blunders and systematic errors can and must be eliminated from a survey, the degree of accuracy of a survey depends on the size of the accidental errors.

Error. Inaccuracy in any one measurement due to the type of equipment used or in the way the equipment was used. The difference between the measurement and the true value is known as the *resultant error.*

Resultant Error in a Measurement. Consists of the individual errors from a variety of sources, some causing the measurement to be too large, others making it too small. Some are positive while some are negative, giving rise to both cumulative and compensating errors, none of which is identifiable but all of which can be fixed within probable limits.

Error cannot be entirely omitted, but conscientious and careful surveyors have and will attempt to minimize error.

Discrepancy. The difference between two measurements of a given quantity, or the difference between the measured value and the known value. A small discrepancy between two measured values indicates that likely there are no mistakes, and accidental errors are small. It does not indicate the magnitude of systematic errors.

A discrepancy may indicate the precision with which the measurements were made. For instance, a discrepancy may show the value the previous surveyor put on direction(s), distance(s), or area(s), compared to your results, so long as you are both measuring between the same points.

Error of Closure. Also known as *closing error.* The misclosure of a survey or closed figure in a mathematical sense. The error of closure divided by the total distance measured is equal to the relative precision of the work.

Standards are set to allow for an acceptable amount of error, based on the relative precision of the type of work performed. Unfortunately, some work is accepted as being in accordance with the standard, whereas due to compensating (accidental) errors the error of closure is misleading and not a true indicator of the amount of error in the work.

Random Errors. Errors that obey no mathematical or natural law other than chance. The magnitudes and algebraic signs of random errors are matters of chance. There is no absolute way to compute or eliminate them. Sometimes called *accidental error* or *uncertainty.* Also known as *compensating errors.*

Index Errors. Arise from imperfections or faulty adjustment of the instrumentation used in making measurements. For example, a tape may be too long or, more likely, too short, affecting each measurement taken. Places where a tape is kinked, repaired, or bent may introduce significant error in longer lines. Any instrument out of adjustment from being in perfect working order will introduce error. Also known as *instrumental errors.*

Personal Errors. Occur through the observer's inability to manipulate or read the instruments exactly. An observer's eyesight may affect readings each time they are taken. At an unusual setup, a person's stance, such that an observation is at an angle, can introduce error.

Natural Errors. Occur from variations in the phenomena of nature, such as temperature, humidity, wind, gravity, refraction and magnetic declination. Changes in temperature can significantly affect steel tapes and in some cases electronic equipment. Changes in magnetic declination and other magnetic disturbances affect compass readings, upon which the majority of early surveys and titles in many regions are based.

Systematic Errors. Occur in the same direction, thereby tending to accumulate so that the total error increases proportionally as the number of measurements increases. The magnitude of the angle does not affect the size of the error, but the length of distance may, such that a line had to be measured in segments, thereby repeating the error each time the measuring device was used and causing errors to be cumulative.

Systematic errors are mechanical errors and result from imperfections in the equipment or the manner in which it is used. Under the same conditions, systematic errors always have the same size and algebraic sign. Although they often introduce serious errors in results, since they generally have no tendency to cancel, they are often difficult to detect.

Accidental Errors. Occur randomly in either direction, thereby tending to cancel one another so that, although the total error does increase as the number of measurements increases, the total error becomes proportionally less and accuracy becomes greater as the number of measurements increases.

Accidental errors are the small, unavoidable errors in observation that an observer cannot detect with the equipment and methods being used. Greater skill, more precise equipment, and better methods will reduce the size and overall effects of accidental errors. However, consistent use of the same equipment in the same way will not result in any improvement. A surveyor using the same equipment and techniques throughout his or her practice is likely to achieve the same type of results, containing the same or similar errors.

Computational Errors. Occur in the computation process. Such errors include gross mistakes (blunders), transposition of figures, simple math mistakes, rounding errors, and so on. A small error in one part of the computational process may result in a large error affecting the end result.

Mistakes. Inaccuracies in any one measurement because some part of the operation is performed improperly. Mistakes are caused by a misunderstanding of the problem, carelessness, or poor judgment. Large mistakes are often known as *blunders*. Blunders are the result of human error and cause a wrong value to be recorded. When a blunder is found, there may be an opportunity for correction. When a legitimate error is found, it must be dealt with in proper fashion. Transposition of numbers is a very common blunder, as is making a mistake of a whole foot, 10 feet, or a whole chain length. Such mistakes are usually the result of miscounting. Blunders can usually be discovered; however, occasionally two blunders will cancel each other out and therefore go unnoticed.

Mistakes, or blunders, can be corrected only if they are discovered. Today's surveyors find themselves with an unexpected problem when they discover mistakes made by one or more previous surveyors. Over time, some of these earlier problems have escalated into serious boundary and title concerns, especially when other unwary surveyors have relied on the mistaken work, believing it to be all right.

Blunders that may not be discovered are such things as staking out the wrong lot in a block or even on the wrong street, misreading a number on a plan, or counting the wrong number of tape lengths in very nearly parallel sides of a traverse.

Certain kinds of mistakes are not blunders. These include mistakes due to lack of judgment or lack of knowledge.

Total Error. The sum of the inaccuracies in a finished operation. Since inaccuracies are either positive or negative, the total error is the algebraic sum of all the errors.

Reliabililty. Measurements taken under the same conditions are equally reliable. A survey where measurements are made at different times warrants a different consideration, since the reliability of the various measurements may not be directly comparable and therefore may be unable to be treated the same. For example, taped distances from two different days where the temperature difference is large may result in two different sets of errors, each of which needs to be considered separately. In the past, some surveyors did not bother to correct for temperature when taping, especially where a relatively high degree of precision was not required. This was especially true for woodlot surveys.

Lower-Precision Surveys. It is human nature to believe, and survey standards reflect this opinion, that high-value property requires more care in making measurements than low-value property. However, this opinion may become a significant problem at some time in the future when the use of property changes and low-value property suddenly becomes high-value property. Retracing lines of a woodlot is difficult in itself, but when it is soon destined to be a multimillion-dollar shopping center, it becomes increasing difficult, and locating correct points becomes exceedingly critical.

In the past, the tendency was to survey woodlots, wetlands, and other low-value land with low-precision equipment and to take only what care was necessary, sacrificing precision to save time and produce a survey at low cost. Retracing such areas today is difficult, particularly in comparison to urban and suburban parcels, which generally were surveyed to a higher precision. Woodlots include sighting problems resulting in short traverse lines, alignment problems because of trees and brush, and a host of other inherent difficulties. They were often surveyed with a compass, with angles measured to one degree or perhaps to 15 minutes of arc, and each observation was subject to local attraction and other influences. In contrast, urban surveys were measured carefully using equipment of higher precision resulting in better closures; survey costs were justifiably higher. The resulting errors from the various types of survey and measurements must be considered in their appropriate context.

Different surveyors operate under different standards, if standards are considered at all. One surveyor might select an acceptable closure of 1 part in 2,500, while another would insist on nothing greater than 1 part in 5,000. The purpose to which the property may have been considered may have changed over the years; subsequent surveys may have demanded a greater precision or smaller error of closure. Investigators must be careful not to compare apples with oranges when comparing two different surveys. The fact that a survey closes mathematically within the desired or required limits is no guarantee against compensating errors.

Measure of Precision. There is no measure for accuracy. Either the property is located or it is not. There is, however, a measure for the precision of a survey figure. Mathematically, a figure that is intended to be a closed figure should end at the same point begun at. Although physically, in the field, this happens, mathematically there is error, so that the beginning point and the ending point, while very close to one another, are not exactly the same. This is because of the error, which can be computed. Standards dictate how much error is acceptable and eventually adjusted out by some method. When the *closing error*, or *error of closure*, is more than acceptable limits, the work must be done over or, if the error is mostly in either one angle or in one line, sometimes it can be isolated and that part corrected to bring the entire work within acceptable limits. Compensating errors, however, cannot be detected this way since there is no gross error of closure. Consider the next example.

Case #30

GOOD CLOSURE WITH 20 FEET OF ERROR

In this case, a 100-acre tract that was nearly exactly rectangular was measured. The error of closure was small, and the error was distributed by an acceptable method. When the parcel was subdivided, gross errors appeared in the mathematics. Upon remeasurement of the perimeter, it was found that two 10-foot errors were made, one each in parallel sides. The errors compensated for each other, resulting in an almost perfect closure. Here we have mathematics, with error, leading us to believe the measurements are acceptable. Again, this survey is precise but not accurate.

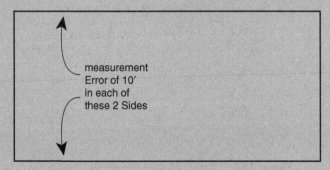

Figure 15.3   Survey of rectangular parcel with parallel sides.

Isolating Traverse Error. In reviewing a past survey, it may be necessary to work backward from the final result to determine raw closure, then attempt to isolate any gross error by one or more of the following methods. Such analysis may lead to a result whereby an original point may be found even though originally it was located with gross error. Or, alternatively, it may demonstrate where a survey subsequent to the original failed to locate one or more original points due to its inherent errors. No survey, original or subsequent, can be mathematically perfect, regardless of whether the plan or the description closes. All a perfect closure demonstrates is that someone has manipulated the numbers by some method to make a perfect figure. Whether they are reliable remains to be seen. If the numbers, whatever their origin, bring the retracement person to the original points, they are sufficient, regardless of how much error they contain. If they fail to bring the surveyor to the original points, further investigation is in order. This type of investigation may come in the form of mathematical analysis.

LOCATING MOST THE LIKELY ANGLE ERROR

Whenever most of the error is in one angle, it can often be found by one of two methods, graphical or computational.

Bisect the Error of Closure Graphically

When the figure has been plotted, the error of closure can be drawn by connecting the ending point with the beginning point. That line is then bisected, and the perpendicular bisector will point to the suspect angle. This can also be done by computation, by

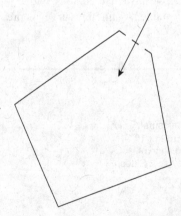

Figure 15.4   Graphical location of angular error.

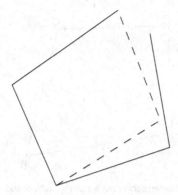

Figure 15.5   Mathematical location of angular error.

halving the error (the inverse between the ending and beginning points) and projecting it into the figure until it intersects an angle or one of the lines. The coordinates for the point of intersection can be compared with the coordinates of the nearest angle point.

Plot in Reverse Direction

Plotting in reverse direction also can be done graphically or computationally. Taking the coordinates of the ending point, then compute (or plot) the figure in reverse direction from the previous computation, reversing the directions and using the same distances. If most of the error is contained in one angle, one pair of coordinates (or one angle point) will match in that the same position, or very close to it, will be found computing from the two different directions. (See Fig. 15.5 above.)

Locating Most Likely Distance Error

If the error is in a distance, and most of it in one length, the line parallel to the direction of the closure error is the most suspect.

TRAVERSE ADJUSTMENTS

The following example illustrates the results expected with nonadjustment of error in comparison with various types of adjustment. The rules or reasons for each are presented, although not everyone followed the rules or knew about them.

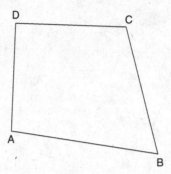

Figure 15.6 Sample traverse for computation and adjustment.

TRAVERSE DATA

STA	Distance	Int. Angle
A		90°-36'-20"
	280.20'	
B		70°-07'-30"
	294.54'	
C		113°-02'-00"
	183.60'	
D		86°-15'-30"
	286.92'	

Table 15.1   Basic traverse data.

Table 15.1 shows a sample traverse for computation. This traverse contained 80 seconds of angular error with an allowable error of 2 minutes. Assuming consistent error throughout, the error was distributed evenly to all four angles. The linear closure was then computed, and an error was found that caused the traverse to not meet the required precision. The error was isolated and remeasured, and the traverse was adjusted by several methods. The results are shown in Tables 15.2 and 15.3.

Traverse Adjustment: A Process of Manipulating Numbers

Figures must be closed mathematically in order to calculate area and, in some cases, to use for additional survey work and computation. Raw data are almost never available to the retracement surveyor, who generally must rely on land descriptions or survey plans. The more angular error is placed in isolated or selected angles, the more the traverse is distorted.

COMPARISON OF COORDINATES SAMPLE TRAVERSE

	No Adjustment		Compass Rule		Transit Rule		Crandall Method		Least Squares	
	North	East	North	East	North	East	North	East	North	East
A	1000.00	1000.00	1000.00	1000.00	1000.00	1000.00	1000.00	1000.00	1000.00	1000.00
B	1000.00	1280.20	999.96	1280.19	1000.00	1280.19	1000.00	1280.19	1000,01	1280.19
C	1276.99	1180.04	1276,91	1180.03	1276.92	1180.03	1276.92	1180.03	1276.93	1180.00
D	1287.05	997.02	1286.94	997.01	1286.98	997.00	1286.99	996.99	1286.96	995.98

Table 15.2   Coordinate values from different adjustment routines.

COMPARISON OF INVERSE SAMPLE TRAVERSE

	SIDE			
Method	AB	BC	CD	DA
No Adjustment				
Length	280.20	294.54	183.30	287.07
Direction	DUE EAST	N 19-52-50 W	N 86-51-10 W	S 00-36-00 E
Compass Rule				
Length	280.19	294.50	183.30	286.96
Direction	S 89-59-31 E	N 19-53-03 W	N 86-51-39 W	S 00-35-56 E
Transit Rule				
Length	280.19	294.48	183.31	286.99
Direction	DUE EAST	N 19-53-09 W	N 86-51-13 W	S 00-35-59 E
Crandall Method				
Length	280.17	294.47	183.32	287.00
Direction	DUE EAST	N 19-52-50 W	N 86-51-10 W	S 00-36-00 E
Least Squares				
Length	280.19	294.49	183.30	286.98
Direction	N 89-59-54 E	N 19-53-22 W	N 86-51-46 W	S 00-36-14 E

Table 15.3   Final adjusted distances by various methods.

Almost anything can be adjusted. There can be gross error in one angle or one line, and it can be distributed throughout the figure, eroding the accuracy of all the values. This, in effect, is what many of the surveyors did in the past if they were at all conscious of their error. They determined the error graphically by plotting the traverse on paper, then forced it to close. Now we do it mathematically.

The method of least squares is capable of weighting values to isolate error to place the most error in problem setups where angles and distances are likely to be. Least squares is founded on probability. The practitioner exercises judgment in where and how much error is placed, and the result is the *most probable* solution, given the data.

Type of Adjustment Used

Most early surveys were not adjusted (or balanced) for error. Therefore, computation of the descriptive information, if complete and if it is a closed figure, will result in some error. From the previous discussion on error isolation, it is clear that closure may be of some value in isolating error to further refine the search area for corner markers.

The very early texts did not contain any discussion of error, or how to deal with error. Closures were graphic when plotting was done, and lines were adjusted as necessary to form a figure that began and ended at the same point.

If the survey was computed and adjusted, there were several techniques available to distribute the error, if it was done. Graduates from engineering schools and those relying on twentieth-century texts may have adjusted their traverse work to distribute acceptable error, resulting in a closed figure. If closure error was small, surveyors may not have bothered to make the final computations.

Many surveying courses taught the use of the compass rule for the adjustment of closed traverses. Some courses also included the transit rule and made a distinction as to which one to use, depending on how the measurements were made. A few schools discussed the Crandall method, and most modern courses include, or require, adjustments to be made by the method of least squares. Prior to the advent of computers, the least squares method was very cumbersome and mostly unused. Today, with sophisticated software in almost all surveying computational programs, and a high-powered computer in every survey office, least squares adjustment is as easy as any of the others.

Three basic conditions can exist:

1. The angular precision is higher than the linear precision.

    This might be expected with a traverse executed in rugged terrain, in which angles are measured directly to seconds and distances are determined by holding the tape horizontally throughout the measurements. The angular precision is expected to be high, while the linear precision expected to be low because it was necessary to break tape.

2. The angular precision is the same as the linear precision.

    This would be typical of a traverse executed in fairly level terrain, in which angles are measured with either a 30-second or a 1-minute transit and the tape is handled by experienced personnel.

3. The angular precision is lower than the linear precision.

    This would likely result from a traverse run over level terrain, in which directions were obtained by compass bearings and the tape is handled by experienced personnel.

Compass Rule   The compass rule assumes that the corrections to be applied to latitudes and departures vary as the length of the courses varies; that is, the longer the line, the greater the error. It also assumes that all the angles are measured with equal precision and the amount of error is independent of the size of the angle.

However, since the compass rule is one of the first rules learned in school and is relatively easy to use, many surveyors used this rule exclusively when traverse balancing. For the most part this use was acceptable, since the precision of the angular and linear measurements were about the same.

Transit Rule   The transit rule was originally intended for surveys where angles were measured with greater precision than the distances, such as in stadia surveys. It is only a rule of thumb, and its use is not widely advised. It meets the assumptions on which it is based only to the extent that each side is parallel to one or the other of coordinate axes and different results are obtained for different orientations of the meridian.

In the past, some software programs were designed using the transit rule. When the computer was instructed to balance the traverse, this is the rule applied, regardless of conditions. More recent programs allow two or more options.

Crandall Method   The Crandall method is valid when errors in linear measurements are considerably greater than those in angular measurements, such as a stadia traverse. The assumption is made that after any small angular error of closure is distributed among the bearings of the traverse, the bearings are without appreciable error, and further adjustments should properly be made to linear measurements only. The method of least squares is applied to the traverse lengths in such a way as to not disturb the bearings.

Least Squares   The least squares method, based on theory of probability, simultaneously adjusts the angular and linear observations to make the sum of the squares of the residuals a minimum. The method is valid for any type of traverse regardless of the relative precision of the angle and distance measurements. It is undoubtedly the best possible method of traverse adjustment but was used very little in the past because of the lengthy computations necessary.

Theory of Probability   To form a judgment of the probable value or the probable precision of a quantity, from which systematic errors have been eliminated, it is necessary to rely on the theory of probability, which deals with accidental errors of a series of either like or related measurements. The theory assumes that: (1) small errors are more numerous than large errors, (2) no very large errors occur, (3) errors are as likely to be positive as negative, and (4) the true value of a quantity is the mean of an infinite number of like observations. In practice, it is impossible either to eliminate systematic errors entirely or to take an infinite number of observations; hence the value of a quantity is never known exactly. But in many cases, investigators may assume that that systematic errors are so far eliminated as to be a negligible factor.

The *most probable value of a quantity* is a mathematical term used to designate that adjusted value which, according to the principles of least squares, has more chances of being correct than has any other. In conclusion, it is a matter of chance, and review of the normal bell curve will show that some values in fact fall outside those that are the most probable. In reality, we are looking for facts, not probabilities, but probability may aid us in determining the facts.

Knowing the type of equipment a previous surveyor used or was likely to have used usually lends insight into what errors might be expected and their magnitude. Each instrument has its idiosyncracies, precision capability, and inherent errors. Knowing this can sometimes aid in determining where particulars errors may have occurred and how to deal with them.

Knowing the type of formal education surveyors had may suggest how they dealt with mathematics. For example, twentieth-century surveyors from engineering curricula would be likely to adjust their traverses by using the Transit Rule. Those who had taken land surveying courses or advanced courses might assess the type of traverse they had and choose a different rule, say the Compass Rule or Crandall Rule. Recent surveyors would be likely to use some form of a least squares program. In using the method of least squares, note that two surveyors might not weight the measurements in a traverse the same. Also consider the type of computational aid a surveyor used, as built-in software for traverse adjustment probably came with the machine, and programs are not consistent in their adjustment routines. Some use one method, others use a different method. Some contain options whereby users can select the type of routine to apply and assign weight to selected measurements.

Assigning weight is a judgment call based on a person's assessment regarding the reliability of a particular measurement. Good, solid setups where everything works well may be weighted heavily, whereas at an unstable location measurements might be considered less reliable than the remainder and therefore given less weight.

APPLYING A RULE OF CONSTRUCTION: REVERSING COURSE

> In solving a problem of this sort, the grand thing is to be able to reason backwards. That is a very useful accomplishment, and a very easy one, but people do not practice it much. In the everyday affairs of life it is more useful to reason forward, and so the other comes to be neglected. There are fifty who can reason synthetically for one who can reason analytically.
>
> —*Sherlock Holmes*
> A Study in Scarlet

Principle of Law: Reversing Course

> The weight of authority is to the effect that the beginning corner of a survey does not control more than any other corner actually well ascertained, and where a disputed or lost line or corner can thereby be established more nearly in conformity with the terms of the instrument and with the intent of the parties as gathered therefrom, it is competent to ascertain such line or corner by first ascertaining

the position of some other bound and tracing the line back from that by reversing the course and distance. Nevertheless, it has been held that the test of reversing in the progress of a survey should only be resorted to when the terminus of a call cannot be ascertained by running forward. Likewise, it has been held that a line will not be reversed for the purpose of showing the termination of a prior line unless the description of the posterior line is more definite than that of the prior line, and a mistake in the prior line can clearly be shown; and that where the beginning corner is known the calls ought not to be reversed except in order to make the survey close. Accordingly, the test cannot be applied where the lines of a survey are not actually run and measured on the ground, and natural monuments and boundaries in a deed must be followed rather than ignored and the courses and distances called for in the description reversed. In no event is it lawful to disregard natural objects called for in the boundary, either as corners or lines, in running the lines backward and in reverse order, to locate a boundary. Due respect must be shown to such objects whichever way the boundaries run, and as much one way as another.

—11 *C.J.S. Boundaries*, § 9a(3)

Case #31

Cornett v. Kentucky River Coal Co.

195 S.W. 149 (Ky. 1917)

REVERSING CALLS IS AS LAWFUL AND PERSUASIVE AS FOLLOWING THEIR ORDER. In this case, the last six calls of the description did not call for any established object. If run by the courses and distances, the patent would not close. There were 13 lines called for in the patent, the first 7 of which were located by marked trees and about the location of which there was no dispute. However, according to the appellant, in running the call between the sixth and seventh corners in the patent, it was found necessary to extend the line approximately 80 poles beyond the distance called for in order to reach the timber that it was agreed marked the seventh corner, and this error in the recorded distance of this line was responsible for the failure of the patent to close. The appellant also claimed that the effect of this error is compensated for and adjusted by extending the distances called for on the courses given in the last two lines of the boundary until they will intersect. This theory was approved by a surveyor of some 12 years' experience, who first surveyed the land for appellee.

A different theory was contended for by the appellee for closing the patent, which was adopted by the lower court and supported by a different surveyor who made the last survey of the land in question for the appellee. His theory was to reverse the last call from the beginning corner and run the course and distance called for in the patent, and the result he accepts as the proper location of the thirteenth corner. He then began at the eighth corner, which is the last marked corner, and ran the courses and distances of the eighth, ninth, tenth, and eleventh lines, the result of which he accepted as establishing the twelfth corner. He then connected the twelfth and thirteenth corners with a straight line, which varies from the course called for of that line by $7^1/_2$ degrees, but which approximates the distances called for on that line, being lengthened almost exactly the same distance as the error in the sixth line. In other words, his theory sacrifices the course called for in one line and attributes to that particular line whatever error that may have been responsible for the failure of the patent to close.

Appellants' theory makes the patent, which called for 500 acres, include 918 acres, and sacrifices the distances called for in the twelfth and thirteenth lines but retains their courses and extends them until they intersect, a distance out of all proportion to the error in the sixth line. He extends the twelfth line from the called-for 806 poles to 1,068 poles and the thirteenth line from the called-for 90 poles to 320 poles.

The appeals court stated:

> It has long been the rule in attempting to locate lost lines to give preference to the courses rather than to the distances, and to close the survey, if possible, by lengthening or shortening the distances rather than changing the courses, but in so doing the error in distances must be reasonable and should be apportioned to all of the lost lines rather than arbitrarily placing it in only some of them. This rule of sacrificing distances to courses is only a general rule, and is subject to many exceptions where, from the evidence of a particular case, some other more satisfactory method of adjusting the error is disclosed. This rule was fist announced in *Beckley v. Bryan*, Ky. Dec. 93, and has been followed and recognized in many cases, but its limitations have been called to attention and it has been departed from about as frequently as it has been followed. A frequent reason given for departing from the general rule is where, by following it, a figure is produced that does not correspond with the original survey and plat upon which the patent is issued, and which always may be looked to in determining the proper mode to be adopted in closing the survey. *Steele v. Taylor*, 3 A.K. Marsh. 226, 13 Am.Dec. 151; *Bruce v. Taylor*, 2 J.J. Marsh. 160; *Harris v. Lavin*, 6 Ky. Law Rep. 304; *Hagins v. Whitaker*, 42 S.W. 751, 19 Ky. Law Rep. 1050; *Gogg v. Lowe*, 80 S.W. 219, 25 Ky. Law Rep. 2176.

> Neither party in this case offered to introduce in evidence the original patent or survey and plat, although they might have been exceedingly helpful in a correct solution of the problem. Both surveyors testified that their respective theories for closing the patent were in accord with an unauthenticated copy of the original survey and plat with which they compared their surveys, but neither was able to say whether or not the copy was correct or not. This evidence was therefore of no value.

> Another reason for departing from the above general rule is where a known error in established lines may be offset by a corresponding change in the lost line opposite, when the course or distance called for in the lost line opposite

the known error, or both, may be changed to close the survey. This exception to the general rule was announced and applied in the case of *Preston Heirs v. Bowmar*, 2 Bibb. 493, decided soon after the *Beckley v. Bryan* opinion was delivered, and has frequently been used since in closing surveys. In that case the survey was closed by changing the course rather than the distance called for, and we think the facts here are sufficiently analogous to call for the application of the same method for closing this survey.

The court continued:

The proven error in the patent is in the sixth line, which forms a part of the irregular northern boundary line of the survey, and the twelfth line is the lost line opposite forming the southern boundary of the patent. To change this twelfth line by the same distance as the error in the sixth line and by $7^{1}/_{2}$ degrees closes the survey so as to include approximately the number of acres called for in the patent, and while the number of acres is not controlling evidence, it has evidential value, as has the fact that the known error is in the opposite line, both of which support the theory of the appellee. Appellant's theory is supported by no evidence whatever, but depends entirely upon the application of a rule of construction manifestly not applicable here because it requires an extension of the opposite line out of all proportion to the known error in the sixth line, and extends the thirteenth line, only 90 poles in length as called for, to nearly four times its prescribed length without any known error in its opposite line. Appellants argue there was a proven error in the length of the second line of 201 poles which, added to the error in the sixth line of 80 poles, makes the total error in the length of the northern boundary lines approximately the same as the extension they propose in the twelfth line, but, there is no competent, if any, evidence of any such error in the second line. The original patent is not in evidence, and this line, as described in the contract between the parties, is 225 poles, and as run is 223 poles, in length. So there is no error proven in its length, and we do not know upon what authority the statement of error in this line is predicated.

There is no hard and fast rule for closing a survey, but such rules as are employed are but rules of construction in aid of an effort to relocate lost lines as they were located in the original survey, which is always the problem for solution, and rules of construction must give way to competent evidence disproving their applicability to a given case.

Two rules of construction often recognized and applicable here are that reversing calls is as lawful and persuasive as following their order (*Pearson v. Baker*, 4 Dana, 323), and that when a party is claiming under a survey where the course or distance must yield without data to determine whether the mistake was in the one or the other, the mode of closing the survey must be adopted which operates most unfavorably to the claiming under it (*Preston Heirs v. Bowmar*, supra, and *Pearson v. Baker*, supra).

Case #32

WORKING BACKWARD
Coburn v. Coxeter
51 N.H. 158 (1871)

Where the termination of one line is indicated only by the course named in the deed, but the termination of the next line is an ascertained monument at a given distance and course from the termination of the preceding line, the precise location of the latter may be determined by reversing the course of the succeeding line, and measuring back the distance called for.

In this case, there was a problem with the second line in the description. It was described as running north $8^3/_4$ degrees east, about 20 rods to the river road; but running that course the line would not strike the river road at all but would run nearly parallel with it, though slightly divergent. Had the deed called for a monument at the road, which could be identified, the line must have been governed by it upon the principles in numerous cases stating that monuments govern over courses and distances, notwithstanding the great departure from the course given in the deed; but no such monument is called for as the termination of that line. The question then is whether the place of intersection with the road can in any way be determined. In a case where a boundary of an extensive character, such as a river or swamp, is called for it has been held that the line must be run to the nearest point in such object.[237]

In this case, even though the deed called for no monument but the river road at the end of the second line, yet at the termination of the third line it does call for a stake and stones at a distance of one rod and a half from the end of the second line, and that stake and stones the evidence tends to identify as the one established by the parties for that corner.

By reversing the course and running back one rod and a half, the place of the termination of the second line is determined, and there is another stake and stones established by the parties as the north-westerly corner of the lands conveyed. That a line may be located in this way by reversing the course from an established monument, is shown by sundry adjudged cases. *Safret v. Hartman*, 7 Jones (Law) N.C. 199, 21 U.S. Dig. 169, sec. 148; *Dobson v. Fenlay*, 8 Jones (Law N.C. 495; *Wilson v. Inloes*, 6 Gill. 121.

If then, the monument at the termination of the third line is established, the terminus of the second line is readily fixed, that is, by measuring back one rod and a half from the end of the third line. In this way the deed is made to take effect, and to grant the land with appears to have been intended by the parties.

---

[237] *Campbell v. Branch*, 4 Jones Law (N.C.) 313 (1857).

Case #33

COMMON SENSE CAN RULE

Ayers v. Watson

11 S.Ct. 201, 137 U.S. 584, 34 L.Ed. 803 (Tex. 1891)

If an insurmountable difficulty is met with in running the lines of a survey of public land in one direction, and all the known calls of the survey are met by running them in the reverse direction, it is only a dictate of common sense to follow the latter course.

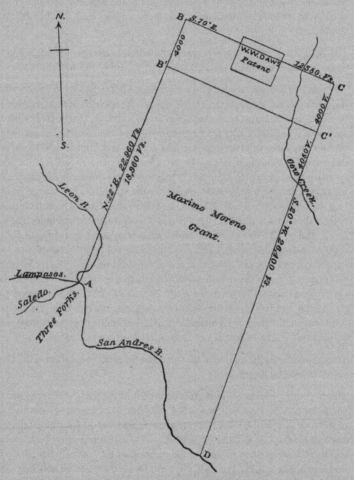

Figure 15.7 Diagram of property in question, *Ayers v. Watson.*

Plaintiff Watson claimed title to one-third of a league of land situated in Bell County, Texas, being a rectangular tract granted by patent of the State of Texas to the heirs of Walter W. Daws, in 1850. The location and boundaries of this tract were not disputed. The defendant, Ayers, claimed title under a grant of the government of Coahuila and Texas to one Maximo Moreno, dated 1833, for a tract containing eleven leagues of land and it was agreed that the defendant owned all right, title and interest created by this grant. This being the older of the two grants, verdict should have been for the defendant if it was shown that the Moreno grant covered the Daws tract owned by the plaintiff. Whether it did not did not is the issue in this case.

The Maximo Moreno grant lies on the north side of the river San Andres. The Daws tract is situated near the north end of the Maximo Moreno grant. The question is whether the north boundary line of the Maximo Moreno grant is situated so far to the north as to include the plaintiff's land, or whether it runs southwardly of it.

The field notes of the Moreno grant, embodied in the grant itself, are in the Spanish language, and, translated into English, are as follows:

"Situated on the left margin of the river San Andres, below the point where the creek called Lampasas enters said river on its opposite margin, and having the lines, limits, boundaries, and landmarks following, to wit: Beginning the survey at a pecan (nogal) fronting the mouth of the aforesaid creek, which pecan serves as a landmark for the first corner, and from which 14 varas to the north 59° west there is a hackberry 24 in. dia., and 15 varas to the south 34° west there is an elm 12 in. dia.; a line was run to the north 22° east 22,960 varas, and planted a stake in the prairie for the second corner; thence another line was run to the south 70° east, at 8000 varas crossed a branch of the creek called Cow Creek, at 10,600 varas crossed the principal branch of said creek, and at 12,580 varas two small hackberries serve as landmark for the third corner; thence another line was run to the south 20° west, and at 3520 varas crossed the said Cow Creek, and at 26,400 varas to a tree (palo) on the aforesaid margin of the river San Andres, which tree is called in English 'box elder,' from which 7 varas to the south 28° west there is a cotton wood with two trunks, and 16 varas to the south 11° east there is an elm 15 in. dia.; thence, following up the river by its meanders, to the beginning point, and comprising a plane area of eleven leagues of land or 275 millions of square varas."

The sketch shows the outline of the tract, and the relative location of the Daws patent owned by the plaintiff:

The beginning corner, A, opposite the mouth of the Lampas Creek, and the southeast corner, D, at the "box elder," or "double cottonwood," on the bank of the river, are well known and conceded points; and the location of the long easterly line, C' D, is fixed by marked trees, concurred in by both parties; and there is no controversy about the position of the westerly line, A B, the first line of the survey. The difficulty is to locate the back, or northerly, line. The defendant, as owner of the Moreno grant, contends for the line from B to C, which includes the greater part of the plaintiff's tract; and the plaintiff contends for the line from B′ to C′, which passes south of his land. If either the northwest, or northeast corner were known, the controversy would be at an end; but they are not fixed by any monuments which the parties agree on. The northwest corner, at the end of the first line in the field notes, was a mere stake set in the prairie, and, of course, soon disappeared. The northeast corner, at the end of the second line, was marked by "two small hackberries;" but no such trees have been found at, or near, the point C, where the north line, run by compass and chain according to the survey, would meet the easterly line. In 1854 one Samuel Bigham, a surveyor, under an order of the District Court of Bell County, surveyed the Maximo Moreno grant, commencing at the beginning corner, A, and following the field notes to the end of the second line, and was unable to find the northeast corner, or the easterly line. Some months afterwards he tried again, and by running across the front of the survey, the distance usually taken for an eleven-league front (13,750 varas), he found the eastern line, marked with blazes, which led him to the southeastern corner of the grant (D), when he found and identified the trees called for in the field notes. From this point, following the line back N. 20° E., he found the line plainly marked with old blazes for 26,400 varas, (the length called for in the field notes,) crossing Big Elm or Cow Creek at the exact distance fro the S.E. corner required by the field notes; and proceeding onward about 560 varas further, on the same course, he found two small hackberries in Cow Creek bottom, at which point, as he testifies, the line gave out. The line passed between these hackberries, and they were each marked on the inside with old blazes facing each other. He took those hackberries to be the identical ones called for in the grant, and fixed upon that point as the northeast corner of the survey. This is the point which the plaintiff claims to be the true northeast corner, and is marked C′ in the sketch. A line run from this point N. 70° W., the reverse of the line called for in the survey, would be the line B′ C′ on the map, and would fall to the south of the plaintiff's land. But B′, the point at which this line would intersect the west line of the survey, would be only about 18,700 varas from the beginning corner, instead of 22,960 varas, as called for in the field notes, or a deficiency of over 4000 varas.

On the other hand, if the field notes are followed, by running the first line from the S.W. corner, N. 22° E., 22,960 varas, and the second line thence S. 70° E., 12,580 varas, the upper line, BC, would be followed, but the distance, 12,580 varas, would fall short of the eastern line at C by about 570 varas, the true distance from B to C being 13,150 varas instead of 12,580. Then, running from C to D, the whole distance is found to be about 30,400 varas instead of 26,400 (as called for in the grant), or about 4000 too much; and the distance from C to Cow Creek is found to be 7500 or 8000 varas, instead of 3520, as called for in the field notes, or 4000 too much. So that the northeast corner of the tract, as fixed by Bigham at the two hackberries, corresponds very nearly with the several distances called for on the east line, but makes the west line 4000 varas too short; whilst the northeast corner, as fixed by running the west line its full length as called for by the field notes, and then running the north line as directed therein, and extending it so as to meet the easterly line, makes the easterly line 4000 varas too long.

The truth is, the original survey must in some parts have been imperfectly executed, or errors must have crept into the field notes. Frank W. Johnson was the surveyor—long well know as principal surveyor of the Austin and Williams colony. His deposition was taken in 1878, and again in 1880, forty-five and forty-seven years after the survey was made. He does not say what time of the year he made the survey, but William Duty, his chain-bearer, says it was in the spring. Both say that it was made in 1833, and was never made but once. Johnson is positive that he followed the courses and distance designated in the field notes of the grant for the first two lines, but that the last line, the easterly one of the tract, though run and marked, was not measured, but only estimated as to length or distance. But the field notes give the distance from the N.E. corner to Cow Creek 3520 varas, and from the N.E. corner to the San Andres River 26,400 varas, which would make the distance from Cow Creek to the San Andres 22,880 varas, which, by subsequent surveys, is found to be precisely accurate. This correspondence for such a long distance (over 12 miles) could hardly have been the result of conjecture; and the evidence of the chain-bearer is, that the easterly line, as well as the westerly and northerly lines, was actually measured by chaining. If this was so, (and it was for the jury to determine whether it was or not,) the judge was entirely right in charging that the footsteps of the original surveyor might be traced backward as well as forward; and that any ascertained monument in the survey might be adopted as a starting point for its recovery. This is always true where the whole survey has been actually run and measured, and ascertained monuments are referred to in it. *Ayers v. Harris*, 64 Texas, 296; *Ayers v. Lancaster*, 64 Texas, 305; *Scott v. Pettigrew*, 72 Texas, 321.

On the question of the true location of the northern boundary line of the Moreno grant, evidence was adduced by both parties. The defendant showed by surveyors who had recently gone over the lines that there were old marked trees in the north line of the survey claimed by him, and that the easterly line was continued to that line by old marked trees extending northerly from the two hackberries discovered by Bigham. The plaintiff, in rebuttal, adduced evidence to show that by blocking these trees the marks and blazes relied on were found to be of comparatively recent origin not more than 18 or 20 years old in 1886.

Duty, the chain-bearer, who was examined several times on the subject, and contradicted himself a good deal, on his last examination, taken by deposition in 1886, testified that the two hackberries found by Bigham, and established by him as the northeast corner, appeared to him (Duty) to be in a location like that where the northeast corner was established in 1833, and that the northeast corner, as claimed by the defendant, is in a location entirely different from that in which said corner was established in the original survey. He also said that the corner was made, not in the prairie, but in the bottom timber, and that he does not think that the corner is a hundred varas from the place claimed by the plaintiff.

The testimony of this witness is not entitled to much weight, but, being corroborated by the existence of the two hackberries discovered by Bigham, and by the distances from that point to Cow Creek and so to the San Andres River, it may be regarded as not so entirely worthless as to be absolutely rejected. The testimony of several other witnesses, including surveyors, was taken to show the situation of the different lines and points named in the grant, and of the condition of the marked trees claimed by the respective parties to be indicative of the true location.

In addition to the two hackberries, relied on by the plaintiff as fixing the position of the N.E. corner and the northerly line of the Moreno survey, he contended that the respective distances of the creeks and water-courses, called for by the field notes on said line, corresponded with the actual distances found on the line run from said hackberries, and did not correspond with the actual distances found on the line claimed by the defendant. To show this more clearly, the plaintiff offered in evidence a certified copy of certain field notes in a field book on file in the General Land Office of Texas, as the original English field notes of the Moreno survey made by Frank W. Johnson. In his deposition, Johnson had testified that his field notes of the survey were made in English, and reported to the empressario, and by him transcribed and translated into Spanish, and thus carried into the title. C.W. Pressler, chief draughtsman of the General Land Office, testified that these field notes were claimed to have been made by Johnson. DeBray, Spanish clerk in the land office, testified that he had heard Johnson claim that this field books was written by him. There was also a map or sketch of old surveys, including the Moreno survey, bound up in an atlas, regarded as the work of Johnson, and which had been in the General Land Office as far back as the witnesses had knowledge of it. Pressler testified that it was claimed by Johnson to have been filed by him, and that he (Pressler) had known it to have been in the land office since December, 1850, and that the words and figures on it resembled Johnson's handwriting. A certified copy of this map, and the said certified copy of the original field notes of the Moreno eleven-league survey, as also a photographic copy of the latter, were admitted in evidence against the objection of the defendant.

The following is a copy of the field notes referred to:

"Sunday, 21st, surveyed for Samuel Sawyer 11 leagues of land, beginning on the N. side of San Andres, opposite the mouth of Lampasas, at a pecan 18 in. diam., bearing N. 59 W.—vs. from a hackberry 24 in. and S. 34 W. 15.2 vs. from an elm 12 in.; thence N. 22 E. 22,960 vs. to the corner, a stake in the prairie; thence S. 70 E. 1690 vrs. to a branch of Cow Creek, 4500 vrs. to 2nd branch, 8000 vrs. to 3rd branch, 11,060 vrs. to Cow Creek, 12,580 vrs. to the corner, two small hackberries; thence S. 20 W. 3520 frs. To Cow Creek; 7500 to N.E. corner of 2nd tract, a stake bearing N. 77 E. 93.3 vrs. from a hackberry 8 in. to Spring branch 23,640 vrs., 23700 vs.; to bottom prairie 24,360 vs.; crossed same branch to the corner, a box elder, 26,400 vs., bearing S. 48 W. 7.2 vs. from a forked cotton wood 48 in., and S. 11 E. 16.4 vrs. from an elm 15 in."

These are evidently the field notes of the same survey that was carried into the grant. It seems that it was made for one Sawyer, and afterwards used for the Moreno grant, which was not issued until October, 1833. Duty, the chain-bearer, says the survey was made in the spring of that years; and the 21st of April came on Sunday in the year 1833. These notes are more full than the field notes in the grant, as they call for four streams crossing the north line, whilst the grant mentions only two of them. The four are as follows: 1690 varas fro the N.W. corner to a branch of Cow Creek; 4500 varas to a second branch; 8000 varas to a third branch; 11,060 varas to the Cow Creek itself. The witness Turner, for many years county surveyor of Bell County, who was employed by the defendant to trace the eastern and northern lines of the Moreno grant in 1880, testifies that by running the north line westerly from the two hackberries the first stream is reached at the distance called for in the field notes; that the distance between the first and second is also right; between the second and third the distance is too great; but between the third and fourth, and between the fourth stream and the N.W. corner (as claimed by the plaintiff), the distances agree with the field notes;—whilst the north line, as claimed by the defendant, crosses only three creeks, and none of them are in any way near the distances called for in any of the field notes. As rivers and streams are natural monuments, entitled to weight in any survey, it is manifest that these English field notes of Johnson must have had an important bearing in the trial.

The map or sketch, as before observed, contained an outline of the Moreno eleven-league tract, and of the streams which traverse it, with notes in Spanish of the courses and distances of the different lines. These notes begin with the easterly line, which is described as "Norte 20° Este, 26,400," [N. 20° E. 26,400]. The north line is partially obliterated, but enough of the notation remains to show that it was measured from east to west. The west line is described as "18,400 Sur 22° Oeste," [i.e. 18,400 S. 22° W.]. This shows that the length of the west line was therein made what it should be to correspOnd with the length of the east line as called for in all the surveys, and, so far as it goes, is evidence of a survey beginning at the southeast corner and running north, and then west, and then south, the reverse of the course which Johnson says he pursued. When it is recollected that his testimony was given forty-five years after the survey was made, and that the field notes, which he undoubtedly had regard, to, may have been written out in reverse order after the outdoor work was done, the fact that this old map or sketch exhibits a survey entirely consistent in all its parts, which the field notes do not, gives it considerable interest and value as independent evidence.

The admission of the field notes and map is one of the errors assigned on the present hearing; and the question of their admissibility will be now considered. These very field notes were admitted in evidence in a recent case in Texas, in an action between the appellant Ayers, as plaintiff, and Harris and others, defendants, and their admission was sanctioned by the Supreme Court of Texas on appeal. *Ayers v. Harris*, 77 Texas, 108. The court, in its opinion, says:

"The evidence, we think, places it beyond doubt that the survey mentioned in the field book as made for Samuel Sawyer was a survey of the same land that was titled to Maximo Moreno, and the only survey that was ever made of it. It cannot be doubted, upon this evidence, that Johnson having made the survey for Sawyer a few months before, adopted it when ordered to make a survey for Moreno, without making a resurvey.

"Very great difficulty existed in ascertaining where the lines of the survey were actually run. Aided by all the evidence that could be secured, and guided by all the rules recognized as being proper to be observed in such cases, repeated trials of the question have been had with conflicting results.

The judge said to the jury:

"Our purpose and your duty is to follow the tracks of the surveyor, so far as we can discover them on the ground with reasonable certainty, and were he cannot be tracked on the ground we have to follow the course and distance he gives, so far as not in conflict with the tracks we can find that he made; and you will constantly bear in mind, in considering the proof in this case, that in fixing the boundaries of al grant the rule require that course shall control distance as given in the calls of the field notes of the survey, and that marked trees, designating a corner or a line on the ground, shall control both course and distance. In order to reconcile or elucidate the calls of a survey in seeking to trace it on the ground the corner called for in the grant as the 'beginning' corner does not control more than any other corner actually well ascertained, nor are we constrained to follow the calls of the grant in the order said calls stand in the field notes there recorded, but are permitted to reverse the calls and trace the lines the other way, and should do so whenever by so doing the land embraced would most nearly harmonize all the calls and the objects of the grant.

"There has been proof given you tending to show where the two small hackberries called for as the intersection of the eastern and north lines of the grant actually stood at a distance from the lower corner on the river corresponding to the length of the eastern line of said grant; and if the proof satisfies you that the two hackberries mentioned in the testimony of the witnesses Sam. And Pat. Bigham were the hackberries called for and marked by the original surveyor as a corner of said grant, in that case a line drawn from the point where said hackberries stood N. 70 W. until it intersects the western line of said grant will bound the eleven-league grant upon the north, and if the Daws one-third of a league is situated wholly north of this line it does not conflict with said eleven-league grant, and you will find for the plaintiff.

"If the proof does not satisfy you that the two hackberries mentioned in the testimony of the wit nesses were the two hackberries called for by the original surveyor to serve as a landmark for corner at the intersection of the back (or north) line with the east line of said grant, and if a consideration of the whole proof satisfies you that the original surveyor began the survey at the 'cottonwood' corner (the S.E. corner) and marked and measured the east line and did not actually trace and measure the west line of said grant, you should follow these footsteps of the surveyor, and from the pint where you find his footsteps stop (for it is not disputed that this line is marked to a greater distance than the distance called for in the grant as the length of this line)—from this point where you find the footsteps stop you will run a line N. 70 W. to the west line of said grant for the north or back line; and if this line so run will fall wholly south of the Daws survey you will find for the plaintiff.

"If from the proof you are not able to fix the place where the two hackberries called for in the grant as a landmark to designate the N.E. corner of the Moreno grant then stood, and the proof does not satisfy you that to reverse the calls and trace the lines the other way would most nearly harmonize all the calls with the footprints left by the surveyor, you will fix the boundaries of the Moreno grant by the courses and distances of the first and second lines of the survey, extending the second line so as to meet the recognized east line, extended on its course to the point of intersection with the extended second or north line; and if the north line so fixed will embrace in the Moreno grant any part of the Daws survey, you will find for the defendant.

In our judgment this charge was justified by the testimony in the cause, and, on the whole, gave a correct view of the questions to be solved. The general rules laid down at the commencement are undoubtedly sound. The judge was also correct in saying, as we have already remarked, that the beginning corner does not control more than any other corner actually ascertained, and that we are not constrained to follow the calls of the grant in the order they stand in the field notes, but may reverse them and trace the lines the other way, whenever by so doing the land embraced would more nearly harmonize all the calls and the objects of the grant.

The court found that the instructions made by the lower court to the jury were proper. The last question related to the following instruction, to wit:

"It would not be proper to reverse the calls of the grant made to Maximo Moreno and to run in reverse course from the southeast corner for the purpose of ascertaining where the northeast corner would be found by the measurement called for in the grant, if the evidence satisfies the jury that the surveyor actually began the survey at the corner called the beginning corner in the field notes and from that corner ran and measured the western and northern lines on the ground."

We think the instruction was properly refused. As already intimated, the judge was right in holding as he did, and in instructing the jury, that the beginning corner of a survey does not control more than any other corner actually well ascertained, and that we are not constrained to follow the calls of the grant in the order said calls stand in the field notes, but are permitted to reverse the calls and trace the lines the other way, and should do so whenever by so doing the land embraced would most nearly harmonize all the calls and the objects of the grant. If an insurmountable difficulty is met with in running the lines in one direction, and is entirety obviated by running them in the reverse direction, and all the known calls of the survey are harmonized by the latter course, it is only a dictate of common sense to follow it.

There is no substitute for the original record. The beginning point is no more important than any other point.

Case #34

REVERSAL OF CALLS OF SURVEY IS PERMISSIBLE IN ATTEMPTING TO LOCATE A LINE

Sweats v. Southern Pine Lumber Co.

361 S.W.2d 214 (Tex. 1962)

It is true, as appellants assert, that a call of natural objects is a call to the highest dignity, but this does not mean that in all circumstances such a call is absolutely controlling. Such a call may be shown to be a mistake like any other call, and this many be shown circumstantially. The cardinal rule is that the footsteps of the surveyor shall, if possible, be followed.

Sweats v. Southern Pine Lumber Company, 361 S.W.2d 214 (1962): Also, in attempting to locate a line it is permissible to reverse the calls. *Ayers v. Lancaster*, 64 Tex. 305. If we do this and begin at point R, the field notes would read "Thence North with the east line of M.L. Ware's Preemption Survey 809 varas to Piney Creek; thence east down said creek 519 varas; thence South 2275 varas." When this is done, you go from R to K to 175 to S. Here it would be the cutoff of Piney that you reach at approximately 809 varas and not the north fork some 500 varas further north. Here you would reject no call.

In ascertaining lines of lands, tracks surveyor so far as discoverable on ground with reasonable certainty should be followed, and, where survey has been actually run and measured and ascertained, footsteps of surveyor may be traced backward as well as forward, and any monument in survey may be adopted as a starting point where difficulty exists in ascertaining lines that were run.

—Ralston v. Dwiggins
225 P. 343; 115 Kan. 842 (Kan. 1924)

Case #35

Richmond Cedar Works v. West

152 Va. 533 (1929)

In the instant case this method [reversal of courses] could be followed without trouble and confirms the plaintiffs' claim. But if an attempt is made to apply such reversed readings to the lines relied upon by the defendants, it would be found to be impossible. It is a matter of common knowledge that surveyors reverse courses and readings in order to locate disputed lines. Not only is this a practical check and rule in the science of surveying, but it stands in law approved by the weight of authority.

As is frequently the case where large tracts of wild land are concerned, the survey does not close. To make it close, one of two things must be done: The 200 chain south line must be extended seventeen chains, or there must be a corresponding extension of the last line of the survey as it is run upon the plat. When either of these things is done, we reach the point of beginning and not otherwise. The plan usually adopted is to follow the lines as they appear on the patent, and to make the necessary changes when the last is reached, rather than make any extension of some intermediate line, but this, is not possible here as a practical proceeding on the ground.

It is an elementary principle in surveying that courses and distances usually give place to recognized monuments and natural landmarks. To illustrate: If this survey were run upon the ground from its point of beginning in a southerly direction, when its south line at the gum had been reached, we would have a call of 200 chains west to the Dismal Swamp Canal. Two hundred chains would not reach that canal by seventeen chains, and its necessary lengthening would be inevitable, not only for the reasons stated, but because its next and westwardly line runs with the canal itself.

In the words of the court:

In order to ascertain and fix upon the ground the lines of a grant, we look first to the grant itself, and follow those lines in the order in which they are there stated. If it is possible to this on the ground with certainty, nothing more is to be said. A grantee takes what the State gives unless some senior right has already vested. If this is not possible, resort must be had to evidence *aliunde.*

Continuing, when we turn north from the 200 chain south line and follow bearings and distances, we find one of these lines described in the patent as running twenty-two degrees east fifty-five chains to a gum in Stewart's line, which is also a line tree in a patent theretofore granted to Hodges and Mills. This line as there described does not come anywhere near the Hodges-Mills grant and leads to an impossible situation. That is to say, you cannot go upon the ground and lay out the Herron land according to the lines of the grant, except in a manner which would be erroneous upon its face. This is admitted in the petition for appeal where it is said: "All parties agree that a call in the grant of 'N. 22 degrees E. 55 chains' is erroneous, and should be read 'N. 22 degrees W. 55 chains'." Hence the necessity for evidence beyond that which the grant gives.

How did this error come about? In the clerk's office of the Circuit Court of Norfolk county is found the surveyor's filed notes on which the patent itself rests. They are a part of the surveyor's record there kept. From them it appears that the lines on the ground were run in an opposite direction from that indicated by the patent; that is to say, the surveyor ran down the east line of the grant and then west to the Dismal Swamp Canal, etc. When copied, courses and distances were reversed, and they were properly reversed in every instance except as to this particular line, so that it alone, of all the lines in the patent, appear as it was originally written in the field notes. This error is also made manifest by an examination of the plat which is attached to, and is a part of the patent. In the map this line bears twenty-two degrees west as it should.

It is perfectly true that surveyor's notes are not competent to contradict patents as issued. *Rousens v. Lawson*, 91 Va. 226, 21 S.E. 347, a leading and constructive case. The grantee takes what the State gives and nothing more. To hold that he might take 2,000 acres of land when the State had given him only 1,000, because a surveyor's notes give him such an increase, is so manifestly unreasonable as not to merit discussion. Extraneous evidence, however, is competent, not to contradict a patent, but to locate and fix its line whenever there is any ambiguity on its face. *citing several cases*. Here, not only is the map attached to the patent not in accord with the patent line, but the patent lines, or certainly one of them, is confessedly wrong, and it is in the state of uncertainty that the surveyor's notes become evidence, not only competent but valuable. *Ayers v. Watson*, 137 U.S. 584, 11 S.Ct. 201, 34 L.Ed. 803. If we follow them as they were run upon the ground, there is no doubt whatever, about the proper location of the eastern boundary. It is certain that this eastern boundary touches the extreme eastern limit of the Hodges-Mills grant. This is not only shown by the "Cassell Picture Map" in evidence, but it is shown by the map with the patent itself, and there is no doubt about the fact that its location with reference to the point of beginning of the Herron survey is fixed. One grant fixes the other, and if they are shifted, they must be shifted together.

It is true that if we follow this last line to the 200 chain south line and follow that line for 200 chains it would fail to reach the canal by seventeen chains, but its extension to that natural boundary would merely be in accord with the common practice of surveyors.

It is a matters of common knowledge that surveyors reverse courses and readings in order to locate disputed lines. Not only is this a practical check and rule in the science of surveying, but it stands in law approved by the weight of authority. *Ayers v. Watson, supra*. This method may be followed without trouble and confirms the plaintiff's claim. If we attempt to apply such reversed readings to the lines relied upon by the defendants, we find it to be impossible, for we must leave the south line of the Hodges-Mills grant at a point several hundred feet west of the eastern extension of these lines as called for in the Herron patent.

Of course, as a surveyor's problem, in this uncharted waste of swamps, it is possible to slip the entire Hodges-Mills patent to the west, but in balancing that possibility, the surveyor's notes, the location of the seventeen acre tract tied to ditches whose identity is established and the manner in which the iron pipe came to be placed as a marker on the line of the Herron survey are all to be remembered and fix its situs with satisfactory certainty.

> *There is no hard and fast rule for closing a survey, but such rules as are employed are but rules of construction in aid of an effort to relocate lost lines as they were located in the original survey, which is always the problem for solution, and rules of construction must give way to competent evidence disproving their applicability to a given case.*[238]

All of the rules of construction of written instruments tend to one end, to discover the real intent of the maker.[239]

---

[238] *Cornett v. Kentucky River Coal Co.*, 195 S.W. 149 (Ky. 1917).

[239] *Holman v. Houston Oil Co. of Texas, et al.*, 152 S.W. 885 (Tex. 1912).

CHAPTER 16

PROFILING

> You do occasionally find a carrion crow among the eagles.
>
> —*Sherlock Holmes*
> "The Adventure of Shoscombe Old Place"

Criminal profiling is a process based on the psychological and statistical analysis of the crime scene, which is used to determine the general characteristics of the most likely suspect for the crime. In property investigation, it is attempting to determine where a particular description originated or where physical evidence came from. In the case of a plat, it is the determination of who the author of the plat is if it is not identified. Determining those answers will likely lead to additional sources of information or provide names of persons who may be consulted.

A profile is based on the premise that a person tends to be guided by personal psychology and will inevitably leave idiosyncratic clues.

One forensics text defined profiling as the identification of certain characteristics of an unknown, unidentified offender based on the way he committed a violent act, and his interactions with the victim(s).[240] In a property investigation, profiling is a matter of drafting style, the choice of certain words or phrases in a land description, the placement of certain types of markers at property corners. Profiling is an art, and is not usually acceptable for courtroom presentation. In fact, it was not conceived for such application. However, knowing something about previous surveyors and the way they performed their work can not only be acceptable in a court of law but also helpful to the court.

---

[240] Stuart H. James and Jon J. Nordby, eds. *Forensic Science. An Introduction to Scientific and Investigative Techniques* (Boca Raton, FL: CRC Press, 2003).

Case #36

Brantly v. Swift
24 Ala. 390 (1854)

> A practical surveyor, who testifies that he is familiar with the peculiar marks
> used by the United States' surveyors in their government surveys, may give
> his opinion, as an expert, whether a particular line was marked by them.

In this case the witness, a retracement surveyor, was asked if he was acquainted
with the manner in which the United States' surveyors marked the lines run by
them. He answered that he was, and that he had seen their marks often, and was
familiar with them. That their marks were peculiar, and made with a hatchet
framed for the purpose, and that no other surveyors marked lines in that way.

The only question in the case on appeal related to the ruling of the court in
reference to the question put to the witness which was objected to by the defendant:
from his knowledge of the manner in which the United States' surveyors usually
marked lines, he believed one of the marked lines earlier referred to in testimony
was a line run and marked by the surveyors of the United States in running
lines. This was objected to and overruled by the lower court, and the defendant
appealed.

The court said that the records shows the witness to be a practical surveyor,
that he had often traced the lines marked by the United States' surveyors and was
familiar with their marks, which were peculiar, differing from the marks or chops
made by other surveyors. Here, stated the court, is a sufficient predicate to allow
the witness, as an expert, to give an opinion, or express his belief, as to whether
the line spoke of was marked by the United States' surveyors. On questions of
science, persons of skill may not only speak as to facts but are allowed to give
their opinions in evidence.

In the case of an earlier surveyor, characteristics of the survey and the evidence left behind may provide clues to the previous surveyor(s). Looking at the
surveyor's behavior (markers set, the way trees are blazed, paint, flagging, traverse points), certain traits and characteristics can be attributed to the practitioner.
Additionally, in the past, most surveyors were fairly limited as to the their region
of practice, so it should be fairly easy to narrow the possibilities down to a few
choices.

Profiling. Knowing the original, or the earlier, surveyor. Some investigators have
a great amount of knowledge about previous surveyors when they have retraced
many of their surveys, they own or have worked in the their firm, or may even
have known the previous surveyor personally.

There once was a case whereby one surveyor testified as to the knowledge of a previous surveyor whose plan was on record. He stated that he did not give any credence to that survey because the surveyor had a poor reputation and his mathematics were mostly undependable. The opposite testimony, which ultimately helped win the case, was that the surveyor was personally known, and while the mathematics may not be reliable, the surveyor had indicated the presence of iron pipes on his plan. Since there were iron pipes called for in the previous deed, and the plan closely matched the description, it could be concluded that the pipes were indeed set and that the surveyor found them at that point in time. Whether his mathematics were or were not reliable was secondary.

Insight may be gained by reading surveyors' field notes, reviewing files, reading the copybook if there is one, or knowing what resources surveyors had at their disposal. A collection of plats will also give a profile of the type of work done by surveyors, whether they set points or a particular type of point, whether they included notes on the plat, what references they used indicating the extent of any research, and the like. The court system places field notes and plats quite high in the hierarchy of conflicting elements. Survey reports can be of tremendous value when they exist, or can be found.

Retracement of an Individual. To successfully retrace an earlier surveyor means learning about the individual and his or her habits. The type of equipment used, measurement techniques, allowances for error, and type of monumentation the person set all factor in to following the individual's footsteps. The successful retracement surveyor learns the previous surveyor's idiosyncrasies, habits, preferences, and procedures.

Serial killers are sometimes caught by using profiling techniques, and crime patterns can often be recognized for certain perpetrators. The same is true for surveyors following certain set procedures and becoming creatures of habit for the sake of efficiency and consistency. Surveyors who have practiced in an area for a period of time become familiar with the work of earlier surveyors they are following. They begin to recognize consistency in the type of monumentation set, the way trees are marked, certain notations in field books and on plans, as well as equipment and procedures used.

The procedures used in a criminal investigation and other types of investigation are often very similar, or even the same, although sometimes they are known by different names. Since they are all part of forensics, of interest to the legal system, they all need to be performed correctly, consistently, objectively, and with concern for keeping within the law and in protecting people's rights. There are rules for each type of investigation, some written and others unwritten. The rules for any investigation include protection and processing the scene, preserving the evidence, and proper procedures for analysis, reporting, and presentation, whether the situation is a major crime scene or a simple lot survey.

Signature. An act that has nothing to do with the process of creating a record or a survey. It is the personal flair that one attaches to a product. While products and the way they are produced change from one to another, a person's signature seldom, if ever, changes. The scribing one attaches to witness evidence, the order in which a description or a reference is written, the drafting technique used, especially when creating a north arrow, are all characteristics that are likely unique to an individual. Learning these, and identifying them in an investigation, can lead to an abundance of additional evidence.

Things to be considered about earlier surveyors may include:

- Education or apprenticeship
- Textbook used
- Preferred equipment
- Computational aids

Education or Apprenticeship   If the surveyors had formal education, what field was it in and what school did they attend? Many engineering schools concentrated on measurements and errors as opposed to evidence. These schools tended to teach the use of higher-precision equipment, transits, theodolites, and steel tapes. Many forestry schools, the source of many surveyors, concentrated more on evidence. Because large tracts of land often were involved, surveys were done with lower-precision equipment, for example, compass and tape.

Many of the earlier surveyors learned through apprenticeships; they worked for, and learned from, other surveyors. This means habits and procedures, whether good or bad, proper or otherwise, tended to be similar, if not identical. Some of these earlier surveyors were highly professional and competent, taking great pride in their work. Their results tend to be very reliable.

Textbooks Used   Knowing the particular textbook surveyors relied on can lend a tremendous amount of insight on the procedures they would have been likely to use and the mathematics they were capable of performing, as the latter often depended on the formulas contained in the textbook. Very early surveyors sometimes compiled their own book, complete with formulas, computations, and drawings. These books were known as *copybooks*.

Ira Allen, Ethan Allen's brother and first Surveyor General of Vermont, wrote on the cover of his book:

> Ira Allen, his book, containing the most useful rules of surveying, wrote with my own hand and agreeable to my invention.

Equipment   Equipment includes not only measuring devices but also computers and calculators. The type of computer may dictate the type of software used, which in turn may dictate the type of processes performed. This can be most critical when dealing with adjustment routines.

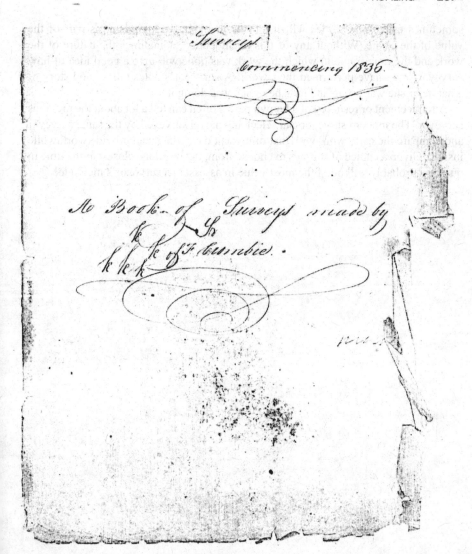

Figure 16.1 The second of two books belonging to Franklin Crombie, the first outlining methods of computation.

Computational Aids The types of computational aids available depends a lot on the time period in question. For instance, almost no one had computers and their associated software until the early 1970s. Before this, what computations were done were performed with a slide rule, a calculator, or longhand. Some surveys were not computed, with the surveyor relying on graphical closure, if at all.

If earlier surveyors are still alive, they can be interviewed to determine their equipment and how they learned to be a surveyor. If the surveyors are deceased, sometimes estate inventories will list maps, books, and equipment as part of the value of the estate. Without any of this documention, sometimes the nature of the work and the time frame in which the work was done will give a good idea of how surveyors may have performed their work. A surveyor's notes tell a good story of what he or she was doing and how he or she was doing it.

A retracement or resurvey of a known parcel often can tell a lot about the previous surveyor. The modern surveyor can select any parcel surveyed by the same surveyor and compare the early work with that of present day, which can provide worthwhile insight. Given a choice of surveys to choose from, the one done closest to the time in question probably will be of the most value in assessing a surveyor's methodology.

CHAPTER 17

INTERVIEWING (INTERROGATION)

It's a wicked thing to tell fibs.

—*Sherlock Holmes*
"The Adventure of the Three Gables"

I tend not to believe people. They lie, the evidence never
lies.

—*Gil Grissom*, CSI: Crime Scene Investigation
"Crate 'n Burial"

The utilization of physical evidence is critical to the
solution of most crime. No longer may police depend on the
confession, as they have done to a large extent in the past.
The eyewitness has never been dependable, as any
experienced investigator or attorney knows quite well. Only
physical evidence is infallible, and then only when it is
properly recognized, studied, and interpreted.[241]

INTERROGATION—INTERVIEW—WITNESSES

Generally a number of witnesses are interrogated or interviewed in the criminal investigation process. Concerning land ownership and location, knowledgeable persons
are contacted to inquire about unrecorded documents, recollections of the location
of boundary markers, and the use of the property. Some may become witnesses if

---

[241]J. I. Thornton (ed.) and P. Kirk, *Crime Investigation*, 2nd ed. (New York: John Wiley & Sons, Inc., 1974).

litigation ensues, to corroborate other testimony or conclusions. Former owners, abutting landowners, visitors, record keepers, and others familiar with a land parcel may be potential sources of valuable information.

Definitions

Interrogate. To examine formally by asking questions.

Interview. A conversation conducted with a person from whom information is sought; also, the record of such a conversation.

Although it may seem like they are the same, it is important to understand that there is a difference between the words "interrogate" and "interview." An interrogation asks questions of an individual who supplies answers, which often can provide or lead to important information.

An interview is a structured, nonconfrontational process. It is a fact-finding discussion, which may lead to other possibilities, clues, or other persons to interview. Both interviews and interrogations need to be handled carefully. While parol evidence is indeed evidence, it may be classed as hearsay, and therefore it must fall under one of the exceptions in order to be used in court.

Interrogation/Interviewing. Either process may involve a variety of persons: a surveyor, another professional or a member of the public, a grantor or a knowledgeable individual. Beware of the person who thinks he or she is an expert but in fact is not.

Case #37

THOSE PIPES ARE NOT THERE, I LOOKED!

A survey investigation once uncovered a sketch indicating a swap of land parcels in order to "square up" the lots line—that is, make them more nearly at right angles with the street line.

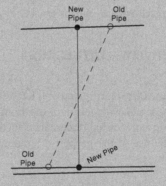

Figure 17.1 New and old evidence; new and old corners.

The two pipes in the rear were obvious, aboveground and in a buffer zone of woods. The two pipes in the front were not visible, so an underground search began using a metal locator. The neighbor appeared on the scene and vehemently stated that the iron pipe could not be found; he and his real estate agent had spent a lot of time searching for it without success. The locator made a sound, and the pipe was uncovered after digging down about a foot. Next the second pipe was searched for, while the neighbor insisted even more strongly that pipe did not exist. In a few seconds the locator sounded, and again the pipe was uncovered after digging down about a foot. The neighbor walked away shaking his head and muttering to himself.

This case demonstrates how sometimes people satisfy themselves that a certin state of facts either exists or does not exist, but could be totally mistaken. The lesson here is to be careful when taking someone else's (anyone else's) word for something, especially if it is a critical piece of information. Satisfy yourself if you are one responsible for it. The neighbor in this case was well intentioned, not malicious, but was totally wrong.

Proper Procedure

In interviewing a person, it is important to follow the proper procedure. Taking careful notes aids the investigator's own memory, and giving a copy to the people interviewed may aid in sealing their statements. Having a witness to assist and hear the conversation is often helpful. The environment can also be an important factor. People tend to be more relaxed and forthcoming when they are comfortable with their surroundings. Neutral ground or the person's home base with often produce better results than a place where a person feels antagonized by the surroundings. People tend to be more helpful when they do not feel pressured.

PAROL EVIDENCE

It is estimated that each year, 4,500 wrongful convictions are based on mistaken eyewitness identification, which generally grows out of faulty memory encoding. It is the most common cause of wrongful convictions—around 65 percent.

—*Katherine Ramsland*
The C.S.I. Effect

Eyewitness Testimony

Eyewitness misidentification occurs even under the best of circumstances.

—*Katherine Ramsland*
The C.S.I. Effect

Numerous studies have been done to test the reliability of people's observations. Once a man ran through a basketball game dressed in a gorilla suit. Some reported they did not even see the gorilla; others remembered seeing something but couldn't describe it. People generally do not remember details very well. When asked, rather than feel or appear stupid, they will sometimes attempt to give an answer. This can be especially true of persons in authority. Human nature indicates that if they do not know something that they feel they should know, they try to come up with an answer that makes sense, at least to them.

Watson. Holmes, you see everything.

Holmes. I see no more than you, but I notice what I see.

Scientific studies have demonstrated that after 48 hours, average people will remember just 10 percent of what they hear, 20 percent of what they see, 40 percent of what they read, but 90 percent of what they do. Confucius said it a long time ago, and it is worth remembering:

> I hear, I forget.
>
> I see, I may remember.
>
> I do, I understand.

Many people tend to remember things by association, especially when perceived in groups of three. Most people will never forget where they were, or what they were doing, when President John F. Kenndy was shot. Most people will never forget the Three Little Pigs; Goldilocks and the Three Bears; faith, hope and charity; blood, sweat, and tears; and many other things that follow the rule of three. Resurrecting that rule in the interview process sometimes refreshes a person's memory. However, it may bring about memory, or a person may insert something to complete the process that he or she did not have knowledge of before.

As explained in *Interpreting Land Records*, people will sometimes read a document according to what they either think it says or what they would like it to say rather than what it says. It is vitally important to read *exactly* what it says, giving meaning to *every* word, and not insert words that do not exist.

Investigators can sometimes inadvertently cue witnesses, who want to please. Eyewitness memory is susceptible to influence and vulnerable to distortion. Rather than asking a person "Did you see an iron pipe at the corner?", a better question might be "What, if anything, did you see at the corner?" The first question implies that there was an iron pipe at the corner. Some witnesses might ask themselves why they did not see such an obvious item. They often do not think the obvious—that they would not see a pipe if there was nothing there, or if something other than a pipe was marking the corner.

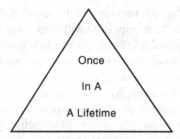

Figure 17.2   The first time people see this, the majority will read it as "once in a lifetime" sometimes they will do so the second and even the third time through.

Accuracy of eyewitness reports relies on the quality of three different perceptual processes:[242]

1. Encoding: converting information for storage
2. Storage: retaining information for short-term use or long-term recall
3. Retrieval: finding and providing the stored information for use

The quality of each of these processes depends on how many interfering factors are present.

Factors that can affect memory encoding include:

How much the information is rehearsed

Whether the information is rearranged into meaningful units or patterns

Whether a person was traumatized or stressed

Whether there is any personal association

Whether someone else made suggestions or used pressure to remember an item or incident a certain way

How divided the person's attention was

Being exposed to new information between storage and retrieval can affect the recollection, even if the information contains errors about the original incident. Exposure to misinformation after an event can lead people to erroneous reports of said information.

When information is encoded, the tendency is to be selective and ignore some aspects. People accept the things that make sense to them. Leading or suggestive comments tend to influence the process and re-form a person's memory, especially if the suggestions are acceptable.

[242] Katherine Ramsland, *The C.S.I. Effect.*

Confidence in one's memory is unrelated to the accuracy of a memory, in part because confidence levels can be manipulated with encouragement. Reseach indicates that memory is flexible, and many influences can cause memories to change.

Memory recall involves a construction of memory, which can be rebuilt two ways: from the original experience and from information to which the person was exposed afterward. However, mistakes made by witnesses, in or out of court, are generally not malicious. The worst issues are attempts by experts to introduce junk science into a proceeding and experts who are simply dishonest enough to lie or to fake the data results. Too many consider themselves hired guns and will testify on behalf of their client, or client's representative, instead of being completely honest and unbiased.

Psychologist Elizabeth Loftus once testified that memory is a reconstructive process in which the mind easily blends fact with fiction. The mind likes to fill in gaps and is susceptible to suggestion; thus a "memory" can be a distortion of actual experience.

Loftus went on to try to prove that techniques of recovered memory therapy could actually plant false memories or at least add details that are not true. She became the premier researcher in the country in showing just how unreliable memory can be—including and especially "recovered" memories—no matter how confident the person who reports it.

Case #38

STRAIGHT-LINE DESCRIPTION OF A CURVED FENCE

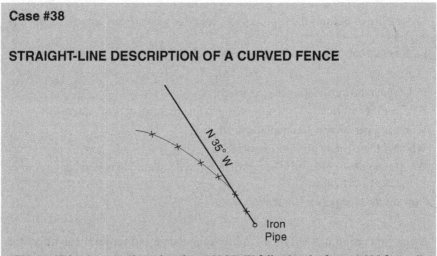

Figure 17.3   "... to an iron pipe, thence N 35° W following the fence, 1,095 feet ..."

The grantor, who had conveyed the parcel in question to his son, was contacted to discuss the conflict in the deed. While he believed that the fence line and the line defined by the bearing were the same, he was taken to the site and showed the difference. It turned out that he when he wrote the description, he was making an assumption and relying on his memory of what the property looked like, both of which were incorrect. If he had had a survey done, the difference would have been apparent.

While a call for a monument would normally control a direction call, extrinsic evidence in this case demonstrated that it would be in conflict with the intent of all parties concerned: the grantor, the grantee, and the abutter who purchased the remainder of the property. Appropriate documents were filed to correct the problem. This case demonstrates how misleading a description can be and how it might lead to a different result (perhaps the wrong result) in the absence of the parol evidence.

> Title to land must not rest upon the fallible memory of
> witnesses when it can be avoided.
>
> —*Reed v. Tacoma Bldg. & Sav. Ass'n*
> 26 P. 252; 2 Wash. 198 (1891)

CHAPTER 18

WOOD EVIDENCE

Where the calls contained within field notes for two pine
witness trees at south corner of survey could still be
identified with reasonable certainty even though the original
witness trees had been destroyed in the 135 years between
the original survey and the boundary dispute, such calls
would be used in locating the footsteps of the original
surveyor.

—*U.S. v. Champion Papers, Inc.,*
361 F.Supp. 141 (D.C. Tex., 1973)

Many of the original surveys and establishment of boundaries involved the setting
and marking of wooden evidence. Trees, where existing, and stakes set where a tree
was not available were among the most popular of all monuments utilized in the
creation and description of land lines. Soon after, fences were erected, and many
times they were constructed partly or entirely of wood. It is the remains of these
pieces of wooden evidence that is sought today. Surprisingly, many of them still
exist, or their location is perpetuated by their remains.

Trees mentioned in descriptions of real estate may be treated as monuments if
properly identified, and they will govern as to the boundaries in preference to courses
and distances. Thus, where a corner is marked by a tree, the call for the tree will
usually control a description of the same corner by a reference to an adjoining tract
of land. Marked trees can not only serve to show with certainty lines of tracts along
the boundaries of which they are located by may also be resorted to, in connection
with field notes and other evidence, to fix the original location of lost monuments.

It will be presumed that any object of perishable nature called for, if not found,
is destroyed or defaced; so that if a tree called for cannot be located it may be
treated as though the call for it had been for a stake or an imaginary point at the

end of the distance. The call for an unmarked tree of a kind which is common in the neighborhood of a place sufficiently described by the other parts of the entry to be fixed with certainty may be considered as an immaterial call.

—12 Am Jur 2d Boundaries, § 70

It is a settled rule of construction that when stakes are mentioned alone in a deed or with other descriptions than that of courses and distances, they are intended to designate imaginary points. A stake when once placed fixes the corner as conclusively as if it were marked by a natural object or a more permanent monument. Owing to the fact that it may be removed or obliterated, its location is frequently more difficult of proof, but if proved, it fixes the corner with the same certainty as where the mark is a permanent object.

—12 Am Jur 2d Boundaries, § 71

Forensic Botany. A combination of several disciplines that results ultimately in their application to matters of law. The botanical aspects of forensic botany include plant taxonomy (nomenclature and identification), plant anatomy (structure), plant growth and behavior, and population dynamics. The forensic aspects require an understanding of what is necessary for botanical evidence to be useful and accepted as evidence in the judicial system. Forensics requires the recognition of pertinent evidence at a site, the appropriate collection and photographing as well as preservation of material, and an understanding of scientific testing methodology. Methods range from simple techniques of macroscopic characteristics to more technical, microscopic analyses. In short, it is the identification, characterization, and use of tangible evidence remaining from the original plant or plant material.

Trees as Monuments.

Trees mentioned in descriptions of real estate may be treated as monuments and given the same effect as other monuments located on the ground. Where a tree is called for as a boundary the terminus of the line is ordinarily the center of the tree unless the grantor's title to the property conveyed extends only to its side. The general rule is that since marked trees remain invariable on the ground they are to govern as to the boundary in preference to courses and distances. Where a corner is marked by a tree, the call for the tree will usually control a description of the same corner by reference to an adjoining tract of land, and for this purpose testimony is admissible that a line of trees was planted so as to correspond to the location of a boundary line. It will be presumed that any object of perishable nature called for in a patent, if not found, is destroyed or defaced. So that if a tree called for cannot be located it may be treated the same as if the call for it had been for a stake or an imaginary point at the end of the distance. Marked trees may not only serve to show with certainty lines of tracts along the boundaries of which they are located but may also be resorted to, in connection with field notes and other evidence, to fix the original location of lost monuments.

—4 Ruling Case Law, Boundaries, § 40.

Tree Blazes

Figure 18.1   The photo shows a half-grown over blaze on an Eastern Hemlock (*Tsuga canadensis*).

Scribe Marks. Trees, stumps, and wood posts, when scribed, may tell a story about the corner itself or a corner it is a witness (accessory) to. Many symbols are standard, such as those listed next, and some are unique. It was common in nonrectangular states for surveyors who marked objects to include their own personal symbol. This symbol can be a valuable tool to lead a retracement person to the surveyor's notes or other records. Figure 18.10 presents an example of personal scribe marks collected for Maine surveyors.

In the rectangular system, certain markings were standard according, to the *Manual of Instruction, Bureau of Land Management:*

B.O.	Bearing object (cliffs and boulders)
B.T.	Bearing tree
E	East
L.M.	Location monument
M.C.	Meander corner
N	North
NE	Northeast
NW	Northwest
R	Range

R.M.	Reference monument
S	Section
S	South
S.C.	Standard corner
SE	Southeast
S.M.C.	Special meander corner
SW	Southwest
T	Township
TR	Tract
W	West
W.C.	Witness corner
W.P.	Witness point
X	Bearing object (cliffs and boulders)
$\frac{1}{4}$	Quarter section
1/16	Sixteenth section

Figure 18.2 This photo contains an American Beech (*Fagus grandifolia*) that has been bark-scribed. The scribing was done on the face of the bark of the tree as opposed to living tissue inside the bark. Smooth-barked trees lend themselves to this type of scribing. The scribing was done in 1949, and the tree was recovered during a survey in 1974.

Where a blaze is scribed and grown over, it is possible to remove the outside layers, revealing the original marks inside the tree. The part removed is a mirror image of the scribing, and can be read by holding the sample in front of a mirror.

Figure 18.3 A cutout section of bark and wood overgrowing scribing. This is a reverse image of the original. Note the discoloration of the wood due to increased pitch and resins as a reaction to the wound. This reaction will result in a denser, stronger wood at this location, which will usually last long after the rest of the tree has deteriorated. See Figures 19.15 and 19.16.

Figure 18.4 Cut open, heavily pitched, Jack Pine (*Pinus banksiana*) bearing tree.

Figure 18.5   Blazed and scribed (now dead) bearing tree. The initials "BT" can be seen in the photograph.

Stakes as Monuments

It is a settled rule of construction that when stakes are mentioned alone in a deed or with other description than that of courses and distances they are intended to designate imaginary points, and therefore a stake called for as located by the side of a street is not such a fixed monument as will control and prevent the operation of the rule that the grantee of land bounded by the side of a street takes to its center. A stake when once placed fixes the corner as conclusively as if marked by a natural object or a more permanent monument. Owing to the fact that it may be removed or obliterated, its location is frequently more difficult of proof; but if proved, it fixes the corner with the same certainty as where the mark is a permanent object. Monuments set at the time of an original survey on the ground, and named or referred to in the plat, are the best evidence of the true line. If there are none such, then stakes set by the surveyor to indicate corners of lots or blocks, or the lines of streets, at the time, or soon thereafter, are the next best evidence. A building or a fence constructed according to stakes set by a surveyor at a time when these were still in their original locations may become a monument after such stakes have been removed or disappeared, and next to the stakes may be the best evidence of the true line. Purchasers of town lots generally have the right to locate their lot lines according to the stake as actually set by the platter of the lots, and no subsequent survey can unsettle such lines. In the event of subsequent controversy the question becomes not whether they were planted by authority, and whether the lots were purchased and taken possession of in reliance upon them. If such was the case, the rule appears to be well established that they must govern, notwithstanding any errors in locating them.

————4 Ruling Case Law, Boundaries, § 41

Figures 18.6 and 18.7 A post set in 1894, found still standing in a retracement done in 1974. This is testament to how long a stake can remain as established under the right conditions.

Figure 18.8   An example of the perpetuation of a corner position, recovered in 1974. The standing post was set around 1935. The larger post lying on the ground still had legible scribing on it, "1874," while the smaller post to the left of it was much older than that.

Figure 18.9   This post and stone pile was recovered during a retracement. The post was set in 1915 and has rotted off at ground level, typical of wood posts as the alternate wet-dry conditions are more conducive to decay than either solely dry or solely wet. An added feature of this post is that it is marked with the surveyor's mark, a "Ø," which led to the discovery of his field notes and ultimately to recovering his other corners.

Collection of Land Surveyors Private Marks or Seals

	Lore Alfred	1846	
	Daniel Barker	1859	
	Charles Vernon Barker	1875	
	Noah Barker	1860	
	T. W. Baldwin		
	H. W. Briggs		
	C. D. Bryant		
	S. T. Buzzell	1890	
	T. B. Buzzell		
	Turner Buswell	1875-1900	
	John H. Burleigh	1900	
	Zebulon Bradley	1833	
	Ed. W. Bateman		N. H.
	Eleazer Coburn	1820	Skowhegan, Maine
	Elmer Crowley	1910	Greenville

	J. A. Lobley	1890	Bangor, Maine
	Caleb Leavitt	1834	
	F. S. Lord		N. H.
	Geo. Moulton		
	Andrew MclIellan		
	William Monroe		Brownville, Maine
	Amaziah D. Murray	1880	The Forks, "
	R. E. Mullaney	1908	Bangor, "
	Roy L. Marston	1910	Skowhegan, "
	McKechnie—also		
	Neal & McKechnie	1814	
	E. McCort Macy	1901	
	T. C. Norriss		
	J. C. Norris	1820	

Figure 18.10 Partial list of surveyor scribe marks. *Source*: From State of Maine, *Report on Publilc Reserved Lots,* State Forestry Department, 1963.

SIGNIFICANT COURT DECISIONS REGARDING TREES AND WOODS

As the law requires every official surveyor to see the survey plainly marked by trees or natural boundaries, the presumption is, that every survey has been thus marked, or bounded, when made, though the abuttals may not now be found.

—*Beckley v. Bryanand Ransdale*
1 Ky. (Ky. Dec.) 91 (1801)

An ancient marked line, corresponding in course with a boundary, and agreeing in ancient appearance with the true boundary of a survey, was *held* to be a part of said boundary.

—*Vance v. Marshall*
6 Ky. (3 Bibb) 148 (1813)

The marking of a tree for the beginning of a location is not competent evidence to prove the corner called for in a grant, unless, by some expression in the grant, it is evident that the tree which it calls for is the one marked in the location.

—*Rutledge v. Buchanan*
Fed. Cas. No. 12,177 (U.S. 1813)

The rules which have been established by the decisions of the courts, for settling questions relative to the boundary of lands, have grown out of the peculiar situation and circumstances of the country; and have been moulded to meet the exigencies of men, and the demands of justice.—

These rules are,

1. That whenever a natural boundary is called for in a patent or deed, the line is to terminate at it, however wide of the course called for, it may be: or however short or beyond the distance specified.

2. Whenever it can be proved that there was a line actually run by the Surveyor, was marked and a corner made, the party claiming under the patent or deed, shall hold accordingly, notwithstanding a mistaken description of the land in the patent or deed.

3. When the lines or courses of an adjoining tract are called for in a deed or patent, the lines shall be extended to them, without regard to distance: Provided those lines and courses be sufficiently established, and no other departure be permitted from the words of the patent or deed, than such as necessity enforces, or a true construction renders necessary.

4. Where there are no natural boundaries called for, no marked trees or corners to be found, nor the places where they once stood ascertained and identified by evidence; or where no lines or courses of an adjacent tract are called for; in all such cases, we are of necessity confined to the courses and distances described in the patent or deed: for however fallacious such guides may be, there are none other left for the location.

—Cherry v. Slade's Administrator
3 Murph (N.C.) 82 (1819)

The second and third corner, called for in the plaintiffs deed, was a *post*. It was in proof that a survey was made before the deed, and the corner posts fixed. The parties had possessed and improved up to a line run from post to post, and the plaintiff had acknowledged that to be his boundary, and the places where the posts stood, which was ascertained, his corners. But finding, after several years, that he had a less quantity of acres than he supposed, and that to run the line the exact course and distance called for, would carry them beyond the place where the posts stood, and give him a number of acres more than lie possessed, brought this Suit. He now claims to regard *the posts* as immaterial and void calls, and to run his lines by course and distance.

Where "a post" is called for in the survey of boundary lines in an improved country between individual holders, it is such a monument as will control course and distance.

—Lessee of Alshire v. J.R. Hulse
5 Ohio (5 Ham) 534 (1832)

A line will be run to a marked tree called for, though it departs from the course called for, and materially varies the shape of the original survey.

—Wash v. Holmes
1 Hill, Law, 12 (S.C., 1833)

In fixing limits, where the original boundary posts can no longer be seen, new posts must be fixed, but placed where the former limits stood, without regard to the title papers.

—*Zeringue v. Harang*
17 La. 349 (La. 1841)

Marked trees, as actually run, must control the line, which courses and distances would indicate.

If marked trees and marked corners are found, distances must be lengthened or shortened, and courses varied so as to conform to those objects.

When there are no natural boundaries called for, no marked trees or courses to be found, nor the places where they once stood, ascertained and identified by the evidence; or where no lines or courses of an adjacent tract are called for, in all such cases, Counts are of necessity confined to the courses and distances described in the grant or deed.

Any natural object, and the more prominent and permanent the object, the more controlling as a *locator,* when distinctly called for and satisfactorily proved, becomes a land-mark not to be rejected, because the certainty which it affords, excludes the probability of mistake.

—*Riley, administratrix, &c. v. Griffin, et al.*
16 Ga. 141 (1854)

It is the duty of a surveyor of public land to run round the land located, and to see that such objects are designated as will clearly identify the locality, and to call for these objects, natural and artificial, in his field-notes of the survey; and when the field-notes assert that the survey has been made, the calls will be presumed to be true until the contrary is proved; and as to lost calls, the presumption will be indulged that they have been destroyed or defaced; and even if it be established that the land was not in fact, surveyed, the patent will not be held void, if the boundaries can be identified by any reasonable evidence.

Natural objects are mountains, lakes, rivers, creeks, rocks, and the like; artificial objects are marked trees, stakes, mounds, &c., constructed by others or the surveyor, and called for in the field-notes, and they should be inserted in the patent.

In all future controversies these calls are to be searched for, and, if found, there can be little room for controversy about the boundaries; if not found, or found out of their places, then the rules of law must control.

The most material and certain calls shall control those which are less material and less certain. A call for a natural object, as a river, a known stream, a spring, or even a marked tree, shall control course and distance.

—*Stafford v. King*
30 Tex. 257 (1867)

When marked trees are called for in an old line, and the trees stand short of the line called for, and the next call is thence with the old line, the line called for is the terminus, and not the trees.

—James v. Brooks
53 Tenn. (6 Heisk.) 150 (1871)

Describing a boundary line as commencing at a tree does not necessarily fix the point as at the center of the tree.

—Stewart v. Patrick
68 N.Y. 450 (1877)

In locating a survey made as far back as the year 1790, it is not necessary that all or any of the corner trees should be proved to be still standing, or be accounted for; the survey can be successfully established by the local description contained in the grant, if it is so given as to separate the land described in the grant from that adjacent.

—Bowers v. Dickinson
30 W.Va. 709 (1888)

Where plaintiff and defendant agreed to settle the boundary of their respective lands, by the terms of which agreement the line was to begin at a certain tree, and run N., 77° W., to a certain other tree, the degree called for was merely descriptive, and must give way to that which will run to the points fixed by the objects named.

—Logan v. Evans
29 S.W. 636 (Ky. 1895)

Courses and distances in a deed give way to boundaries found on the ground, or supplied by proof of their former existence when the marks and monuments are gone; and this is true of the grant by an individual and of official surveys.

—Rozelle v. Lewis
37 Pa. Super. Ct. 563 (1908)

A "monument" is any physical object on ground which helps to establish location of line called for; it may be either natural or artificial, and may be a tree, stone, stake, pipe, or the like.

—Delphy v. Savage
177 A.2d 249, 227 Md. 37 (1961)

Evidence sustained finding that defendant's predecessor in title acquiesced in boundary line between adjoining timberlands as established by old blazed line.

—Amey v. Hall
181 A.2d 69, 123 Vt. 62 (1962)

Post is an "artificial monument."

—U.S. v. Gallas
269 F.Supp. 141 (D.C. Md. 1967)

Although proof showed that bois d'arc tree, which was mentioned in both deeds as being southern terminus of plaintiffs' eastern boundary and northern terminus of defendants' eastern boundary, had several branches from a single stump or rot, Chancellor properly held that center of tree, and not one particular branch thereof, was dividing point in eastern boundaries of both lots.

—Morrison v. Jones
430 S.W.2d 668, 58 Tenn. App. 333, appeal after remand,
458 S.W.2d 434 (1968)

Wooden stakes placed by surveyor on the land to mark corners of lots or the intersection of boundaries and measuring lines constitute "monuments."

—Sellman v. Schaaf
269 NE.2d 60, 26 Ohio App.2d 35 (1971)

Where the calls contained within field notes for two pine witness trees at south corner of survey could still be identified with reasonable certainty even though the original witness trees had been destroyed in the 135 years between the original survey and the boundary dispute, such calls would be used in locating the footsteps of the original surveyor.

—U.S. v. Champion Papers, Inc.
361 F.Supp. 141 (D.C. Tex. 1973)

TREES

Colloquial Tree Names

One of the biggest problems people have in reading records, or in searching for evidence on the ground, is that of tree names. Most people identified trees as they knew them, with little regard for whether the name chosen was the correct one or that the next person looking for that tree would know it by that name. As a result, there are some pretty strange tree names scattered thoughout the records, and some of them can be very misleading. Many colloquial names have been sorted out and identified, and many appear in Appendix VIII.

All living things have one established scientific name made up of three parts, a *genus* (or generic name) and a *species* (or specific) name, followed by the *name of the author* or discoverer of that animal or plant:

Tsuga canadensis (L.) Carr.

All living things have *one* accepted common name, established by the International Committee for Biological Nomenclature, but may have many colloquial names, depending on the region or who is talking about it:

Eastern Hemlock
aka
Spruce Pine
Hemlock Pine
Hemlock Spruce
White Hemlock

Trees marked and used as evidence were usually identified by the surveyor or farmer who measured the boundaries, marked the lines, or drafted the description. There are two very important considerations when dealing with tree names:

1. People identify a tree by what they know it as, which may not be proper, or the same name you know it by.
2. As a surveyor, you need to be very careful how *you* name a tree. Try to call it by its correct name. If you don't know, don't guess.

Many hours have been wasted looking for the wrong tree mentioned in field notes, on a plat, or in a deed.

Tree Identification

There are several good tree identification manuals available, especially for the individual who is not well versed in tree identification. Appendix IX provides a brief summary of important characteristics for a number of common tree species.

AERIAL PHOTOGRAPHS

Tree and Object Identification

The extent to which tree species can be recognized on aerial photographs is largely dependent on the scale of the photography as well as its method and quality. On photographs at very large scales, such as 1 inch = 50 feet (1:600), most species can be recognized. At scales of 1 inch = 200 feet (1:2400) or 250 feet (1:3000), small and medium branches are still visible, and individual crowns can be clearly distinguished, but these scales are no longer suited for exact species identification. At 1 inch = 660 feet (1:7920), individual trees can still be separated, except when growing in dense stands, but it is not always possible to describe crown shape. At 1 inch = 1320 feet

(1:15,840), crown shape can still be seen by tree shadow for trees growing in the open or for very large trees.

For identifying tree species, the main factors are the crown image, stand pattern, topographic location, and species associations. In addition, shadows, branching patterns, photographic texture, and tonal contrast can be important factors.

One of the biggest advantages of species identification comes from knowledge of the ecological and silvicultural characteristics of species. Many tree species are characteristic of a certain habitat, and in practice physiography and other features of the area examined are very important keys to the species found there. For instance, white pine will grow over a great range of sites, but hardwoods in a swampy location are more likely to be elm or ash than beech or oak.

Tone

On infrared photography, conifers usually photograph dark gray to black, while deciduous trees photograph in varying degrees of white to gray.

Crown Characteristics

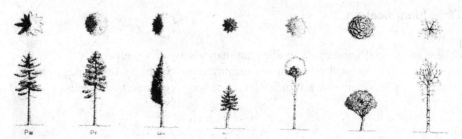

Figure 18.11 Schematic comparison among white pine, red pine, eastern hemlock, jack pine, aspen, red oak, and a dead tree. *Source*: From Victor G. Zsilinszky, *Photographic Interpretation of Tree Species in Ontario* (1966). *Source*: Ontario Ministry of Natural Resources. Copyright: 2007 Queens Printer Ontario.

WOOD

When it is not possible to identify the species of tree, often the object can be identified as a wood specimen. This is applicable if the leaves or other obvious characteristics are not available, the tree has died, there is only a small part of it remaining, or wood is the last resort. After trees have deteriorated, their remains maybe identifiable.

Wood Identification

Technical manuals are available to aid in the identification of wood, which is a bit more complex than identifying trees. Appendix XII provides a simple, general means for wood identification along with a set of key characteristics to aid in the identification of common species.

REMAINS

At any corner position, the surface of the ground should be carefully examined for remains of a decayed wooden post or for discoloration of the soil. In heavy clay soil, the dimensions of the corner stake and decayed wood fragments remain. In very wet soils or in water, the bottom of the original corner post is as perfectly preserved as the day it was placed.

Stump Holes

Basic clues of tree recovery are the remains of the tree trunk, stump, and the roots. For instance, bark on the roots is the last to decay in species such as yellow birch, white pine, red pine, and cedar. For sugar maple, very black wood flakes off in scales. If the species is hemlock, the stump may be found below the raised root system of a yellow birch because a decayed hemlock stump is a choice site for sprouting yellow birch seeds. Hemlock, upon decaying, leaves a reddish-brown stain in the soil, and hemlock knots are found along the decayed fallen trunk. The size and direction of the knots are a clue to the stump end of the tree. Basswood or red maple sprouts around a decayed stump or stump hole will indicate the species and the location of the original tree, as both species sprout from the decayed stump. White pine is soft and light in color and has a distinct smell. Red pine has a distinct smell, is harder to chop than white pine, and is resistant to ground fires. Cedar and tamarack are found in or near a swamp. Cedar is usually detected by the fragrant odor of the wood and the shredlike bark. Tamarack is highly rot resistant and decays by outer layers of wood.

Microscopic Analysis

Many wood samples can be identified from the gross anatomy, or macrocharacteristics. Frequently a 10-power hand lens is all that is necessary. However, with some difficult species and with very small wood fragments, microscopic identification is necessary. If the investigator is not well trained in this area, it is important to consult an expert or send the sample to an organization that offers an identification service. The Forest Products Laboratory in Madison, Wisconsin (www.fpl.fs.fed.us) is a choice, but there are others.

Decayed Wood

Fungi that invade wood tissue are broadly classified as either stain fungi or decay fungi. Stain fungi simply invade the cell structure and cause discoloration, primarily in the sapwood, but they do not significantly soften or weaken the wood. Consequently, wood cell structure and anatomical detail remain intact.

Decay fungi, however, cause eventual breakdown of cell structure. Advanced decay produces drastic changes in many gross features. Useful characteristics such as color, evenness of grain, ray fleck, and parenchyma arrangement may lose their original macroscopic appearance. Features such as resin canals and rays, although present, may be more difficult to detect. In cases such as these, it is usually necessary to rely more on microscopic features.

Charcoal

Charcoal is common in burned-over areas and, unlike other forms of wood, is little affected by fungi and other wood-destroying organisms. Because of this fact, it may survive in the ground indefinitely. Charcoal identification is possible because the anatomical features of the wood remain essentially intact through the carbonizing process, although some of the fine detail may disintegrate. Because charcoal identification is more difficult than normal wood identification, it is best to have a good working familiarity with wood structure and some experience with the identification procedure for normal wood before attempting to analyze charcoal.

In hardwood species, routine details such as ring porosity, pore arrangement, parenchyma arrangement, and ray widths are readily determined, so species such as oaks, ash, elm, hickory, chestnut, and beech are most easily recognized. In softwood charcoal, texture, evenness of grain, earlywood/latewood transition, and the size and number of resin canals can be assessed, and an unknown wood can at least be narrowed to a group of possibilities.

Simple Method of Distinguishing Hardwood from Softwood

While this test will not determine the specific tree species of a sample, it can place the sample in a broad group. The test is reliable for wood even in the most advanced stages of decay.

First place the sample in a small dish and add 1 or 2 drops of potassium permanganate (strength: 1 gram in 100 ml), and leave set for two or three minutes, then pour off the excess. Next add 2 drops of 12 percent hydrochloric acid (strength: 12 ml raised to 100 ml), and let stand until sample is decolorized. Next add 1 or 2 milliliters of concentrated ammonium hydroxide. A hardwood sample will turn bright red to violet, while a softwood sample will turn coffee brown to dull brown.

An alternate method with ordinary household chemicals is to place sample in a dish, add ordinary undiluted household bleach, and let set for 10 seconds. Then add diluted hydrochloric acid (equal volumes of HCl and water) and let set for 10 seconds. Then add concentrated ammonium hydroxide and again let set for 10 seconds. A hardwood sample will turn a red-purple color, while a softwood sample will turn a yellow brown.

These tests should be performed only while wearing rubber gloves and safety goggles and in a well-ventilated area, preferably out of doors.

DENDROCHRONOLOGY

The annulations of the trees for the purpose of proving the early existence of the surveys, is resorted to. This species of evidence may often be entitled to great weight, as that which arises from the progress of nature itself, without danger of corruption.

—*Bulor's Heirs v. James McCawley*
10 Ky (3 A.K. Marsh. 573) (1821)

TIME AND CYCLES

Trees contain some of nature's most accurate evidence of the past. Their growth layers, appearing as rings in the cross section of the tree trunk, record evidence of floods, droughts, insect attacks, lightning strikes, and even earthquakes.

Each year, a tree grows. The new growth is called a tree ring. How much the tree grows depends on a number of factors, such as how much water is available. Because the amount of water available to the tree varies from year to year, scientists can use tree ring patterns to reconstruct regional patterns of drought and climatic change. This field of study, known as dendrochronology, was begun in the early 1900s by American astronomer Andrew Ellicott Douglass.

Modern dendrochronologists seldom cut down a tree to analyze its rings. Instead, core samples are extracted using a borer that is screwed into the tree and pulled out,

Figure 18.12   Extracting an increment core from a Table-mountain Pine (*Pinus pungens*). *Source*: Photo courtesy of Henri D. Grissino-Mayer.

bringing with it a straw-size sample of wood about 4 millimeters in diameter. The hole in the tree is then sealed to prevent disease.

Dendrochronology, or tree ring dating, is defined as the study of the chronological sequence of annual growth rings in trees. Usually what is of interest is the calendar date concerning a wood or charcoal specimen.

The dendrochronologist is concerned with the study of chronological sequence of tree rings. Dendrochronology is made possible by the fact that in many trees, the annual rings visible in cross section, rather than all looking alike, exhibit characteristic patterns. Four conditions are necessary for these patterns to be usable in dating a specimen.

1. Wood used for dating purposes must add only one ring for each growing season, hence the term "annual ring."
2. Although total seasonal growth is the result of many interacting factors, such as genetics and environment, only one environmental factor must dominate in limiting the growth. In some areas it is precipitation; in others, temperature.
3. This growth-limiting climatic factor must vary in intensity from year to year, and the resulting annual rings must faithfully reflect such variation in their

width. The recognizable sequence of wide and narrow rings makes possible *cross dating*, or the matching of ring patterns in one specimen with corresponding ring patterns in another.

4. The variable environmental growth-limiting factor must be uniformly effective over a large geographical area. This way composite chronologies can be compiled on a regional basis.

A tree grows by increasing in height (apical growth) and by increasing in diameter (radial growth). The annual ring is divided into two parts, earlywood and latewood. It is the sharp contrast between the last-formed latewood cells of one growing season and the first-formed earlywood cells of the following season that delineates the boundary of an annual ring. Because of the sharp contrast between the two cell types, annual rings can be seen in most cross sections with magnification.

> By translating the story told by tree rings, we have pushed back the horizons of history in the United States for nearly eight centuries before Columbus reached the shores of the New World, and we have established in our Southwest a chronology for that period more accurate than if human hands had written down the major events as they occurred.
>
> —A. E. Douglas

The term "dendrochronology" is made up of these Greek root words:

- *ology*: the study of
- *chronos*: time or, more specifically, events in past time
- *dendros*: using trees or, more specifically, the growth rings of trees

Terminology

Annual ring. The layer of wood, consisting of many cells, formed around a tree stem during a single growing season. It is seen most easily in cross section.

Trees that grow in situations where growth stops or slows during a portion of the year will form annual rings which can be read to determine tree age and rate of growth. The science of dendrochronology studies tree rings to infer knowledge about past climatic conditions. This is based on the fact that trees will form wider annual rings during seasons when growing conditions are favorable and narrow rings when not.

Annual rings are highly visible in tree species that form less dense wood during favorable growing conditions early in the season and denser wood during less favorable conditions later in the year. In some tree species this differentiation does not occur, and annuals rings are difficult to see. In tropical species growth never, or seldom, ceases, and annual rings may not be apparent.

Bark. The outer layer of the stems, limbs and twigs of woody plants. Often bark is characteristic of the species and can be used for identification.

Cross section. The wood surface exposed when a tree stem is cut horizontally and the majority of the cells are cut transversely.

Diameter tape. Usually a steel or cloth tape graduated with numerals that are 3.1416 inches apart. When placed around a tree at d. b. h (diameter, breast-high, which is considered to be 4.5 feet up from the ground), the tree's diameter can be read directly in inches. The same result could be obtained by using a standard measuring tape and dividing the reading by 3.1416.

Diffuse porous. Wood of a hardwood species in which the vessel diameter remains approximately constant throughout the annual ring.

Earlywood. The less dense part of the growth ring. It is made up of cells having thinner walls, a greater radial diameter, and shorter length than those formed later in the year. Also known as *springwood*.

Hardwood. "Hardwood" as opposed to "softwood" is a relative term. Hardwoods are generally defined as the woods of deciduous trees, that is, trees that shed their leaves in the winter. However, some hardwoods do not shed their leaves. Moreover, some hardwoods are softer than some softwoods.

   To confound the situation, the group is divided into hard hardwoods, such as oak, ash, and hickory, and soft hardwoods, such as elm, cottonwood, willow, and soft maple. In Iowa and the Midwest United States, hardwoods are those species that lose their leaves on an annual basis and softwoods are evergreens.

Heartwood. The inner part of the stem, which, in the growing tree, no longer contains living cells. It is generally darker than sapwood, though the boundary is not always distinct.

Increment borer. A T-shaped tool consisting of a bit, a handle, and an extractor that is used to measure the age or growth rate of a tree. The bit is hollow and, when turned into the tree, cuts a pencil-shaped piece of wood showing the growth rings. By counting the number of rings in the inch of wood closest to the bark, a statement can be made about the increase in tree diameter in the last X years. By drilling into the center of the tree and counting annual rings, the tree's age can be determined.

Latewood. The denser part of the growth ring. It is made up of cells having thicker walls, smaller radial diameter, and generally longer than those formed earlier in the growing season. Also known as *summerwood*.

Parenchyma. Tissue composed of cells that are usually brick-shaped, have simple pits, and frequently only a primary wall. They are mainly used to store and distribute food material.

Ray. A ribbonlike group of cells, usually parenchyma, extending radially in the wood and bark.

Resin duct. A space in which resin has accumulated between the cells.

Ring porous. Wood of a hardwood species in which the vessel diameter is considerably larger in the earlywood than in the latewood.

Sapling. A young tree that has grown beyond the seedling stage. When a tree has grown to a diameter of 3.5 inches at a point 4.5 feet above the ground, it is no longer a sapling, having become a small pole.

Sapwood. Wood immediately inside the cambium of the living tree, containing living cells and reserve materials, and in which most of the upward water movement takes place.

Sensitivity. The growth of trees can be affected by slope gradient, sun, wind, soil properties, temperature, and snow accumulation. The more a tree's rate of growth has been limited by environmental factors, the more variation in ring-to-ring growth there is likely to be present. This variation is referred to as *sensitivity,* and the lack of ring variability is called *complacency.* Trees showing sensitive rings are those affected by conditions such as slope gradient, poor soils, and lack of moisture. Those showing complacent rings generally have constant climatic conditions, such as a high water table, good soil, or protected locations.

Softwood. Generally considered to be the wood of conifers, although the wood of some conifers is harder than that of some hardwoods. See the definition of *hardwood* for a further explanation.

Springwood. That part of an annual ring formed early in the growing season, a period of more rapid growth. Walls of wood cells are thin and wood formed is less dense. Also known as *earlywood.*

Summerwood. That part of an annual ring formed during the summer when growth has slowed. It is more dense than springwood, having thicker cell walls. Also known as *latewood.*

Tree ring. A layer of wood cells produced by a tree or shrub in one year, usually consisting of thin-walled cells formed early in the growing season (called earlywood or springwood) and thicker-walled cells produced later in the growing season (called latewood or summerwood). The beginning of earlywood formation and the end of latewood formation form one annual ring, which usually extends around the entire circumference of the tree.

Tree ring chronology. A series of measured tree ring properties, such as tree ring width or maximum latewood density, that has been converted to dimensionless indices through the process of standardization. A tree ring chronology therefore represents departures of growth for any one year compared with average growth.

Tyloses. An expansion of a parenchyma cell through a pit into a neighboring vessel; the tyloses may fill the invaded cell cavity partially or completely, reducing or preventing passage of liquids or gases.

Vessel. A series of cells extending longitudinally in the stem, the ends of the cells having fused together to form a long tube.

FACTORS AFFECTING TREE RING GROWTH

- Slope gradient
- Soil properties
- Temperature
- Wind
- Sun
- Snow accumulation

PRINCIPLES OF DENDROCHRONOLOGY

Uniformitarian Principle

The uniformitarian principle links current environmental variability evident in tree ring growth to past environmental variability and uses it to predict future environmental change.

The principle of uniformitarianism was not realized until 1785, when a Scotsman, James Hutton, presented two papers entitled "Theory of the Earth with Proof and Illustrations." Hutton recognized the cyclical nature of geological changes and the way in which ordinary processes, operating over long time intervals, can effect great changes. The fact that natural processes of today also would have operated in the past led to the statement: *"The present is the key to the past."*

Uniformitarianism, however, does not apply just to physical processes. In dendrochronology, the principle states that physical and biological processes linking current environmental processes with current patterns of tree growth must have been in operation in the past. This does not necessarily mean that the conditions are exactly the same but that similar kinds of influences affected similar kinds of processes.

Dendrochronology also adds a new twist to the principle *"The past is the key to the future."*

> Study the past to divine the future.
>
> —Confucius, ca. 500 BC

> In history lies all the secrets.
>
> —Winston Churchill

In other words, by knowing environmental conditions that operated in the past (by analyzing such conditions in tree rings), we can better predict and/or manage such environmental conditions in the future.

Limiting Factor Principle

According to the limiting factor principle, the fastest rate that plant processes or growth can occur is equal to the greatest limiting factor.

As stated by Fritts in 1976, "A biological process, such as growth, cannot proceed faster than is allowed by the most limiting factor. If a factor changes so that it is no longer limiting, the rate of plant processes will increase until some other factor becomes limiting" (*Tree Rings and Climate*. New York: Academic Press). For example, precipitation is often the most limiting factor to plant growth in arid and semiarid areas. In these regions, tree growth cannot proceed faster than that allowed by the amount of precipitation, causing the width of the rings (i.e., the volume of wood produced) to be a function of precipitation. In some locations, rainfall is not the most limiting factor. For example, in the higher latitudes, temperature is often the most limiting factor that affects tree growth rates. In addition, the factor that is most limiting is often acted upon by other nonclimatic factors. While precipitation may be limiting in semiarid regions, the effects of the low precipitation amounts may be compounded by well-drained (e.g. sandy) soils.

This principle is important to dendrochronology because ring widths can be cross-dated only if one or more environmental factor becomes critically limiting, persists sufficiently long, and acts over a wide enough geographic area to cause ring widths or other features to vary the same way in many trees.

Principle of Aggregate Tree Growth

According to the principle of aggregate tree growth, any individual tree growth series can be broken down into an aggregate of environmental factors, both natural and human, that affected the patterns of tree growth over time.

This principle states that any individual tree growth series can be decomposed into an aggregate of environmental factors, both human and natural, that affected the patterns of tree growth over time. For example, tree ring growth in any one year is a function of an aggregate of factors:

- Age-related growth trend due to normal physiological aging processes
- Climate that occurred during that year
- Occurrence of disturbance factors within the forest stand (e.g., a blow-down of trees)
- Occurrence of disturbance factors from outside the forest stand (e.g., an insect outbreak that defoliates the trees, causing growth reduction)
- Random (error) processes not accounted for by these other processes

Therefore, to maximize the desired environmental signal being studied, the other factors should be minimized. For example, to maximize the climate signal, the age-related trend should be removed, and trees and sites should be selected to min-imize the possibility of internal and external ecological processes affecting tree growth.

Principle of Ecological Amplitude

The principle of ecological amplitude states that a tree species will be more sensitive to environmental factors at the latitudinal and elevational limits of its range. This is important because tree species most useful to dendrochronology are often found near the margins of their natural range. For example, lodgepole pine (*Pinus contorta*) is among the most widely distributed of all pine species in North America, growing in a diverse range of habitats. Therefore, lodgepole pine has a wide ecological amplitude. Conversely, yellow cedar trees (*Chamaecyparis nootkatensis*) grow only on the Pacific coast of British Columbia and Alaska. Therefore, this species has a narrow ecological amplitude.

Often the growth of trees near arid forest limits is most affected by drought, while the growth of trees near the upper elevational or high latitudinal forest limits is most affected by low temperatures.

Principle of Site Selection

The principle of site selection states that sites useful to dendrochronology can be identified and selected based on criteria that will produce tree ring series sensitive to the environmental variable being examined. For example, trees that are especially responsive to drought conditions can usually be found where rainfall is limiting, such as rocky outcrops or on mountain ridgecrests. Therefore, a dendrochronologist inter-ested in past drought conditions would purposely sample trees growing in locations known to be water-limited. Sampling trees growing in low-elevation, mesic (wet) sites would not produce tree ring series especially sensitive to rainfall deficits.

Principle of Crossdating

Matching patterns in tree ring width, density, and other characteristics across several tree ring series allows for the identification of the year in which the growth ring was formed.

The principle of crossdating states that matching patterns in ring widths or other ring characteristics (such as ring density pattern) among several tree ring series allow the identification of the exact year in which each ring was formed (after Kaennel and Schweingruber, *Multilingual Glossary of Denrochronology*; Berne, Switzerland. Paul Haupt Publisher, 1995). For example, the construction of a building can be dated by matching the tree ring patterns of wood taken from the structure with tree ring patterns from living trees. Crossdating is considered the fundamental principle of dendrochronology. Without the precision given by crossdating, the dating of tree rings would be nothing more than simple ring counting.

Crossdating is perhaps the *most crucial procedure* in tree ring analysis. It is of vital importance in assuring that each ring width/climatic value/density value/isotopic measurement/and so on is placed in its proper time sequence.

Often tree ring and climatic studies count only rings. Chronologies resulting from such studies can contain errors due to counting, mistaken identification of ring features, and ring absence. A surprising number of absent sets or false rings occur, even for sites that are relatively temperate. Through cross-dating, these problems can be ameliorated.

Trees can be crossdated in five ways:

1. *Directly from wood.* This is the classic method performed by A. E. Douglass and his students. Dendrochronologists with many years of experience in a particular area carry the patterns of hundreds of years of ring width variations in their head and can compare an observed pattern with this mental reference.

2. *Skeleton plots.* This is the graphic representation of those rings considered important in cross-dating. In the case of the stress-grown conifers, for which the method is normally most suited, these indicator years are the narrow rings. The method is explained in detail by Stokes and Smiley (1968). In the last decade, the method has been expanded to facilitate dendroecological studies. Referred to as "event dating," the procedure graphically maps event years, suppression and release, and other variables seen on the wood itself. For an in-depth explanation, see:

    Schweingruber F. H., D. Eckstein, F. Serre-Bachet, O. U. and Bräker. "Identification, Presentation and Interpretation of Event Years and Pointer Years in Dendrochronology," *Dendrochronologia* 8 (1990): 9–38.

3. *Cross-dating using measured ring widths.* The usual procedure is to plot measured ring widths in a linear scale representing the years of growth for the x-axis and either a linear or logarithmic y-axis for the width of each ring. The dating is carried out on a light box by superimposing one or more graphs.

4. *Computers and cross-dating.* Although the human brain is very efficient at cross-dating, the process lacks the objectivity demanded of a scientific discipline. Since the mid-1960s, a number of computer programs for tree ring research have been written, which has now culminated in the ITRDB Program Library.

5. *Statistical tests*

Principle of Replication

The principle of replication states that the environmental signal being investigated can be maximized, and the amount of "noise" minimized, by sampling more than one stem radius per tree and more than one tree per site. Obtaining more than one increment core per tree reduces the amount of intratree variability (in other words, the amount of nondesirable environmental signal peculiar to only one tree). Obtaining numerous trees from one site and perhaps several sites in the region ensures that the amount of "noise" (environmental factors not being studied, such as air pollution) is minimized.

PROCEDURAL CONSIDERATIONS

Tree rings are never identical, but the patterns are similar, assuming you are looking in the same geographic area. When the climate is particularly moist, it will produce wider rings; and in the dry years, narrow rings. Due to severe weather, trees may not produce a ring every year. To ensure that they are counting accurately, scientists have developed a cross-check system that uses nearby resources to verify the data. By looking at a species with a known sequence of growth, they can look for matching patterns in the unknown and perhaps see the past more clearly.

In order for this to be a reliable method for dating, four factors must be present:

1. The species studied must produce only one ring per growing season or year.
2. Only one dominant environmental factor can be the cause of hindered or increased growth.
3. The dominant environmental factor should vary each year so the changes can be seen clearly in every ring.
4. The environmental factor must affect a large geographic area so tests can be compared easily.

SUBFIELDS OF DENDROCHRONOLOGY

Dendroarchaeology (Dendrohistory). The system of scientific methods used to determine the exact time span of a period during which timber has been felled, transported, processed, and used for construction.

Dendrochemistry. The use of tree rings to document trends in chemical constituents of tree rings (whether organic or heavy or trace metal concentrations).

Dendroclimatology. The reconstruction and study of past and present climate, including temperature, precipitation, and/or moisture availability. The science that uses tree rings to study and reconstruct the past and present climate of an area. Example: analyzing ring widths of trees to determine how much rainfall fell per year long before weather records were kept.

This is the oldest and probably also the most widely practiced subfield of dendrochronology. It is concerned with measuring the widths, densities, or other characteristics of annual rings in selected trees at a certain site with the aim of producing an averaged tree ring chronology for that site. This chronology enables the study of past and present climatic conditions and possible paleoclimatic reconstructions.

Dendroecology. The study of past and present forest ecological form and function, including topics such as fire and/or insect dynamics and forest composition history. The science that uses tree rings to study factors that affect the earth's ecosystems. Example: analyzing the effects of air pollution on tree growth by studying changes in ring widths over time.

Dendrogeomorphology. The application of dated tree rings to the investigation and study of present landforms and past geomorphic processes, including volcanic eruptions, earthquakes, avalanches, glacial movement, and landslides.

Dendrogeomorphology concerns the timing of recent geomorphic and hydrologic events that can be established, where they have interrupted or disturbed the growth of coniferous trees. Landslides generally result in tree mortality, especially along the lateral and terminal margins and on the lower parts of landslides, where the substrate tends to be severely disturbed. The year of mortality can be determined by cross-dating of ring width signatures from the dead and living trees. Some trees survive, although generally under poor growing conditions and tilted. On the upper valley sides, slump blocks and the associated soil and vegetation commonly remain intact, despite the considerable displacement of bedrock. The failure is deep below the surface, over a curved plane, causing the block to rotate and the trees to tilt backward toward the scarp face. Titled trees produce reaction wood and grow asymmetrically, so that the trunk curves upward to resume a vertical growth. Dating of landslides is thus "based on changes in the amount and eccentricity of tree growth." Dendrogeomorphology is a reliable means of determining the age of recent slope failures, where a single event, such as the rotational failure of a slump block, dominates the signal. Delayed growth response to disturbance and errors in tree ring counting caused by missing or false rings, prevent exact results.

The disturbance of trees by floods produces similar growth responses, plus the scarring of cambium from the impact of boulders and logs carried in the flow. In general, there is better agreement among the samples from flood-affected trees than with landsliding, since they occupy level ground, where overbank stream discharge is the only geophysical disturbance. There are meteorological data, photographs, and written documentation for the three most recent floods. Whereas the use of these methods is limited by the relatively short life spans of most tree species, evidence of earlier floods and landslides should exist in dead wood. Reconstructing geophysical events from dead wood is more involved, however, since the logs are not in situ and cross-dating is required to determine the years of anomalous growth.

Dendroglaciology. The application of dated tree rings to the study of glacier-related process, such as glacial fluctuations.

Dendrohydrology. Related to dendroclimatology, it is the utilization of tree ring data for the reconstruction and study of past and present hydrologic phenomena, including water supply, river flow, and flood frequencies, and the variations in stream flows, lake levels, and flood history.

Dendropyrochronology. The science that uses tree rings to date and study past and present changes in forest fires. Example: dating the fire scars left in tree rings to determine how often fires occurred in the past. Originally termed *pyroden-drochronology*.

Archaeology. The dating and study of behavior of past human cultures.

Stable Isotopes. The study of stable isotopes of carbon, oxygen, and/or hydrogen of tree rings and application of results in various climatological, ecological, or hydrological contexts.

Dendrochemistry-Environmental Studies. Analysis of inorganic elements in tree rings and interpretation of chemical changes in the environment through time.

Basic Techniques. Improving methods of dendrochronology, including use of high-technology methods such as X-ray densitometry and image analysis.

Quantitative Methods. Basic research on how best to quantitatively analyze tree ring data.*

CREATING A TIME LINE

> A **tree** has grown only once and ultimately its **ring** pattern can only fit at one place in **time**
>
> —Ballie, M. G. L. (1982) Tree-ring data and Archaeology
> (Chicago: University of Chicago Press)

Taking a Sample

Increment cores can be taken at any point on the stem (bole) of the tree, the result of which will be the age at that point. Standard samples for forest inventory, for instance, are taken at what is know as dbh, the diameter at breast height, or 4.5 feet from ground level. For most trees, this point is above the root swell of the base of the tree. Also, for most trees, it will take 6 to 10 years for a tree to reach that height, so some figure must be added to the reading to determine the total age of the tree.

An increment borer is a precision instrument and must be handled with care. The end of the bit is very sharp and must be kept sharp in order to extract a good sample. After a core has been taken, it is a good practice to immediately return the bit to its holder, the handle.

*Much of the information in this list of definitions was adapted from the Ultimate Tree-Ring Web Pages by Henri D. Grissino-Mayer (http://web.utk.edu/~grissino).

A time line extends a chronology based on living trees farther back in time through cross-dating.

How Crossdating Works (Read figure from right to left)

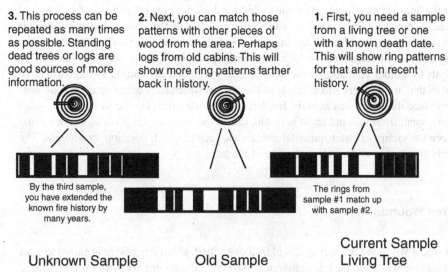

3. This process can be repeated as many times as possible. Standing dead trees or logs are good sources of more information.

2. Next, you can match those patterns with other pieces of wood from the area. Perhaps logs from old cabins. This will show more ring patterns farther back in history.

1. First, you need a sample from a living tree or one with a known death date. This will show ring patterns for that area in recent history.

By the third sample, you have extended the known fire history by many years.

The rings from sample #1 match up with sample #2.

Unknown Sample

Old Sample

Current Sample Living Tree

Figure 18.13 Determining the age and date of an unknown sample (left), by comparing it with an old sample (middle) and a current sample (right). *Source*: Adapted from Jody Lyle, "Making Maps Out of Tree Rings" US National Park Service.

To get a good sample, one in which the rings can be easily seen and counted, demands a clean start to the boring. The borer should be kept steady and the handle should be turned consistently. Bore into the tree as far as wanted, to the wound if aging a wound, to the center if aging the tree. The center should be approximately the center as eyeballed, unless the tree is eccentric due to growing on a slope or at an angle, or has been severely wind-blown. When such is the case, the offset center must be accounted for.

Once the boring is complete, very carefully remove the extractor with a slow, steady pull, keeping one hand under it in case of breakage of the core. Once core has been removed, it should be placed in a wooden holder or at least a drinking straw, and labeled. If desired, the rings can be counted on the spot, and the core returned to the hole in the tree. If difficulty in counting is an issue, the sample is returned to the lab and counted under different lighting, increased magnification, or with the aid of enhancements.

Cores are sometimes made easier to read by rubbing them with soft chalk, carefully taking off a smooth slice with a razor blade or sharp knife, or by using various stains and varnishes.

Aging a Tree

Some trees need to be examined for age, to determine whether the tree found is old enough to have been the tree called for in a description. Just because there is a tree at the approximate location of one called for does not mean that it is in fact the right tree. Not only does it need to satisfy the position of the corner, but it also needs to be old enough to have been in place when the description was first written. This is another important reason to trace a description back to its origin; otherwise an analysis cannot be made on the tree. Size is not a good indication, as there are large trees that are not very old because of the site on which they occur or if they are near anyplace that is or was heavily fertilized, such as a farm. By the same token, some very small trees turn out to be very old, due to the site they are growing on or having been the victim of environmental influences, such as wind, flooding, or drought. The only true test is to determine the age of the tree.

Tree Wounds

A tree is wounded whenever it is blazed or scribed, when fence wire is nailed to it, or otherwise when the bark is penetrated. Affecting the outer bark will not leave a scar inside, but when living wood tissue is injured, it reacts as any living thing does. Not only is a scar left, but usually there also is a discoloration due to the invasion of bacteria and other types of microorganisms. This is of great benefit to the survey investigator, since it allows the aging of survey marks and the installation of fence wire.

Dating a Blaze

The date that a tree was blazed or otherwise wounded may be found within a year or two by extracting a core near the location of the blaze. If the blaze is entirely grown over, some would take the core at the position of the blaze. Other investigators would prefer to take samples above and below the site of the blaze, whether it is grown over or not, If the blaze was painted, taking the sample exactly at the site of the blaze will sometimes yield small flakes of the paint.

Dating the Installation of Wire

The installation of wire onto a tree may generally be dated within a year or two by extracting a core close enough to the wire to capture the staining. It is best to take at least two samples, one above and one below the wire whenever possible.

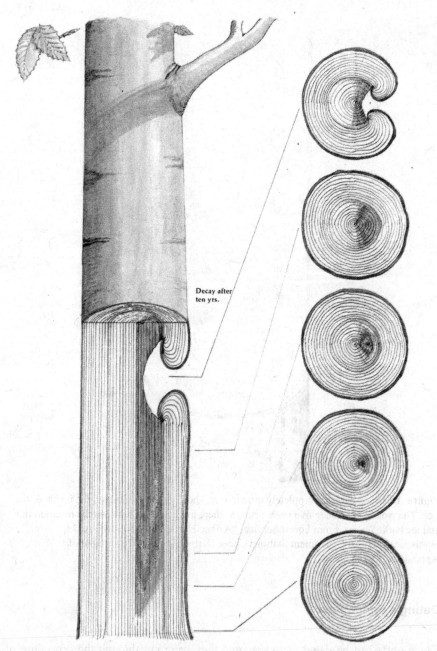

Decay after
ten yrs.

Figure 18.14   A partially grown over wound showing cross sections at various points on
the stem and the staining within. This wound was made 10 years prior, showing 10 years of
clear growth rings from the staining outward to the bark. In this case, rot has entered
through the wound and the interior of the tree has begun to deteriorate. *Source*: From Forest
Service, "A Tree Hurts, Too," NE-INF-16-73, Northeastern Forest Experiment Station
(Upper Darby, PA: U.S. Department of Agriculture, 1973).

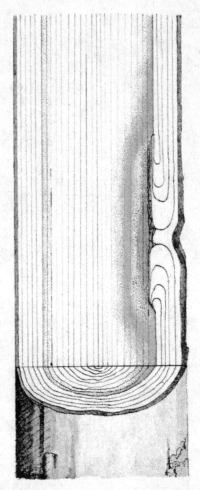

Figure 18.15 A wound completely grown over, showing the staining on the inside of the tree. The wound was made five years prior, as there are five rings between the discoloration and the bark. *Source*: From Forest Service, "A Tree Hurts, Too," NE-INF-16-73, Northeastern Forest Experiment Station (Upper Darby, PA: U.S. Department of Agriculture, 1973).

Dating a Fence Post

Fence posts can be dated as to the time they were cut through the procedure of crossdating. Comparison with a known time line, a match can be made. However, posts can be cut one year, and the fence put up later. Also, without extensive sampling and comparison, there is no guarantee that a particular fence post was not a

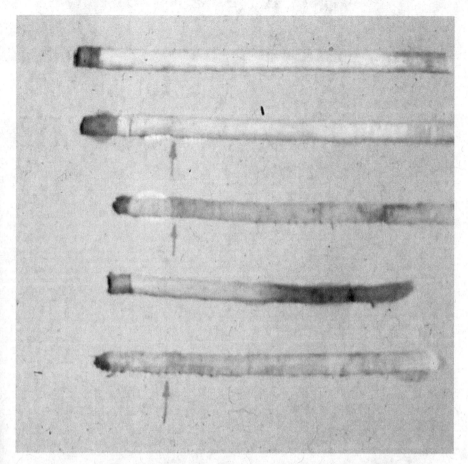

Figure 18.16   Stains on increment cores indicating when the trees were blazed. These were all taken from one blazed line and were consistent at 12± years.

replacement. Dating the installation or erection of a fence by this method is approximate at best.

Dating a Corner Post

Without a scribed date, a post can also be dated by the method of cross-dating. It might be expected that most posts were installed at the time they were cut, especially if cut at or near the location of the corner. In addition, if the tree that the post was taken from can be located, an additional check may be available.

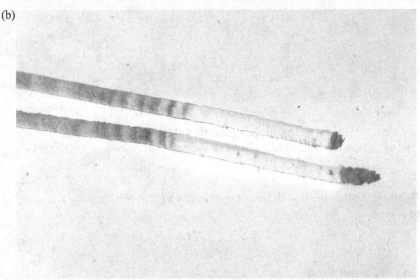

Figures 18.17 and 18.18 This large pine (a) shows scarring in the form of lines on the outer bark. There are also remnants of wire leading out of the tree and onto the ground, where the wire is covered over with pine needles and other debris. Cores taken from the tree, above and below the wire (b), demonstrate the difference in diameter by one being longer than the other, but also exhibit the same ring *pattern*. Counting inward from the bark (the dark-colored material on the right-hand end) toward the center of the tree, the staining indicates the discoloration of the wood due to the wound.

Dating a Building

Building timbers can be aged by cross-dating. However, there is no guarantee that the building was erected at the same time the timber was cut. Again, unless several samples are taken, there is no way to tell if a single timber is a replacement or perhaps a part of later construction, such as building expansion.

Durability of Wood Posts

A number of native North American woods are known heartwood durability. The USDA Forest Products Laboratory *Wood Handbook* lists heartwood of the species listed in Table 18.1 as "resistant or very resistant" to decay.

The older farmers putting up miles of fencing had this to say: Black locust fence posts will last *one year longer than a rock*. Aspen posts need to be kept track of. When setting one, stick a wooden match in the ground beside it; when you return, the match will still be there.

Knowing the resistance of woods provides two things: the ability to select a wood that will be the most durable, and when something was set, the ability to have an idea how long it might last under certain conditions and what we might expect to find.

```
Baldcypress (old growth only)
Black locust[1]
Post oak
Catalpa
Mesquite
White oak
Cedars
Red mulberry[1]
Osage orange[1]
Black cherry
Bur oak
Redwood
Chestnut
Chestnut oak
Sassafras
Arizona cypress Gambel oak
Black walnut
Junipers Oregon white oak
Pacific yew
```

[1]Exceptionally high decay resistance.

Table 18.1   Durability of Wood Posts

Species	Expected life span if untreated	Comments
Osage-orange (hedge)	35+	Best native post species. Does not need treatment. "Outlasts the hole."
Black Locust	20+	Used for railroad ties and posts. Good strength. Holds staples well. Does not need treatment.
Eastern Redcedar, Juniper	20+	Readily available. Heartwood is very decay resistant. Does not need treatment.
Honeylocust	15+	Good strength. Little shrinkage. Best if treated.
Hickory	15+	Very hard, moderate decay resistance. Best if treated.
Catalpa	15+	Good species to plant for posts. Treating will extend life.
Mulberry	15+	Easily grown for posts. Best when treated.
Bur Oak	10+	Slow growing. Treating will extend life.
Black Walnut	10	Good short-term post. Best if treated.
Hackberry	<10	Holds staples well. Short life span as post. Little shrinkage.
Green Ash	<10	Holds staples well, good strength. Needs treatment.
Ponderosa Pine	<5	Decays rapidly if untreated. Use only if treated.
Cottonwood	<5	Poor choice for use as post.

Table 18.2   Decay resistance of selected trees

Figure 18.19 Two photographs of fence remains at the location where one was called for in an 1854 description.

Figure 18.20   Two photographs of fence remains at the location where one was called for in an 1854 description.

An 1854 deed gave calls for directions and distances, abutters and physical monuments. The last line in the description was "all calls as the fences now stand." Figures 18.19 and 18.20 show the remaining evidence found during the retracement. Figure 18.19 shows a section of board fence, while Figure 18.20 shows a fence post with square-cut nails protruding. While it is not known whether this fence was in existence when the description was written, it compares reasonably well with the other calls in the description.

CHAPTER 19

FENCES

> Stone walls of ancient vintage offer strong evidence of
> boundary lines.
>
> —*Hawley v. Macdonald*
> 7 Conn. Supp. 516 (1940)

ANCIENT FENCES

Law of Fences

A fence has been defined as "a hedge, structure, or partition, erected for the purpose of inclosing a piece of land, or to divide a piece of land into distinct portions, or to separate two contiguous estates."[243] Put another way, in layman's terms, a fence is for the purpose of keeping something in, keeping something out, or to mark a property boundary.

> Ancient fences used by a surveyor in his attempt to reproduce an old survey are strong evidence of the location of the original lines and, if they have been standing for many years, should be taken as indicating such lines as against the evidence of a survey which ignores such fences and is based upon an assumed starting point. It is said that a long-established fence is better evidence of actual boundaries settled by practical location than any survey made after the monuments of the original survey have disappeared. Accordingly, a fence erected on a surveyed line shortly after the land has been surveyed may serve as a monument to control courses and distances or a subsequent survey after the stakes set out at the time of the original survey have disappeared.[244]

[243] *Black's Law Dictionary.*
[244] 12 Am Jur 2d, Boundaries, § 71. Stakes and fences.

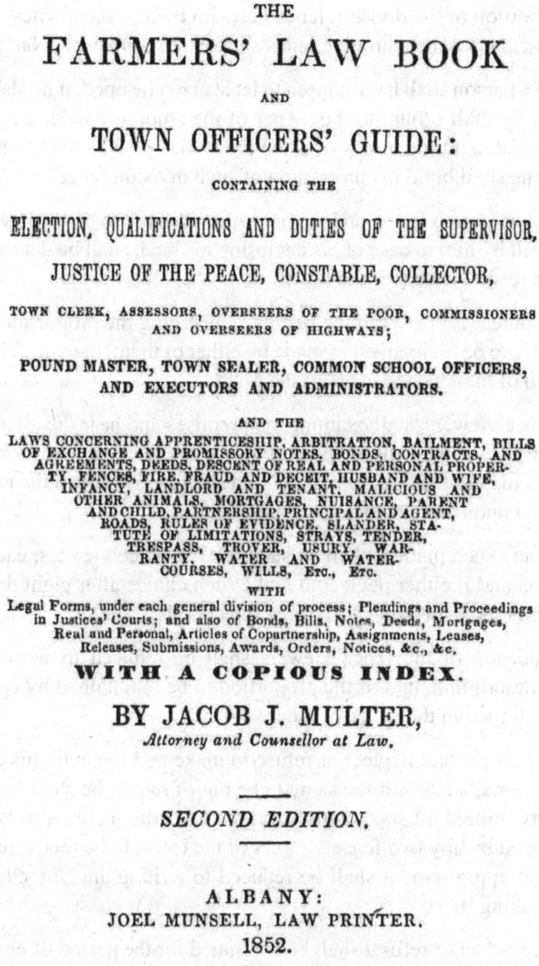

THE

FARMERS' LAW BOOK

AND

TOWN OFFICERS' GUIDE:

CONTAINING THE

ELECTION, QUALIFICATIONS AND DUTIES OF THE SUPERVISOR,

JUSTICE OF THE PEACE, CONSTABLE, COLLECTOR,

TOWN CLERK, ASSESSORS, OVERSEERS OF THE POOR, COMMISSIONERS
AND OVERSEERS OF HIGHWAYS;

POUND MASTER, TOWN SEALER, COMMON SCHOOL OFFICERS,
AND EXECUTORS AND ADMINISTRATORS.

AND THE

LAWS CONCERNING APPRENTICESHIP, ARBITRATION, BAILMENT, BILLS
OF EXCHANGE AND PROMISSORY NOTES, BONDS, CONTRACTS, AND
AGREEMENTS, DEEDS, DESCENT OF REAL AND PERSONAL PROPER-
TY, FENCES, FIRE, FRAUD AND DECEIT, HUSBAND AND WIFE,
INFANCY, LANDLORD AND TENANT, MALICIOUS AND
OTHER ANIMALS, MORTGAGES, NUISANCE, PARENT
AND CHILD, PARTNERSHIP, PRINCIPAL AND AGENT,
ROADS, RULES OF EVIDENCE, SLANDER, STA-
TUTE OF LIMITATIONS, STRAYS, TENDER,
TRESPASS, TROVER, USURY, WAR-
RANTY, WATER AND WATER-
COURSES, WILLS, ETC., ETC.

WITH

Legal Forms, under each general division of process; Pleadings and Proceedings
in Justices' Courts; and also of Bonds, Bills, Notes, Deeds, Mortgages,
Real and Personal, Articles of Copartnership, Assignments, Leases,
Releases, Submissions, Awards, Orders, Notices, &c., &c.

WITH A COPIOUS INDEX.

BY JACOB J. MULTER,

Attorney and Counsellor at Law.

SECOND EDITION.

ALBANY:
JOEL MUNSELL, LAW PRINTER.
1852.

Figure 19.1 Title page from the second edition of *The Farmers' Law Book and Town Officer's Guide* by Jacob J. Multer, published in 1852.

Local Rules and Regulations

The treatise that follows, excerpted from *The Farmers' Law Book and Town Officer's Guide* by Jacob J. Multer, like so many in the mid-nineteenth century, contained the current law regarding fences.

CHAPTER XI. FENCES—LAWS CONCERING.

The provisions of the statute in relation to fences are as follows: Where two or more persons shall have lands adjoining, each of them shall make and maintain a just proportion of the division fence between them, except where the owner or owners of either of the adjoining lands shall choose to let such land be open.

Where a person shall have chosen to let his land lie open, if he shall afterwards enclose it, he shall refund to the owner of the adjoining land, a just proportion of the value at that time of any division fence that shall have been made by such owner, or he shall build his proportion of such division fence.

The value of such fence, and the proportion thereof to be paid by such person, or to be built by him in case of his enclosing his land, shall be determined by any two of the fence viewers of the town.

If a dispute arise between the parties, concerning the proportion or particular part of fence to be maintained or made by either of them, the same shall be settled by any two of the fence viewers of the town.

The fence viewers shall examine the premises and hear the allegations of the parties. In case of their disagreement, they shall select another fence viewer to act with them, and the decision of any two shall be final upon the parties to such dispute, and upon all parties holding under them.

When any such matters shall be submitted to fence viewers, each party shall choose one, and if either neglect to make such choice after eight days' notice to make such choice, the other party may select both.

The decision of the fence viewers shall be reduced to writing, and shall contain a description, and of the proportion to be maintained by each, and shall be forthwith filed in the office of the town clerk.

If any person shall neglect or refuse to make and maintain his proportion of such fence, or shall permit the same to be out of repair, he shall be liable to pay to the party injured all such damages as shall accrue thereby, to be ascertained and appraised by any two fence viewers of the town, to be recovered, with costs of suit. The appraisement shall be reduced to writing and signed by the fence viewers making it.

If such neglect or refusal shall be continued for the period of one month after request in writing, to make or repair such fence, the party injured may make or repair the same, at the expense of the party so neglecting or refusing, to be recovered from him with costs of suit.

The following is the form of said notice:

To———

Take notice that you are required to make or repair your proportion of the division fence along the eastern boundary of my land, [describe it so that it can be understood,] within one month from the service of this notice.

—Yours, &c.

Dated Milford, April 6, 1852.

John Platt

If any person who shall have his proportion of a division fence, shall be disposed to remove his fence, and suffer his lands to lie open, he may, at any time between the first day of November in any year, and the first day of April following, but at no other time, give ten days' notice to the owner or occupant of the adjoining land, of his intention to apply to the fence viewers of the town, for permission to remove his fence; and if at the time specified in such notice any two of such fence viewers, to be selected as aforesaid, shall determine that such fence may with propriety be removed, he may then remove the same.

If any such fence shall be removed without such notice and permission, the party removing the same shall pay to the party injured, all such damages as he may sustain thereby, to be recovered with costs of suit.

Whenever a division fence shall be injured or destroyed by floods or other casualty, the person bound to make and repair such fence or any part thereof, shall within ten days after he shall be thereunto required by any person interested therein, make or repair said fence. Such requisition shall be in writing, and signed by the party making it.

If such person shall refuse or neglect to make and repair his proportion of such fence, for the space of ten days after such request, the party injured may make or repair the same, at the expense of the party so refusing or neglecting, to be recovered from him with costs of suit. Witnesses may be subpœnaed, sworn and examined by either of the fence viewers, and either may issue subpœnas for that purpose.

Whenever the electors of any town shall have made any rule or regulation, prescribing what shall be deemed a sufficient fence in such town, any person who shall thereafter neglect to keep such a fence according to such rule or regulation, shall be precluded from recovering compensation in any manner, for damages done by any beast lawfully going at large on the highways, that may enter on any lands of such person, not fenced in conformity to said rule or regulation, or for entering through any defective fence.

When the sufficiency of a fence shall come in question in any suit, it shall be presumed to have been sufficient until the contrary be established.

The above are all the statutory provisions in relation to fences. The proportion that each is to make of a division fence, often becomes and is established by custom and length of time that each has kept the same.

Fence viewers may be dispensed with in the case where a party is trespassed upon through defect of his neighbor's fences, or the part which he is bound to repair, and by his cattle. Such action may immediately be brought.

Where the town make no rule or regulations in relation to cattle running at large, or fences, then one is not bound to fence along the road as far as street cattle are concerned, for they have no business there, and every man is bound to keep his cattle on his own land. Defect of fences of course is not in question in relation to such fences or cattle. And in the case of line fences, one is only bound to fence against his neighbor's cattle, and not against those not lawfully in his neighbor's lot. For further particulars in relation to fences, the reader is referred to trespass. Where the town makes no rule or regulation in relation to the height or sufficiency of fences, men of age and experience are to testify in relation to the sufficiency of the fences when questioned or disputed in any action.

But it appears from a recent decision that the town has no right to make rules and regulations in regard to fences, and to permit cattle to run at large. That being the case, the common law rule prevails, which is, that every man (except in the case of line fences), is bound to keep his cattle on his own land; and that a party is bound only to make fences to keep his own cattle in, and not others out of his premises. After all there is some doubt whether the law clearly sustains the above decision. The latest decision on that subject contradicts the above, but the court was divided.

FENCE VIEWERS

Not long ago, and in some towns still existing, an important job in every American town was that of the fence-viewer. Fence viewers decided the necessity and sufficiency of all the fences in their neighborhood. They settled disputes between landowners, and they were liable (by fine) for the neglect of fences within their jurisdiction.

Fence viewers had a deputy and assistants, two of whom carried a Gunter's chain for measuring acreage and fence mileage.

Case #39

FENCE VIEWERS: JUDICIAL OFFICERS?

Edgerton et al. v. Moore

28 Conn. 600 (1859)

Under the statute (Rev. Stat., tit. 15, § 5,) which provides that, where any person shall, after notice from the fence viewers, have neglected to repair his part of a divisional fence, and, upon such neglect, the same shall have been by the adjoining owner, the fence viewers shall estimate the value of such repairs and make a certificate thereof, and that the party making the repairs shall have a right to recover double their value from the delinquent party, it is not necessary that the fence viewers should give notice to such delinquent party of the time of their meeting to estimate the value of the repairs.

Fence viewers are not judicial officers.

The cases from the Massachusetts and Maine reports, which hold that notice should be given by the fence viewers in such cases, do not affect our argument. 1st. The courts of those states have adopted radically different views from those of our own state, which regard to the character and functions of those officers. They treat them as judicial officers, before whom a trial or judicial inquiry is to be had. *Lamb* v. *Hicks*, 11 Met. 496. *Scott* v. *Dickinson*; 14 Pick., 276. *Harris* v. *Sturdivant*, 29 Maine, 366. But our court, in *Fox* v. *Beebe*, supra, takes an entirely different view of the their character, and classes them with inspectors of provisions and other like officers, who decide by direct examination and not by the testimony of witnesses, and before whom there is no formal hearing or trial, as before auditors or committees in chancery. 2d. But there is an important difference between the statutes of those states and our own. That of Maine is a transcript of that of Massachusetts, and both require that the fence viewers, before proceeding to adjudge the old fence insufficient, "shall give due notice to each party." The fact that notice is required of the time of their inspection of the old fence, would seem to imply that a like notice is to be given of the time of their inspection of the new one. But our statute contains no such requirement, and therefore there is no room for the same inference here.

1. The proceedings of fence viewers in estimating the value of the repairs made under the statute referred to, are of a judicial nature. They involve the necessity of an inquiry into the amount of labor performed by the party making the repairs, its value, and the quantity and value of the materials furnished by him. And their decision is final and conclusive upon the other party, the statute allowing him no appeal. Sanborn v. Fellows, 2 Foster, 473.
2. Whenever a court, or any person exercising a legal authority, is to act judicially, or to exercise discretion in a matter affecting the rights of another, the party to be affected is to have reasonable notice of the time and place where such act is to be done, to the end that he may be heard in defense of those rights. Several citations.

In the case of *Fox* v. *Beebe*, 24 Conn. 271, it was held that fence viewers are not judicial officers, that their functions are more analogous to those of appraisers and inspectors, and other boards of that character, than of judges or of courts, and that no notice whatever need be given of their first examination of the defective fence until after it has been made. We think the doctrine of the court in that case should be adopted in the case now before us. The character of the office is indicated by its name, and its duties are, in the section under which these proceedings were had, so distinctly specified as to leave no room to doubt but that fence viewers are expected to proceed in a summary manner, upon their own view, and not upon the testimony of others. They are to be, in the first place, "called on" by the aggrieved party, "to view" the defective fence, and if, upon such view, they find it insufficient, they are to give notice in writing of such insufficiency to the party whose duty it is to keep the fence in repair, and when the repairs have been made by the complainant, they are to be "judged" by the fence viewers complete, and the fence viewers are to "estimate the value of," or appraise, such repairs, not to ascertain their cost by evidence or by calculation. Fence viewers are generally selected on account of their presumed fitness for the place, by reason of their familiar acquaintance with the subject of their examinations, and they constitute a domestic tribunal of great utility in preserving the peace and harmony of neighborhoods; a tribunal always at hand, readily accessible, economical in its action, and prompt in its conclusions; while its awards, being conclusive only as to the insufficiency of the fence when first examined, and the sufficiency and value of the repairs actually made, involve but little property, determine no question of title, and fix on future liability. Its proceedings being regulated by the express provisions of our statute, the validity of such proceedings must be tested by those provisions, rather than by the rules of the common law regarding proceedings in courts of justice, or by the determinations of courts in states whose statutes are unlike our own.

In some states, fence viewers were not judicial officers, and, contrary to the belief of some, did not and do not decide boundary issues. They decided *fence* issues: who was responsible for maintenance, whether a fence was "good and sufficient" and in proper order. In other states, fence viewers did, or do, have limited judicial power. Consulting the statutes existing at the time and the relevant court decisions, will assist in deciding the value of a fence viewer's activities and resulting report. Regardless, fence viewers' reports can be very helpful. Since the position was either an elected or appointed municipal position, the records of fence viewers generally are found in the files of the municipality.

According to a review of court decisions, fence viewers were found in most states east of the Mississippi River. See Appendix VII for a list of decisions.

Case #40

Shaw v. Gilfillan

22 vt. 565 (1850)

A fence-viewer decision had been reached in this dispute, and the case was appealed. One party claimed that the decision was in the nature of a judgment. The court stated:

> This doctrine would invest fence viewers with a much higher and more responsible duty, than they have generally been supposed to possess, and much beyond the terms of the statute defining and prescribing their official duty. In cases like the present the fence viewers are only authorized, by statute, to determine the proportion, or part, of the fence, which each adjoining owner shall

make, or maintain. If disputes arise between the occupants of adjoining lands, as to their ownership, or their boundary lines, these are to be settled by some other tribunal than the fence viewers;—they are authorized to divide fences, and the statute declares their judgment in that respect conclusive; but they have no authority to settle the rights of different claimants to landed property, or to establish disputed boundaries. Neither party, therefore, is concluded, by the decision of the fence viewers, from contesting the question of ownership in himself, or his adversary, or the location of their boundaries.

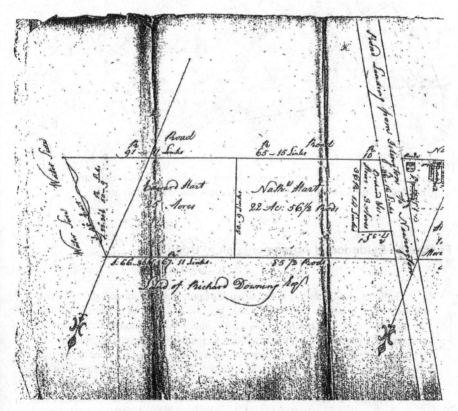

Figure 19.2 Example of a fence-viewer's map.

Forestry-type maps such as the one shown in Figure 19.3 often depict features of interest to an investigator. In this example, white pine blister rust control maps were produced under a cooperative state and federal program, beginning in the 1920s. One value is that they document a particular feature at a point in time. Some later maps were made using aerial photography. Features shown are fences, roads and trails, cemeteries, and a few other miscellaneous items.

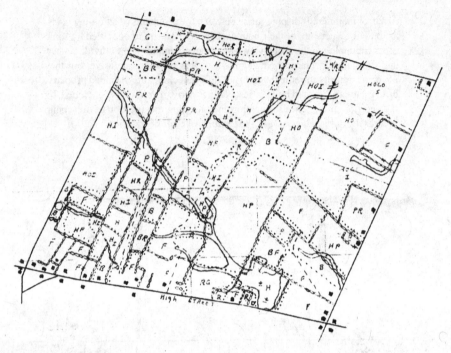

Figure 19.3 Example of a blister rust control map, one type of forest reconnaissance map. Evidence in the form of fences and stone walls is abundant in this area.

Fences as Evidence of Agreements as to Boundaries

The location of a fence, in the absence of proof of a continued and exclusive adverse possession for twenty years, may be evidence of an agreement and acquiescence in the line thereof as the true line. A presumption that an agreement formerly was made as to the location of a boundary line may arise from the fact that one or both the adjoining owners have definitely defined such line by erecting a fence or other monument on it and that both have treated the same as fixing the boundary between them for such length of time that neither ought to be allowed to deny the correctness of its location. Two sections of a fence, situated on the same straight line, but not adjoining have been considered sufficient evidence as to the existence of an agreement between parties settling such line.

—*Smith v. Hosmer*
7 N.H. 436, 28 Am. Dec. 1835

Ancient fences, used by a surveyor in his attempt to reproduce an old survey, are strong evidence of the location of the original lines, and if they have been standing for many years, should be taken as indicating such lines, as against the evidence of a survey which ignored such fences and was based upon an assumed starting point. A fence erected on a surveyed line shortly after the land has been surveyed may serve as a monument to control courses and distances or a subsequent survey, after the stakes of the original survey have disappeared.

—4 Ruling Case Law, Boundaries, § 70

Acquiescence in Fences as Boundaries

Under the general principles of the law of adverse possession it is essential, in order that possession may be considered as being adverse, that there should be an intention to claim title. Where the intention is to claim only up to the true line, wherever it may be, the necessary element of an intent to claim title adversely is absent. Accordingly, where lands are divided by a fence which their owners suppose to be the true line, each claiming only to the true line wherever that may be, they are not bound by the supposed line, and must conform to the true line when it is ascertained; but where a person takes and holds possession of land up to a fence, and claims to be the owner up to it, his possession will be adverse, although he may believe the fence to be on the true line, when in fact it has been erroneously located. If a fence has been recognized by adjoining owners of land as on the true line for more than twenty years, both of them may be conclusively estopped from denying that it is on the true line, whether it was originally established on the true line or not. Apart from the question of adverse possession the erection of a fence may be evidence of the location of a boundary line which it was intended to make, and acquiescence in it for a reasonable length of time may become binding on the adjacent landowners. Yet a fence may be maintained between adjoining proprietors for the sake of convenience merely, and without intention of thereby fixing boundaries, and, therefore, will not be given that effect. A property owner is not entitled to rely on the erection of a fence by a neighbor as the establishment of a boundary line, where the fence exited only a comparatively short period of time, and the neighbor was honestly mistaken in erecting it, while the property owner had the means of knowing that it was not on the true line.

—4 Ruling Case Law, Boundaries, § 71

SIGNIFICANT COURT DECISIONS ON FENCES

In the absence of natural boundaries or monuments, and of monuments or stakes set in the course of the original survey, the lines of ancient fences and long-continued occupation of adjacent lots and blocks in the same plat, if evidently intended to mark the true lines of such lots and blocks, have greater probative force than mere measurements of courses and distances.

—*Galesville v. Parker*
83 N.W. 646, 107 Wis. 363 (1900)

Case #41

CHIEF JUSTICE THOMAS M. COOLEY #1

Stewart v. Carleton

31 Mich. 270 (1875)

Where a fence has been treated and acquiesced in as the correct boundary line between adjacent owners for fifteen years, the boundary ought not to be disturbed, even if there were some variance from the true line; but where such fence was the line actually agreed upon, and a deed giving distances had been accepted under assurances that it did not vary such boundary line, the grantor in such deed, and all claiming under him, except an honest purchaser without notice, would be estopped from setting up any different boundary.

The entire lot of the complainant in this case was a parcel of land formerly owned by Beard and Haynes, and the easterly end sloped downward to a piece of somewhat lower ground, concerning which the difficulty seemed to have arisen. In January, 1856, Beard and Haynes agreed to sell to the Port Huron & Milwaukee Railroad Company, the eastern part of a larger tract of land which they then owned (and of which the premises in dispute were a part), and in the contract the western boundary of the railroad tract was to be bounded "*by the slope of the hill.*" A fence was built then or previously, which ran along the base of the slope, and which complainant claimed was the line of the tract in question. In July, 1856, Beard and Haynes, at the request of the company, and in pursuance of the contract, executed a deed, which they were assured by the company's agent and engineer was in conformity with the contract, whereby the western boundary was fixed at four hundred and twenty feet from the easterly line of section fifteen, of which the tract was a part. The reason given for mentioning distances was, that the company preferred an exact measurement; and the parties, when they conveyed, were assured that the actual survey put the line where they had agreed it should be, at the fence, or a trifle east of it.

This fence continued undisturbed, and coincided in direction with the line of railroad lands bought of adjoining owners.

In August, 1856, the land west of the fence was conveyed to complainant's grantors, and in the deed was described, not by metes and bounds, but by adjoining property, and the easterly line was defined as "*the line of lands sold by said Beard and Haynes to the Port Huron & Milwaukee Railroad Company.*"

The court stated:

"Upon the testimony in the case the facts do not seem to us to be left in any doubt. There can be no question but that all parties assumed and acted on the assumption, that the fence was the boundary, and that no one ever questioned it until defendant made his purchase. The railroad company from whom he bought did not suppose there was any controversy about lines, and had no actual knowledge on the subject at all. The line had been acquiesced in as properly located, for fifteen years before defendant purchased.

The landmarks which have been recognized and acted on so long, ought not to be disturbed, even if there had been some variance from the lines intended. But here the line so fixed was the line actually agreed upon; and if the deed varied from the contract it was by mutual mistake or fraud; and we do not think any fraud was designed. If it had been, the equity would be still stronger."

Case #42

CHIEF JUSTICE THOMAS M. COOLEY #2
Diehl v. Zanger
39 Mich. 601 (1878)

In defense to an action of ejectment based upon an alleged mistake in the original survey, evidence is admissible that the existing boundaries had been defined for more than twenty years by buildings, fences, and harmonious occupancy.

A re-survey, made after the monuments of the original survey have disappeared, is for the purpose of determining where they were, and not were they ought to have been.

A long-established fence is better evidence of actual boundaries settled by practical location than any survey made after the monuments of the original survey have disappeared.

Long practical acquiescence in a boundary, between the parties concerned, may constitute such an agreement on it as to be conclusive, even if it had been erroneously located.

The controversy concerns part of a lot on a plat of a subdivision of a part of outlot 182, Rivard farm, surveyed by Thomas Campau in 1850, and recorded in 1851. There are forty-eight of these lots in the subdivision. Whether they have all been sold off and improved by the purchasers we are not informed, but it appears from the record that many of them have been. It also appears that there has been a practical location of a street on one side the plat and of other streets across it, and also of the lot lines. The lot the boundary of which is in dispute in this case has been fenced in for twenty years by fences on the supposed lines, and it does not appear that the lines have been disputed until recently. The adjoining lots have also been claimed, occupied and improved according to the practical location of the lines.

This litigation grows out of a new survey recently made by the city surveyor. This officer after searching for the original stakes and finding none, has proceeded to take measurements according to the original plat, and to drive stakes of his own. According to this survey the practical location of the whole plat is wrong, and all the lines should be moved between four and five feet to the east. The surveyor testified with positiveness and apparently without the least hesitation that "the fences and building on all the lots are not correctly located" and there is of course an opportunity for forty-eight suits at law and probably many more than that.

"The [following] surveyor should have directed his attention to the ascertainment of the actual location of the original landmarks set by [the original surveyor], and if those were discovered they must govern. If they are no longer discoverable, the question is where they were located; and upon that question the best possible evidence is usually found in the practical location of the lines, made at a time when the original monuments were presumably in existence and probably well known. *Stewart v. Carleton*, 31 Mich., 270. As between old boundary fences, and any survey made after the monuments have disappeared, the fences are by far the better evidence of what the lines of a lot actually are."

Fences not referred to in a deed cannot control the distances stated in the deed.

The evidence offered to show that the fences were a boundary or monument which the jury could consider as sufficient to control the distances stated in the deed was properly excluded. The monuments which control courses and distances are those to which the conveyance itself refers. A reference to the adjoining land of the grantor as a boundary cannot be treated as describing a monument intended to control the dimensions stated because of the existence of a fence, which is not mentioned in the deed.

—Kashman v. Parsons
39 A. 179, 70 Conn. 295 (1898)

Fences and monuments established shortly after lots were surveyed and platted are better evidence of the true boundary lines than a survey made after the stakes of the original survey have disappeared.

No stakes of the original survey were found. The westerly fences of complainant's lot and other fences and monuments were established not long after the survey and plat were made. It was said in Carpenter v. Monks, 81 Mich. 103, 45 N.W. 477, "Fences of long standing, erected upon what parties have called the true line, and up to which they have improved and cultivated, are better evidence of the true line than surveys made after the monuments have disappeared." This feature of the case is ruled by that case, and also by Flynn v. Glenny, 51 Mich. 580, 17 N.W. 65, and other cases therein cited. We are also of the opinion that the

Lynns and Mr. Woolsey established a boundary line by agreement, and thereafter acquiesced in it, and so did the parties to this litigation, until the complainant believed that the original survey and plat would give him more land. Such an agreement is binding, although the line established is not the original one. Brown v. Bowerman, 134 Mich. 695, 97 N.W. 352; Manistee Mfg. Co. v. Cogswell, 103 Mich. 604, 61 N.W. 884; F.H. Wolfe Brick Co. v. Lonyo, 132 Mich. 162, 93 N.W. 251.

—*Breakey v. Woolsey et al.*
112 N.W. 719 (Mich. 1907)

A deed calling for a fence which had been built five years as a boundary could give no title beyond it.

Kinman and Tool owned adjoining lots in the town of Lusby, Kentucky. Kinman brought this suit against Tool, charging that he had fenced in a strip off his lot 20 feet wide. Tool filed an answer, controverting the petition, and pleading champerty and adverse possession. On a trial of the case, there was a verdict and judgment in favor of Kinman for a strip 16 1/2 feet wide. Tool appealed.

The proof shows that the strip in controversy lies within Tool's fence, and that the fence was built where it now stands about six years ago. The deed to Kinman was made on January 25, 1909, or about five years after the fence was built. The call of his deed for the line in controversy is as follows: "Running back 60 feet with Dr. Sparks' (now John Tool's) garden fence." Kinman's deed having been made about five years after the fence was built, and calling for Tool's fence, he took by the deed not title to the land beyond the fence. If this land does not belong to Tool, the title is still in Kinman's vendors, and to recover in ejectment the plaintiff must rely upon the strength of his own title, and not upon the weakness of his adversary's. Not only so, but when the deed to Kinman was made, Tool was in possession of his lot, claiming adversely up to the fence, having the land actually inclosed and in occupation. The deed to Kinman would be champertous as to all land within the fence.

It is said that the deed is not champertous as to this strip, because it was made pursuant to a contract entered into before the fence was built. Since nothing was agreed on, such a supposed contract was not in existence, and therefore did not apply.

—*Tool v. Kinman*
130 S.W. 1073, 140 Ky 208 (1910)

A fence claimed by defendants in ejectment for many years to mark the boundary between their lands and plaintiffs' is a "monument."

The building of a fence does not conclude adjoining landowners as to the boundary line, if it was built for temporary purposes without intent to make it the permanent boundary, or was the result of mistake or fraud.

A witness, one J.C. Boyd, a civil engineer and surveyor of long experience, testified for the defendants. He stated, "there were practically two surveys of the entire city as a whole; the first in laying out of the city originally by Sutter; subsequently, in 1878, a re-surveying by L.S. Bassett, then city engineer, who attempted to adjust the apparent difference in the property ownership to the line of the streets, for the purpose of street improvement largely, and for the purpose of rectifying errors that were shown to have been made by private surveys of private property, lots and so forth. It had been the custom theretofore to make surveys from the established monuments, from buildings, taking the nearest building as being a corner, approximately correct, and run out from that"; that the Sutter survey was made in 1848 or 1849 and the Bassett survey in 1878; that these two surveys do not agree, that Bassett attempted to adjust differences through the city, that his policy "was to run a line for several blocks as long as possible, the longer the better, and adjust it with existing conditions"; that there is no way of determining whether this recent survey "conforms to the original survey, the Sutter survey"; that he considered that there was physical evidence that there was a line fence established by agreement between the coterminous owners and "the only monument that we could accept of the old Sutter survey would could accept of the old Sutter survey would be the consideration of old buildings"; furthermore, that the Bassett survey "most assuredly is not the accepted survey at this time for the measurement of lots in the city of Sacramento" and "that considered in the light of the original survey, I do not think there is any part of lot 2—as originally surveyed—in this inclosure," declaring this was his professional opinion from a physical examination of the entire block including the said fence and "other collateral evidence."

There was no objection made that this was not a matter of expert testimony. Beside, the witness proceeded to state certain facts tending at least to support his conclusion. The most important of these, probably, was in reference to the fence which has been claimed for so many years as marking the boundary line. The fence itself is a monument, visible and obtrusive, which has existed for forty years or more, that under the peculiar circumstances of this case is quite persuasive in favor of the claim of defendants. It is a fair presumption that this fence was originally place upon the true line as then recognized and understood, and it is proper to assume, in the absence of evidence to the contrary, that when the fence was built, either that the contiguous owners had knowledge and information of the lines of the Sutter survey and acted accordingly, or that the true line was uncertain and by agreement it was fixed and marked by said fence. It is not surprising that this fence seemed so important to Boyd and to the trial judge. Its existence for so many years, the recognition accorded it as the true boundary, the acquiescence of the respective owners in the location and the improvements made accordingly were rightfully regarded as important if not decisive considerations in the determination of a line otherwise obscure and uncertain. It is true that no one testified directly that the boundary was actually located on the ground or that there was uncertainty in reference to it, but such may be fairly inferred form the various facts disclosed, and we may invoke the principle announced in Schwab v. Donovan, 165 Cal. 360, and cases therein cited. This must be especially true since there is not clear and satisfactory showing here that the legal title is different from what is indicate by the fence.

Of course, it is true, as said in Dierssen v. Nelson, 138 Cal. 398, 71 Pac. 456, that—

"the building of a fence does not always conclude the parties as to the boundary line; it may have been built for mere temporary purposes and with no intent to

*make it the permanent boundary, and it may have been the result of a clear
mistake or fraud*; but nothing of that kind appears in the case at bar."

—Perich et al. v. Maurer et al.
155 P. 471, 29 Cal. App. 293 (1915)

Where starting points cannot be found, purchaser can rely on stakes planted
or fences built which have long been recognized as indicating lot lines, and a
subsequent survey establishing a starting point cannot unsettle such boundaries.

Mistakes made in the measurement of original surveys may not be corrected
to disturb boundaries thus fixed.

—Pere Marquette Ry. Co. v. Tower Motor Truck Co.
192 N.W. 634, 222 Mich. 190 (1923)

Where starting points, such as section corners or quarter posts, cannot be found,
purchasers have a right to rely on stakes planted or fences built which have long
been recognized as indicating lot lines, and no subsequent survey establishing a
starting point should be allowed to unsettle such boundaries.

—Gregory v. Detroit Grand Haven & Milwaukee Railway Co.,
277 Mich. 317 (1936)

Evidence that landowner and his predecessors occupied land for more than 53
years claiming under deed to well-defined boundary, including **picket fence**
maintained thereon for part of such period, and exercised act of ownership con-
tinuously during such time and gave adjoining landowner permission to cultivate
disputed land, held to establish that landowner and predecessors were in adverse,
open, continuous possession of land to such boundary so as to acquire title thereto
by adverse possession.

Possession of land to well-defined boundary which is open, notorious, adverse,
and continuous for period of 15 years or more is sufficient to sustain claim of
title by adverse possession.

Where parties on each side of division fence have recognized fence as true
line for 15 years or more, reliance on paper title and tracing of paper title back to
commonwealth is unnecessary.

Division line between lands of adjoining owners which is agreed on and
thereafter recognized and treated as true division line for 15 years will be regarded
to be so located by court.

Landowners who acquiesced in location of boundary line called for in deeds
of adjoining owners and their predecessors in title for more than 30 years and their
successors in title *held* not entitled subsequently to claim as against adjoining
owner that such line was not line of deeds or patents if properly located.

—Lewallen v. Mays
265 Ky. 1, 95 SW.2d 1125 (1936)

Stone walls of ancient vintage offer strong evidence of boundary lines.

—Hawley v. Macdonald
7 Conn. Supp. 516 (1940)

In quiet title action, whether a *fence between adjoining lands* had become the boundary line by acquiescence or implied agreement, although the fence was not on boundary line as shown by survey was a question of fact.

In this case, appellant's contention was not that the old fence followed the true half section line according to any survey, but that the fence became the boundary line by acquiescence or implied agreement. This presented a plain question of fact. There was no evidence of agreement or acquiescence other than the conceded fact that the parties pastured their stock up to the fence from a time as early as 1910. Contra is the evidence that both parties, and their predecessors treated the fence as a cattle guard only, that both parties, and their predecessors, recognized and publicly stated that they were in doubt as to the true boundary line, and both agreed that a survey ought to be made to determine the boundary between their respective holdings.

The essence of the doctrine of a boundary being fixed by acquiescence is the implied agreement between the parties that a fence or other structure represents the true boundary and that the acquiescence of the parties, or their failure to object, supports an inference of such agreement. Essentially the doctrine is a mixture of implied agreement and estoppel. See Board of Trustees v. Miller, 54 Cal.App. 102, 105, 201 P. 952; Vowinckel v. N. Clark & Sons, 217 Cal. 258, 260, 18 P.2d 58; Roberts v. Brae, 5 Cal. 258, 260, 18 P.2d 698; 8 Am. Jur. pp. 802, 806.

But it is settled law that, to give rise to such an inference, the fence or monument must have been accepted or acquiesced in "as a boundary line," and not "as a barrier." Roberts v. Brae, supra, 5 Cal.2d at page 360, 54 P.2d at page 700; Phelan v. Drescher, 92 Cal.App. 393, 397, 268 P. 465.

During the trial the able trial judge made a personal inspection of the fence and the premises surrounding it. At the conclusion of the trial he filed an opinion which clearly states the reasons for the subsequent judgment. We quote in part:

"Now, there are certain fundamental principles that must be kept in mind. The plaintiff in this cause has the burden of proof and must prove his case by the preponderance of the evidence, so that his evidence, when weighed with the defendant's evidence, has the more convincing force and from which it results that the greater probability is in the plaintiff's favor. If the evidence is equally balanced the decision must go to the defendant.

"After considering all the evidence, the Court finds that the true boundary line between the land of the respective parties is as established by and as stated by Mr. Boling on the witness stand.

"There is no question but that the fence was allowed to remain in its present position fro over 34 years after the defendant acquired the north half of the Section. On the witness stand he testified to conversations with plaintiff's predecessor in interest, which if true, were admissions against interest and of itself would establish that the fence was known by the parties not to be on the true line and was built and maintained where it was by the plaintiff and his predecessor in interest at the place on the line and for the reasons stated by the defendant in his brief. As the defendant was a party in interest and the person making the alleged statements was dead, the Court viewed his testimony with great caution, but after viewing the fence, its location and direction, and the manner in which the defendant testified, the Court can not say that it was not true. On the other hand, the Court feels that the greater probability is that it is true, and it is corroborated by the direction of the fence, its deviations and the contour of the ground over which it is built. It clearly appears that if it had been constructed along the true line, it would have been constructed in parts over very difficult terrain on which to construct a fence. That is particularly true of the westerly end of the fence. The land is rough grazing land in the hills and some pretty high and rugged hills at that. It was of little value at the time the fence was constructed and is not of great value even now. In fact, the Court feels it is hardly worth more than the value of the attorneys' fees reasonably earned by the counsel in this case.

"There is no question but that the long acquiescence in a boundary fence line is presumptive evidence of an agreement that the fence was constructed on what the parties thought was the true line and constituted an agreed boundary line, arising out of an uncertainty, but the Court can not say that presumption was not overcome by the evidence of Eade and the physical facts."

—Copley v. Eade
184 P.2d 698, 81 Cal. App. 592 (1947)

Where stakes placed by original surveyor of platted property were still present when plaintiff purchased tract with reference thereto and *plaintiff erected fence* in accordance therewith, which *fence was still present* when defendants purchased adjoining property, plaintiffs were entitled to land included within boundaries of old stakes and subsequent fence.

Mr. Neeley's testimony was definite and he was careful to have the line pointed out to him and he and his wife, and others, stated positively that the fence was put in between two survey stakes. It could have only been put there for one purpose,—to mark the line—of course, to keep in stock, too, but primarily to mark the line. It remained there—was there when the present owners bought the property to the south.

—Neeley v. Maurer
195 P.2d 628, 31 Wash.2d 153 (1948)

In action to quiet title to a strip of land between plaintiff's and defendants' *land which was within a partial fence* built by defendant and claimed by him under a deed, evidence was sufficient to show that defendant was holding adversely to the world the entire strip of land *along a line indicated by the fence* and that plaintiff and his grantees knew of such adverse possession.

The record contains sufficient proof to show that Spainhower, ever since the time he erected the fence upon assuming possession under a contract of sale and after he acquired title under a deed, which was recorded, held the land in controversy adversely to the world and that these facts were known both to appellant and his predecessors in title.

While it is true that the fence did not extend completely across the division line between the two properties, it did extend over half the distance and was, as the trial court found, sufficient to establish the direction of this boundary. In Rader v. Howell, 246 Ky. 261, 54 S.W.2d 914, while construing K.S. 210, now KRS 372.070, we said:

"... possession under a recorded deed extends to the outside boundaries as defined in the deed, regardless of inclosure of only a portion, and such possession is adverse so as to render a deed thereto champertous. Lanham V. Huff, 228 Ky. 139, 14 S.W.2d 402;..."

See also Edwards v. Clark, 261 Ky. 749, 88 S.W.2d 914.

—Cox v. Spainhower
249 WS.W.2d 719 (Ky. 1952)

The circuit court's finding of location of land, excepted as graveyard from tract conveyed by deed describing excepted land as half acre *"under a barbed wire fence"*, by following *woven wire fence built along line of old barbed wire fence* on one side of cemetery and fixing other lines in relation thereto by metes and bounds enclosing nearly half acre, was proper.

The expressed intention of the original grantor in defining the exception is the area "now under a barbed wire fence." It does appear, however, that at that time the cemetery was not completely enclosed by a barbed wire fence, but there were two strands of such wire on three sides of it and perhaps a rail or board fence on the fourth side. The particular line is that on the north of the cemetery and south of the church lot. Its location determines the controversy, for it must be regarded to have been the line intended by the grantor, Eli Goins, in his deed of 1906 which carved out the exception which has been carried through subsequent conveyances. And any doubt as to the lines must be resolved against the Goins heirs. Sargent v. Trustees of Christian Church, 252 Ky. 57, 66 S.W.2d 5; Chaney v. Chaney, 300 Ky. 382, 189 S.W.2d 268.

It is well established in the evidence that later a woven wire fence was built, at least around the south line of the cemetery, exactly along the same lines of the old wire fence. Men who erected it testified that the new posts were put in the holes from which the rotted posts were removed. Using this as the basic line, the court, following the fence, located the other lines of the church lot in relation thereto. The judgment also locates the cemetery as being that area "confined to the present woven wire fence and to the present metes and bounds of said enclosed tract of land," the particular metes and bounds being given.

—Goins v. Beech Bottom Baptist Church
231 S.W.2d 23, 313 Ky. 287 (1950)

Brick foundation and *line of rotted fence posts*, the existence of which was testified to by witnesses and by photographs, weenot fictitious landmarks, and decree which established boundary line was not improper because based on these landmarks.

It was clear that the court was not using a fictitious landmark when the decree mentioned the brick foundation of the barn at the Gilbert corner and the old line of fence posts to South Street as monuments in determining the true boundary. In addition to the evidence of the witnesses, two pictures showing the fences were before the appeals court and one of them showed the brick foundation of the barn.

—Parkman et al. v. Ludlum
69 S.2d 434 (Ala., 1953)

Where circumstances warrant, fences may be considered as "monuments."

The true location of the survey of a tract is a question of fact. It is an old and well established principle that in boundary disputes monuments control over courses and distances. Where circumstances have seemed to warrant, fences have been considered as monuments as will appear from the following language in Keiper v. Dunn, 207 Cal. 643, 279 P. 772, quoting with approval from Perich v. Maurer, 29 Cal.App. 293, 155 P. 471:

"The fence itself is a monument, visible and obtrusive, which has existed for forty years or more, that under the peculiar circumstances of this case is quite persuasive in favor of the claim of defendants. It is a fair presumption that this fence was originally placed upon the true line as then recognized and understood, and it is proper to assume, in the absence of evidence to the contrary, that when the fence was built, either that the contiguous owners had knowledge and information of the lies of the Sutter survey and acted accordingly, or that the true line was uncertain and by agreement it was fixed and marked by said fence. It is not surprising that this fence seemed so important to Boyd [a licensed surveyor] and to the trial judge. Its existence for so many years, the recognition accorded it as the true boundary, the acquiescence of the respective owners in the location and the improvements made accordingly, were rightfully regarded as important, if not decisive, considerations in the determination of a line otherwise obscure and uncertain."

—W.B. Rodgers, et al v. Roseville Gold Dredging Co.
286 P.2d 536 (Cal. 1955)

Old fence line set up at time when there was no boundary dispute could not be used as line from which to measure boundary where deeds involved, which referred to fences which marked or where "supposed to mark" boundaries, were clear and unambiguous as to proper boundary marked by steel stake and where no acquiescence of fence line as boundary by predecessors of present parties in title was shown.

The general rule is that in determining boundaries resort is to be had, first, to natural objects of landmarks, because of their permanent character, next to artificial monuments or marks, then to boundary lines of adjacent owners, and then to courses and distances. Pritchard v. Rebori, 135 Tenn. 328, 186 S.W. 121. When all of these tests are applied to the facts of the case at bar, it is clear that complainant must prevail, and that defendants' contention must fail. In Windborn v. Guinn, 7 Tenn. App. 60, our own Court, speaking through Senter, J., held in a boundary dispute case that a fence, temporary in nature, erected at a time when there was not dispute over the boundary line, which made no attempt to follow the proper boundaries, does not constitute an agreed boundary.

—Minor v. Belk
360 S.W.2d 477 (1962)

Where there had been *long standing acquiescence in existing fences* and boundary lines as dividing boundary between two farms, this became the true boundary line, notwithstanding recent survey which differed little from preexisting fences and boundary lines between the properties.

As a matter of law the parties and their predecessors had acquiesced in the existing fences and boundary lines, as the boundary line between their respective properties rather than a recent survey which differed little from the preexisting fences and boundary lines between these two large farms, that respondent was accordingly the owner in fee of the disputed property and that appellant had, therefore, wrongfully changed the boundary line.

On the instant record we concur in the trial court's determination that, as a matter of law, there was such a long standing acquiescence in the existing fences and boundary lines as the dividing boundary between the two farms that notwithstanding any survey determination this became the true boundary line (*Baldwin v. Brown,* 16 N.Y. 359; *Fisher v. MacVean,* 25 A.D.2d 575, 266 N.Y.S.2d 951; *Van Dusen v. Lomonaco,* 24 Misc.2d 878, 204 N.Y.S.2d 778; see 6 N.Y. Jur., Boundaries §§ 79–82).

—Konchar v. Leichtman et al.
35 A.2d 890, 315 N.Y.S. 2d 888 (1970)

For corners to be lost, they must be so completely lost that they cannot be replaced by reference to any existing data or other sources of information, and before courses and distances can determine boundary, all means for ascertaining location of the lost monuments must first be exhausted.

Means to be used to locate lost monuments or corners include collateral evidence such as *boundary fences* that have been maintained, which should not be disregarded by surveyor, and artificial monuments such as roads, poles, and improvements may not be ignored; surveyor should also consider information from owners and former residents of property in the area.

The authorities recognize that for corners to be lost "[t]hey must be so completely lost that they cannot be replaced by reference to any existing data or other sources of information." Mason v. Braught, 146 N.W. 687, 33 S.D. 559. Before courses and distances can determine the boundary, all means for ascertaining the location of the lost monuments must first be exhausted. Buckley v. Laird, 493 P.2d 1070 (Mont.); Clark, Surveying and Boundaries § 335, at 365 (Grimes ed. 1959).

The means to be used include collateral evidence such as boundary fences that have been maintained, and they should not be disregarded by the surveyor. Wilson v. Stork, 171 Wis. 561, 177 N.W. 878. Artificial monuments such as roads, poles, fences and improvements may not be ignored. Buckley v. Laird, supra, 493 P.2d at 1073; Dittrich v. Ubl, 216 Minn. 396, 13 N.W.2d 384. And the surveyor should consider information from owners and former residents of property in the area. See Buckley v. Laird, supra, 493 P.2d at 1073–1076. "It is so much more satisfactory to so locate the corner than regard it as 'lost' and locate by 'proportionate' measurement." Clark, supra § 335 at 365.

—U.S. v. Doyle
468 F.2d 633 (Colo. 1972)

The purpose of the inquiry in boundary dispute action is to locate and follow the footsteps of the original surveyor.

The purpose of rules of priority of calls in original survey is to aid the court in finding best evidence of what the original surveyor actually did on the ground; in the event the footsteps of the original surveyor can be more accurately traced or his intention more accurately ascertained by following a call of lower order, then the rules of priority of calls are inapplicable.

The intention of the parties to boundary dispute is considered to be essentially the same as that of the surveyor.

The surveyor's intention is to be ascertained by scrutinizing what he actually did in making the survey as reflected by his field notes and the attending totality of circumstances of the survey.

The various calls contained within the surveyor's field notes must be harmonized and as few calls as possible disregarded so that the calls which result in the least conflicts in the total survey are given precedence.

In establishing the existence and location of original witness trees in order to establish boundary of national forest, the government would be required to introduce sufficient evidence to carry the case beyond coincidence and into the realm of reasonable certainty.

A survey must be construed in accord with presumption that the original surveyor actually surveyed all the lines, ran the course and distances and marked the boundaries as called for in the field notes unless later surveyors, following the original surveyor's footsteps, demonstrate that the original surveyor's calls constitute a mistake or are otherwise incorrect.

Where the calls contained within field notes for two pine witness trees at south corner of survey could still be identified with reasonable certainty even though the original witness trees had been destroyed in the 135 years between the original survey and the boundary dispute, such calls would be used in locating the footsteps of the original surveyor.

—U.S. v. Champion Papers, Inc.
361 F.Supp. 141 (D.C. Tex. 1973)

Remains of wire fence and stone wall were properly regarded by master, in boundary dispute, as evidence of location of disputed boundary line.

The master [in the lower court] found that "there were certain monuments on the ground that do indicate an old established boundary between land presently owned by plaintiffs and land presently owned by defendants. This line and the monuments are on . . . a surveyor's plan of the property of the defendants." The master properly regarded the remains of the wire fence and the stone wall as evidence of the location of the boundary line. Knight v. Coleman, 19 N.H. 118 (1848). Similar inferences were drawn in Frew v. Dasch, 115 N.H. 274, 339 A.2d 18, 20 (1975), where the court relied on the course of certain hedges to establish a boundary line. The trial court approved the master's finding and entered its decree accordingly. The plaintiffs contend that the monuments referred to are parts of an old fence intended to restrain cows rather than to mark the boundary. This assertion is based on the testimony of plaintiffs' first two witnesses, which is confused on the point. Their evidence tends more to refute than to support plaintiffs' contention.

—Starvish v. Farley
347 A.2d 175, (N.H. 1975)

References, within deeds, to artificial boundaries such as fences, roads, streets, and land lot lines are evidence of the points which land owners, past or present, have had in mind in their contractual dealings.

For purposes of determining boundary line, boundary fence referred to in deed prevailed over distances called for in the deed, whether overstated or understated.

In action in which owner of property adjacent to plaintiff's property counterclaimed for decree locating the property line between parties' properties, issue whether certain fence was located in same place as original referred in deed conveying property to plaintiff was for jury.

Monuments have been defined as permanent landmarks established for the purpose of indicating boundaries. *Thompson v. Hill*, 137 Ga. 308, 313, 73 S.E. 640 (1911). Artificial boundaries as applied to this case includes fences, roads, streets, and land lot lines. They are evidence of the points which land owners, past or present, had in mind in their contractual dealings with one another. Pindar, Georgia Real Estate Law, § 13-4. All monuments, whether natural or artificial, are deemed superior to courses and distances. *Cherokee Ochre Co. v. Ga. Peruvian Ochre Co.*, 162 Ga. 620, 134 S.E. 616 (1926). Also see *Brantly v. Huff*, 62 Ga. 532, 536 (1879). The superiority of monuments over metes and bounds is limited to those which are referred to in the deed itself. Therefore, the location of the boundary fence called for by plaintiff's deed on the north side, separating his land from the land formerly owned by Broadfield, and now defendant, would be the true line called for by plaintiff's deed. This monument prevails over the distances called for in the deed, whether overstated or understated. Perhaps this is better illustrated where the deed says the north line measures 1,098 feet from Stanley Street to Myddleton Avenue. All readily agree the line runs from street to street irrespective of the footage recited in the deed. The fence is no less an artificial monument than the streets.

Plaintiff introduced evidence that the latest fence was located in the same place as the original fence. Although this was disputed by the defendant, the issue should have been submitted to a jury.

—*Lyons v. Bassford*
249 S.E.2d 255, 242 Ga., 466 (1978)

There was ample proof in record to support findings of referee, adopted by trial court that *fence and hedgerow* constituted east-west boundary in action to settle boundary line dispute to a strip of land between respective farms of parties.

The court found "ample proof" in the record to support the findings of the referee adopted by the court to settle a boundary line dispute to a strip of land between the respective farms of the parties. The record indicated that the parties and their predecessors treated the fence line and hedgerow as the east-west boundary between the lots until the present dispute. There is no evidence that the parties treated this as other than the boundary line until 1968 when this dispute arose. A practical location of a boundary line acquiesced in for a long series of years will not be disturbed (*Baldwin v. Brown*, 16 N.Y. 359, 362; *Reed v. Farr*, 35 N.Y. 113, 116–117; *Sherman v. Kane*, 86 N.Y. 57, 73–74; see also *Allen v. Cross*, 64 A.D.2d 288, 292, 409 N.Y.S. 2d 865).

—*Domin v. Walters*
79 A.D.2d 1086, 435 N.Y.S. 2d 823 (1981)

Absent evidence of contrary intent by parties to deed, order of preference governing inconsistent land description is: first natural monuments of landmarks; then artificial monuments and established lines, marked or surveyed; then adjacent boundaries or lines of adjoining tracts; then calls for courses and distances; and finally designation of quantity.

Recorded distance prevailed over inconsistent measurement of acreage, and thus surveyor was negligent in preferring quantity designation over natural monument and recorded distances in fixing boundaries.

At trial, the cause of the disagreement in this case between the surveyors became apparent. Huffman's surveyor, had disregarded the distance called for by the 1907 deed for two reasons. First, measuring that distance from the pipe in the chestnut stump brought him to a point which established a boundary line which did not agree with the alignment of an old fence he found across Route 693. Although the fence was not mentioned in any description of record, and had "bow" in it, it showed signs of such antiquity that the surveyor concluded that it had been accepted as a boundary between the parent tract and its abutter. By extending the line of the fence across the road to the point of its intersection with the reciprocal of the course to the pipe in the chestnut stump as called for in the 1907 deed, he arrived at his conclusion concerning the 1907 beginning point.

The appeals court stated, "the extension of the fence line on which Huffman relied, being neither marked nor surveyed, and being unrelated to any monument described of record, fails to meet the criteria of "artificial monuments and established lines, marked or surveyed," the second category of the relative importance of conflicting deed elements. The recorded distance should have prevailed.

—Spainhour v. Huffman & Associates, Ltd.
377 S.E.2d 615 (Va. 1989)

Approximate location of range line in boundary dispute could be determined based in part on original plan of town, despite evidence of usage and of marker erected on east line; town plans all had common feature of range lines running parallel to southern boundary of town, and evidence, including survey locating range line west of point zero, aerial photographs and range line between ranges established that range lines were parallel.

The land in dispute in this case is a triangular shaped parcel containing approximately 36 acres. Gammon owns lot 1 range 6, which is directly south of the defendants' property, lot 1 range 7, in the Town of Bethel. The deed in both parties' chain of title refer to the common boundary only as the range line between range 6 and range 7. Thus the range line is the boundary.

The expert witnesses agree that a 1949 survey marker, located at the southwest corner of defendants' property and the northwest corner of Gammon's property, marks the location of the range line on the western side of the parcels (point $\emptyset$). The present controversy concerns the course of the range line east of point $\emptyset$ as it traverses lot 1. Gammon claims that the range line bears approximately east 10° north from point $\emptyset$ to the Bethel-Rumford town line. The range line asserted by Gammon has the same bearing as the range line of the adjacent lot and the southern boundary of the Town of Bethel, and is parallel to the range line between range 7 and 8. Defendants claim the range line bears approximately east 10° south from point $\emptyset$ to a post marked 1909 on the Bethel-Rumford town line.

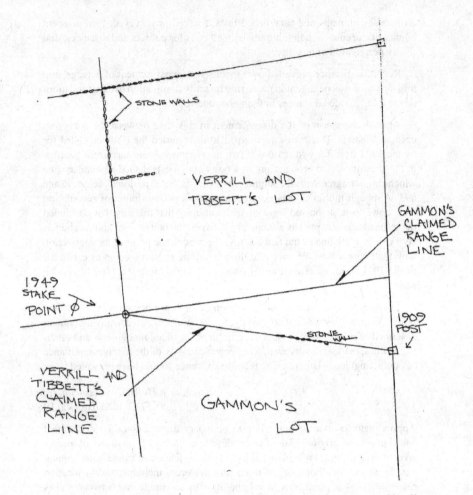

Figure 19.4 Diagram accompanying text of the case of *Gammon v. Verrill*.

Because the evidence provided neither original surveys nor original monuments the court relied solely on usage in locating the range line.

The case does not involve a dispute as to the legal boundary in a deed; rather, it involves the factual question of where the boundary is located.

The original plan plan of the Town of Bethel is not drawn to scale, but it lays off the range lines. Several town plans were introduced into evidence, none of them identical, but they all have one feature in common: the range lines are parallel to the southern boundary of the town. The range line, as located by the court, deviates 21° from the bearing of the southern boundary of the town. The court erroneously discounted the usefulness of the original plan so far as that plan established that the range lines are parallel to, and on the same bearing as, the southerly boundary of the town.

The southerly boundary of the Town of Bethel, as established by the Massachusetts General Court in 1796, was east 20° north. The declination for Oxford County has increased 8° 15′ west since 1796. Thus the southern boundary of Bethel today bears east 11° 45′ North. Gammon's surveyor testified that the course of the range line is east 11° north. Correcting the southerly boundary of Bethel for the change in magnetic declination, the bearing of the surveyor's range line is within 1° of the bearing of the town's southerly boundary.

It is clear from this record that the range line must be approximately parallel to the south line of the town. The evidence including the survey locting the range line west of point Ø, aerial photographs and defendant's own exhibit showing the range line between range 7 and 8 all establish that the range lines are parallel. Moreover, both parties' surveyors testified that the range lines in the Town of Bethel are parallel to its southern boundary. Evidence of usage as well as evidence of the marker erected on the east line in 1909 are insufficient, as a matter of law, to overcome the undisputed evidence of the bearing of the range line. Applying that bearing from point Ø, the range line can be located without resort to evidence of usage. This would establish the range line where Gammon asserts it is located.

—*Gammon v. Verrill, et al.*
600 A.2d 832 (Me. 1991)

Rock wall could not, as matter of law, locate boundary where all of the evidence, including aerial photographs, showed that rock wall generally deviated from bearing of township lot lines by 15 degrees.

The parties did not dispute the legal boundary; they disputed the location of that boundary and in such a case a factual finding will not be disturbed on appeal unless it is clearly erroneous.

Because all of the evidence, including aerial photographs, show the rock wall generally deviates from the bearing of the township lot lines by 15 degrees, that wall cannot, as a matter of law, correctly locate the boundary. See *Gammon v. Verrill*, 600 A.2d 832 (Me., 1991).

—*DuPont et. al. v. Randall*
648 A.2d 437 (Me. 1994)

Stone wall is strong evidence of boundary line.

A "monument," when used in describing land, has been defined as "any physical object on the ground which helps to establish the location of the line called for and term 'monument,' when used with reference to boundaries, indicates a permanent object which may be either a natural or artificial one .... Natural monuments include such natural objects as mountains, streams, rivers, creeks, springs, trees ..... Artificial objects and monuments consist of marked lines, stakes, rocks, fences, buildings and similar matters marked or placed on the ground by the hand of man." 12 Am.Jur.2d Boundaries, § 4, p. 549; 4 Tiffany, Real Property (3d Ed.1975) § 993, p. 193; 3 American Law of Property, (Casner Ed.1952) § 12.105; see *Delphey v. Savage*, 227 Md. 373, 374–75, 177 A.2d 249 (1962). It has been said that "a stone wall is strong evidence of a boundary line. *Roberti v. Atwater*, 43 Conn. 540, 546 [1876]"; *Pendleton v. Macdonald, Highway Commissioner*, 6 Conn. Sup. 5, 7 (1938); see *Wallingford Rod & Gun Club, Inc.*

v. Nearing, 10 Conn. Sup. 414, 116 A.2d 517 (1955). One court has said that a monument, when used in describing land, is "any physical object on the ground which helps to establish the location of the line called for," whether it be natural or artificial. *Delphey v. Savage*, supra, at 378, 177 A.2d 249. That court noted that, just as in contracts or wills, the intention of the parties governs the interpretation of deeds and that it is for that reason "that monuments named in deeds are given precedence over courses and distances, because the parties can see the tree, stone, stake, pipe or whatever it may be, which is referred to in the deed, but would require equipment and expert assistance to find a course and distance." Id. "[T]he physical disappearance of a monument does not terminate its status as a boundary marker, provided that its former location can be ascertained through extrinsic evidence." *Bailey v. Look*, 432 A.2d 1271, 1274 (Me.1981); see *Theriault v. Murray*, 588 A.2d 720, 722 (Me. 1991); *Seely v. Hand*, 119 N.H. 303, 402 A.2d 162 (1979); 6 G. Thompson, Real Property (1962) § 3042.

—Koennicke v. Maiorano
642 A.2d 1046, 45 Conn. App. 1 (1996)

FENCES AND HIGHWAYS

Fences are not original evidence of street lines; but it is only in the absence of original measurements to show their location, that it would be important to ascertain the boundaries of the streets as actually opened and used.

—Winchester v. Payne
10 Cal. App. 501 (1909)

The above quote would seem to be the guiding light in the retracement of highway right-of-way boundaries. The approach is the same as for any boundary, which has been stated by at least one court (see below-easement boundary is like any other boundary), reproduce the establishment of the title, if that fails, rely on physical evidence.

Evidence reviewed, in a suit by a landowner to restrain the removal of his fence from its location in the public highway, and held sufficient to show that the fence was not on the true line, and was not located in conformity with the highway, as shown in field notes of the survey on which it was originally established.

—Pine v. Reynolds
187 Iowa 379 (1919)

It was held that the east line of the street was where the original surveyor placed it, not where it should be according to resurveys or subsequent surveys; that subsequent surveys are worse than useless; they only serve to confuse, unless they agree with the original survey.

—Johnson v. Westrick
200 Wis. 405 (1930)[245]

---

[245] See Chapter 22 for a discussion of this case.

If fences or walls on opposite sides of a wrought road may be found to have been designed to mark the limits of use for highway purposes, a varying distance between them will not destroy their evidentiary value in properly locating the lines of the highway; and, in so far as walls are found to have been built on either side, they may be evidence tending to show the location of the side lines along that part of the highway where there are no walls.

—Hoban v. Bucklin
88 NH 73 (1936)

Border of right-of-way is "boundary line" like any other.

—Manufacturer's National Bank of Detroit v. Erie
County Road Commission
63 Ohio St.3d 318; 587 N.E.2d 819 (Ohio, 1992)[246]

These cases and their significance are presented in detail in Chapter 21.

FOR FURTHER REFERENCE

170 ALR 1144 Comment: Fence as Factor in Fixing Location of Boundary Line.

Types of Fences

Fences can be constructed of a variety of materials: Stone, stumps, sod, brush, wood, and wire are the most common. The first fences were made of native materials, stone and wood, with wire coming much later. Barbed wire was first patented in 1867, and to date there are approximately 1,000 different patents, with many variations, mostly from the late nineteenth century. Woven wire fences date to about 1883.[247] Even wire fences have to be supported by wooden posts or trees, which can be very useful evidence.

Brush Fence.   Brush fences, a type of primitive fence, were once common in the southern states in forested regions. Stakes were erected and brush piled between them to create a barrier. Unless such fences are maintained, once deteriorated, little is left as evidence except perhaps a very slight elevation in the landscape.

Stump Fence.   Stump fences generally were constructed along roads and boundaries as farmers pulled stumps from pastures and placed them along the outside, with their top ends inside of the field. The gaps between where the stumps could not be rolled close together were filled with brushwood. After deterioration, they tend to

[246] See Chapter 22 for a discussion of this case.

[247] For interesting accounts of different types of fencing, refer to Eric Sloane's "Americana" books.

leave behind the dirt and stones that were part of the root mass, so sometimes a slight ridge remains where the line of stumps once was.

Hedge Fence.   Before barbed wire, the Osage orange tree (*Maclura pomifera*) was sold throughout the Midwest as a livestock hedge fence. In four years it was said to grow to be *horse-high, bull-strong, and hog-tight*. There was once estimated to be over 250,000 miles of Osage orange hedgerows. Many are long gone; they are easy places to cut the trees for fence posts when the farmer no longer needs them to contain livestock. However, since Osage orange is an extremely durable wood, remants of trees and posts tend to remain a long time.

Wicker Fence.   Wicker fences were used where other material was too costly or not easily obtained. A wicker fence is constructed of stakes and willows and will last from 10 to 15 years. They were relatively common in the far West. Many were built on a small embankment of earth from one to two feet high, so that slight remains are sometimes found.

Sod Fence.   Sod is cut with a special implement and piled similar to a rock wall. It was always considered a double barrier against prairie fires, since the usual construction was a wide strip cleared of sods, with the fence itself standing in the middle of it. The usual sod fence was about three feet high with stakes driven along its summit and two strands of barbed wire attached to the stakes.

Board Fence.   Board fences were just what they sound like—fence made of boards. Posts were erected, then boards nailed to the posts, creating a barrier.

Figure 19.5 An example of a board fence.

Rail Fence.   Many rail fences were contstructed of chestnut, oak, cedar, or juniper, or sometimes of original-growth heart pine. These woods are quite durable, and such fences tend to last from 50 to 100 years, so that material of this sort will remain one or two generations. Remnants of many of these fences are quite common.

Figure 19.6   Stone fence post with slot for rails.
Photo courtesy of George Butts.

Figure 19.7   Barway in fence used for a gate. Rails are slid in and out.
Photo courtesy of George Butts.

   In 1883 the Iowa Agricultural Report stated that the United States had 6 million miles of wood fence at a cost of $325 a mile. The dollar then was much more valuable than today and amounts to nearly $2 *billion* in today's dollars (the same as the national debt for 1883), which provides evidence regarding just how important the simple fence was in the overall picture of the nation.

Figures 19.8 and 19.9   A typical rail fence (a) with close-up showing the location of "heel stones" under the intersections of the rails. Often after the rails have deteriorated or have been removed, the stones remain in a zigzag configuration, indicating where the fence once stood.

Figure 19.10   Section of moss- and lichen-covered stone fence found in a wooded area that was once cleared.

Stone Fence.   Also know as rock fences, rock walls, and stone walls, stone fences were very labor intensive but last until dismantled and taken away. If fact, because of the weight of a stone wall, many settle into the ground. Even after a wall has been removed, remants, such as bedstones, likely remain. When that is the case, a bit of digging will uncover the location of the fence.

No one knows how many miles of stone wall exist in New York and New England today, but in 1871 the Department of Agriculture issued a report entitled "Statistics of Fences in the United States." According to that report, approximately one-third of Connecticut's fences were made of stone, amounting to 20,505 miles of stone wall—enough to extend almost once around the equator. Most of Rhode Island's 14,030 miles of fencing were stone, as were nearly half of Massachusetts' 32,960 miles. In New York, 18 percent of the fences were stone, some 95,364 miles, more miles than the entire coastline of the United States. By estimation, it would have taken 1,000 men working 365 days a year about 59 years to build all the stone walls in Connecticut and 15,000 men 243 years to build the 252,539 miles of walls in New England and New York.

In other parts of the United States, stone was also used as a fencing material, though not nearly as extensively as in New England and New York. In the 1871 survey, stone walls were reported in Maryland, North Carolina, South Carolina, West Virginia, Tennessee, Ohio, Michigan, Wisconsin, Kansas, Pennsylvania, and New Jersey.

Figure 19.11   Barway in a stone fence.

Walls may indicate something about the land use. For cultivation, stones had to be removed from the fields, and they were either left in piles at various locations around the area or piled along the periphery. Many stone piles were later used to build walls around pastures, fields, or the perimeter of the property by itinerant wall-builders. Stone walls made very poor fences for sheep, so where raising sheep was of concern, likely different types of fence will be found.

Thorson[248] explains the nature of today's fence remnants by comparing field sizes when clearing. Clearing fields greater than one or two acres took great effort, and it did not take a farmer long to learn how much longer it took to clear an eight-acre field in comparison to eight one-acre fields. Using a computer model and assuming a square field, hauling rocks one at a time, and assuming 100 rocks per acre, Thorson calculated that clearing an eight-acre field would require 58 miles of walking, whereas eight one-acre fields would require less than 20 miles.

Thorson[249] also explains that it is possible to age a wall under certain circumstances by examining lichen cover, the character of the stones, and whether it has been rebuilt. The purpose of a wall can sometimes be concluded from its width and height and by searching for human tool marks. Land descriptions can sometimes aid when they include the character of land being transferred as pasture, field, or woodland. The author also characterizes walls by the way they were laid out and their composition. Either of the references listed may be of assistance in evaluating a stone fence.

[248] *Stone by Stone.*
[249] *Exploring Stone Walls.*

Wire Fence.

Before Wire   Since the beginning of time, humans have constructed barriers from natural materials adjacent to the barrier site. These materials were mostly wood from trees, stone, thorny brush, and mud. When settlers arrived on the Great Plains of America, they found these materials in short supply, and a demand for a more economical type of fencing was created.

Smooth Wire Development.   Dating back to AD 400, the process of pulling hot, bloom iron through dies in a drawing plate produced short lengths of various sizes of smooth wire. By 1870, good-quality smooth wire was readily available in all sizes and lengths. Stockmen used the smooth wire in fencing but found it was not a dependable deterrent to livestock passage.

Wire fences were known in Pennsylvania as early as 1816. An "Account of Wire Fencing" was read at the Philadelphia Agricultural Society, and it spoke of "living trees connected with rails of wire." Cost comparisons demonstrated that, at that time, there was a cash savings of $1,329 per 100 acres enclosed.

Fence Wire.   There are many types of fence wire. Besides plain, woven, and barbed wire, fences are known by a variety of names: sheep fence, pig wire, chicken wire, Page wire, woven wire, and so on. Knowing the use of the land may be a clue as to what type of fence was there. For example, barbed wire would not be used for enclosing sheep, since the animals would continually get caught on the barbs.

Barbed Wire.   The development of barbed wire began about 1863, when several individuals experimented with the creation of fences that could be classed as barbed wire. In 1873, when a type of wire with barbs was exhibited at the county fair in DeKalb, Illinois, three men took immediate interest. Each had his own idea on how to improve on the original idea. Each separately invented his own wire and started his own business, and eventually they all came together. By the 1880s, barbed wire production was under way. With the development of the Midwest, farmers needed fencing, and millions of miles of wire fence were produced, with most patents being registered during that time.

Native Americans called barbed wire Diablo Reata, "the devil's rope."

Preservation and Collection of Barbed Wire.   There are over 530 patented barbed wires, approximately 2,000 variations, and over 2,000 patented barbed wire tools to collect. There is also an active society that meets annually, and several museums.

Identification and Dating of Barbed Wire   It is possible to identify wire, but not to date it from the idenfication alone. Most patents were registered in the 1880s, and sometimes fence wire was kept in a barn for years before being installed. Methods of approximating the date erection and installation of wire fencing are discussed in Chapter 18.

Figure 19.12   Collage of barbed wire images.
Reprinted courtesy of the Devil's Rope Museum www.barbwiremuseum.com.

Picket Fence and Other Types of Decorative Fencing.   Through the years, especially around the immediate yard of a dwelling, residents have installed decorative fencing. Some had a purpose of keeping things or out. Most of these fences were erected without regard for property lines because they were for a different purpose. Before selecting any fence as evidence of a property line, the investigator must evaluate the history and purpose of the fence.

FENCES AS BOUNDARY EVIDENCE

Ancient fences used by a surveyor in his attempt to reproduce an old survey are strong evidence of the location of the original lines and, if they have been standing for many years, should be taken as indicating such lines as against the evidence of a survey which ignores such fences and is based upon an assumed starting point. It is said that a long-established fence is better evidence of actual boundaries

Figure 19.13   Remnants of barbed wire fence protruding from a White Ash (*Fraxinus americana*).

settled by practical location than any survey made after the monuments of the original survey have disappeared. Accordingly, a fence erected on a surveyed line shortly after the land has been surveyed may serve as a monument to control courses and distances or a subsequent survey after the stakes set out at the time of the original survey have disappeared.[250]

Retracing Fences

Calling for a fence in a land description is not enough. The type of fence should be noted, and whether it is continuous along the entire length of the line. Knowing the

Figure 19.14   This retracement resulted in finding the remains of three fences: rail fencing, a stone wall, and posts with barbed wire. Likely this fence began as a rail fence, stones were piled underneath it as the area was cleared, then wire erected later to confine animals.

[250] 12 Am Jur 2d, Boundaries, § 71. *Stakes and fences.*

Aging a Fence

Figures 19.15 and 19.16   A deteriorated white pine (*Pinus strobus*) stump with remnants of barbed wire through it. The wood that the wire is attached to, being denser and more durable due to the reaction of the wound when the wire was attached, is still in good condition and aids in protecting the evidence of the old fence and its location.

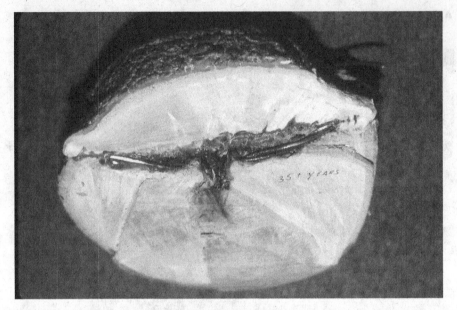

Figure 19.17 This gray birch (*Betula populifolia*) had to be cut down in order to determine how many years ago the fence was attached. The wood adjacent to the wire was so hard as to be almost not able to be cut or sanded, but no rings could be read there. Counting inward from the bark to the edge of the reaction wood yielded a count of 35 years.

area and the use made of it in the past can lend insight as to possible fences. For example, since most barbed wire was not invented until late in the nineteenth century, earlier calls for fences would have to be for fences constructed of other materials. However, fence lines may be upgraded over time, beginning with a wooden fence and ending with a wire fence.

The retracement individual should keep an open mind, because any type of fence can be encountered. In addition, sometimes fences are removed when there is no more use for them, so an early call for a fence may only mean that there was a fence in place at that point in time, with no guarantee that one still is there. Sometimes, however, some remnants may be found, such as short pieces of wire in the ground or grown into trees.

A FINAL WORD

A fence may be erected for any combination of the following reasons.

To keep things in

To keep things out

To mark a property boundary

—Paul Bigelow,
Vermont Land Surveyor and Sage

Don't ever take a fence down, until you know why it was put up.

—Robert Frost (1874–1963)

CHAPTER 20

OTHER TYPES OF MONUMENTS AND MARKERS

> We've got to search back to our last known safe landmark. I
> can't say exactly where, but I think it's back there at the
> start of the Industrial Revolution, we began applying energy
> in vast amounts to tools with which we began tearing the
> environment apart.
>
> —*David R. Brower*
> (1912–2000)

Almost anything can be used or tied into locate a land parcel. Not all descriptions, especially those in the past, were located by surveyors. Landowners would reference a property to about any imaginable object, permanent or otherwise. Objects such as fence posts, building corners, street corners, utility poles, trees and rocks, even stream intersections can be found in documents. Some of these are more subject to change than others. Items to beware of that are not related to boundaries but are readily found on the landscape are grave markers for pets, gate posts, horseshoe pins, and things of that nature. If something doesn't look right, it probably isn't right, and if something is suspect, check it out—it may turn out not to be what it seems.

Stones: Stake and Stone or Stake and Stones?

Pay close attention to words in a description, whether they call for a stake and stone (singular) or a stake and stones (plural). After the stake has deteriorated or otherwise disappeared, it will make a big difference whether the search is for a single stone or a group of stones. Also keep in mind that if the stone(s) is very old, it may not be readily visible, or may be buried.

Figures 20.1 and 20.2 The stone on the left, marked with a "B," is a typical boundary stone; the stone on the right, marked "B 46," is a turnpike marker, or milestone, indicating that it is 46 miles to Boston from that point.

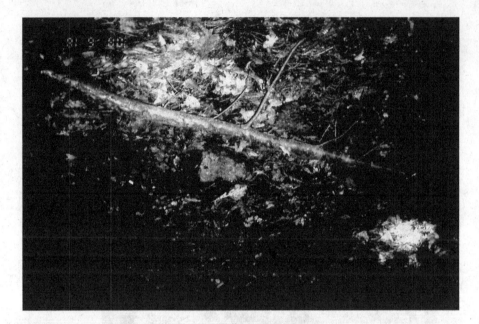

Figures 20.3 and 20.4   The stone on the top was buried under leaves, duff, and organic matter. It was set around 1875, recovered in 2005, and found through the use of measurements from other monumentation. The pile of stones on the bottom was set in 1796 and recovered in 1984. Its location was found under the leaves by using corrected bearings and distances from other, known monuments.

Figures 20.5 and 20.6   Two stones found through the use of directions and distances from known, or suspected, corners. The stone on the top is unmarked and is a natural field stone. The stone on the bottom is a cut (processed) stone with the initial "B" carved in it. This particular stone was broken off, had fallen over, and was found under about a foot of organic debris. It is believed that the growth of the tree it was set next to caused it to break. The base of the stone can be seen to the right of the stone itself.

Figure 20.7   A natural stone marked with the initial "K," which signified "Knowles," one of the early landowners. This stone was called for in the original land description (1892 deed) and was found by following the course of an ancient fence line, as directions and distances in this and abutting descriptions were nonexistent.

Metal Objects: Pin, Pipe, or Post?

The designation of "i.p." in a document can have at least three different meanings. If marking, be clear; if retracing, keep an open mind.

Figure 20.8   An iron pin (reinforcing iron/rebar) set at the base of a tree.

Figure 20.9   This pipe was found with a magnetic locator.

Ferrous locators or magnetic locators will locate iron or anything that produces a magnetic field. They will not locate aluminum, brass, or any nonferrous metals. A different type of locator is needed for nonferrous metals.

Case #43

IRON PINS VERSUS IRON PIPES

Litigation ensued from a disagreement over an iron pipe, since it did not quite fit the distances recited in any of the deeds and was the only pipe anywhere in the vicinity, all of the other markers being iron pins. A search of the records of the original surveyor, field notes and plats, indicated that he set iron pipes. Testimony was that since iron pipes were set by the surveyor, the particular pipe in question was likely an original, and the pins were replacements. Subsequent testimony from a former owner verified that considerable surveying had taken place over the years, that the pins were indeed replacements and therefore not original. Since there was no evidence to indicate the contrary, the location of iron pipe was accepted by the court as being the best evidence of an original location.

Stakes

Stakes are usually made of wood but can be iron. Some people call iron pins and iron rods "iron stakes."

Other Markers

A number of other metal objects have been found or called for as corner markers. Gun barrels are not infrequent and may be iron, steel, or brass. Pieces of television antennae, pipes, and other types of aluminum objects are occasionally found. Mostly used as a supplement or as a witness rather than as the corner marker itself are items like bricks, charcoal, glass and pottery. Frequently these latter items will be placed with the corner marker as a permanent memorial should the corner marker itself disappear.

If something is found other than what is called for in the title
document, it may not be original, not part of the record, is secondary
at best and may be suspect.

Case #44

CORNER OUTSIDE HOUSE, SURVEY INSIDE

A corner mentioned in a deed was tied with a direction and a distance to a house corner. There was considerable fill at the site for the house construction, and digging failed to produce the corner. The question was made whether the house itself had ever changed, and it was found that it had been expanded. Determining that one of the interior room corners was very close to one of the original outside corners, the traverse was extended through the front door and to the corner in a back bedroom. Computing the new position of the corner, an iron pipe was found six feet down in the fill.

CHAPTER 21

HIGHWAYS, ROADS, AND STREETS

> I have long been interested in landscape history, and when
> younger and more robust I used to do much tramping of the
> English landscape in search of ancient field systems, drove
> roads, indications of prehistoric settlement.
>
> —*Penelope Lively* (1933–)

IMPORTANCE OF ROAD LOCATION AND RIGHT-OF-WAY DEFINITION

Since a right-of-way line is a boundary like any other boundary,[251] and since location is the responsibility of only a land surveyor,[252] road lines are part of the survey process. The older the road, generally the more difficult it is to locate. In addition, older records tend to be less definitive than later ones, and less likely to be based on an accurate survey. Nevertheless, a location problem exists, and when a road is an issue, such as for access, or is used as a monument, its location becomes exceedingly important.

There are five considerations with any road:

1. *Status*

   In other words, who owns what? Is the road or right-of-way *owned* by a person or entity, or is it merely a right of passage and maintenance on someone else's land?

2. *Type*

   What is its character: road, trail, or some combination?

---

[251] *Manufacturer's National Bank of Detroit v. Erie County Road Commission*, 63 Ohio St. 3d 318; 587 N.E.2d 819 (1992).

[252] *Rivers v. Lozeau*, Fla. App. 5 Dist., 539 So.2d 1147 (Fla. 1989).

3. *Location*

Where, on the surface of the earth, is the road or right-of-way located? Does its description, written and/or platted, coincide with the physical location?

4. *Width*

What is the extent of the road or right-of-way? This may be fixed or variable. What evidence indicates its limits?

5. *Encumbrances*

What effect, if any, do encroachments have on the road or right of way, or its extent?

Types of Roads

There are many types of roads, and it is important to fully understand exacting which category is being dealt with, or investigated.

Highways
    Federal
    State
    County
    Municipal
    Private

Figure 21.1   Toll gate on a turnpike.

Trail
  Dirt (unpaved)
  Paved
Special categories
  Stage road
  Turnpike
  Roads to mills
  Roads to ferry landings
  Winter roads

With any road, there are several elements of concern:

Center line
Right-of-way line
  Specified
  Unspecified
Kinds of records that exist to support creation of the road
  Eminent domain
  Statutory layout
  Dedication and acceptance (expressed or implied)
  Legislative act
  Prescriptive way
  Mentioned in deed or other document
  Public or private

Case #45

WHICH ROAD?

Sproul v. Foye

55 Me. 162 (1867)

In this case, a parcel of land was conveyed, bounding on one side by "the new county road leading from Wiscasset to Dresden." Some 40 or 50 rods of the road were built outside of the original location by the authorities, and the question was whether the line of location or the road as actually built was to be regarded as the true boundary of the land conveyed. The deed bounded the land "by the new county road. . . ." At the time the road had been open and used for public travel about three years. The question was whether the parties meant the road as located or the road as actually built. The court stated that "when a road is referred to in a deed as one of the boundaries of land conveyed, we should ordinarily suppose that something more than a mere location was meant. A road is a way actually used in passing from one place to another. A mere survey or location of a route for a road

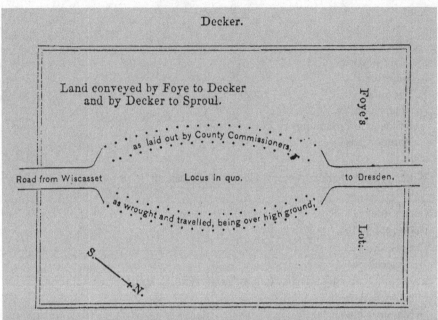

Decker.

Land conveyed by Foye to Decker
and by Decker to Sproul.

Foye's

as laid out by County Commissioners,

Road from Wiscasset   Locus in quo.   to Dresden.

as wrought and travelled, being over high ground;

Lot.

S.

N.

Figure 21.2 Diagram from case of *Sproul v. Frye.*

is not a road. A mere location for a road falls short of a road as much as a house lot falls short of a house." Once again, the correct meaning or interpretation of the words selected is the key to answering the question.

Case #46

ROAD CANNOT BE A BOUNDARY IF IT DOES NOT EXIST

A land parcel was subdivided in stages, and not all of it divided into lots. There were two roads running through the subdivision, and one of the deeds, with a description by abutters and no dimensions, called for being bounded southerly by the road. The question arose as to which road satisfied the description, the land between them amounting to several acres. Unlike the previous example, there were in fact two constructed roads that could be driven on.

Tracing the deed back in time, the origin of the description was determined and the evidence collection began. Trees were sampled along both roads in an attempt to determine if one road was older than the other. Aerial photos were examined to see if the roads could be dated. Neither of these investigations produced an answer. Finally the surveyor's records were located and reviewed. The file contained working drawings of the stages of the subdivision and documented that at the time of the creating deed, only one road existed, thereby defining the boundary of the description.

Case #47

ACCIDENT RECONSTRUCTION
State v. Gallagher
101 N.H. 335 (1960)
REFERENCE 77 ALR. 2D 1167

> Under statute defining a "way" as any public highway, street, avenue, road, alley, park or parkway, or any private way laid out under authority of statute, a private park or a private way not laid out under authority of statute does not constitute a "way."

An accident taking place outside of the right-of-way of a highway, even though on pavement connected to a public way, is not an accident on a public way. There may be a question of jurisdiction, and there may be question as to which laws regarding accidents apply, and who, if a question arises, may be liable.

Case #48

LOCATION OF RIGHT-OF-WAY LINE
Racine v. Emerson
85 Wis. 80, 55 N.W. 177 (1893)
This argument was between a recent survey of a street line and the original survey of the street line from which improvements were based on. In analyzing the problem, the court stated that the ruling question in the case was "Where is the east line of the street in front of the lot in question, according to the original plat? 'It is not, "where is such line according to any subsequent survey or plat?" It continued:

> All resurveys or subsequent surveys are of no effect except to determine that question. A resurvey that changes lines and distances and purports to correct inaccuracies or mistakes in the old plat is not competent evidence in the case. There are only two questions [in this case]: (1) where is the true line fixed by the original plat? (2) is the fence in question on that line? A resurvey that changes or corrects the old survey and plat can never determine the first question. A resurvey must agree with the old survey and plat to be of any use in determining it.

A subsequent survey was ordered by the city in this case to "correct the old plat, to straighten the streets, and make a better plat than the old one." "Resurveysfor the lawful purpose of determining the lines of an old survey and plat are generally very unreliable as evidence of the true lines. The fact, generally known and quite apparent in the records of courts, is that two consecutive surveys by different surveyors seldom, if ever, agree; and the greater number of surveys, the greater number of differences and disagreements will occur. When two surveys disagree, the correct one cannot be determined by still another survey. It follows that resurveys are of very little use in a case as this, except to confuse it.

The court noted in this case that it was shown by old settlers that the lot owners set out shade trees, built fences, and located their buildings with reference to stakes set on the original survey, and this testimony was considered better evidence of the location of the street line than any resurveys. Continuing its discussion, the court stated:

> It is fortunate that this evidence is yet in existence. The time will soon come when it will have been lost by the destruction of all monuments, natural or artificial, and by the death of the old inhabitants. Then resort must be had to evidence of lesser degree to establish ancient boundaries, and long-continued occupation with respect to unchanged lines, and reputation, even, may be the best evidence available. In any case of disputed boundary the testimony, or even the acts, of the surveyor who originally established it, and who pointed out the stakes set by himself to mark the line so many years ago, accompanied by continued use and occupation in recognition of such line, is not only proper, but strong, evidence that such was the true line, and better evidence than a new survey made more than forty years afterwards, which changes such line.

Case #49

RIGHT-OF-WAY LINE IS LIKE ANY OTHER BOUNDARY

Manufacturer's National Bank of Detroit v. Erie County Road Commission

Ohio St. 3d 318; 587 N.E. 2d 819 (1992)

A motorist stopped at a road intersection only to find the view of the other road blocked by a cornfield, which was growing into the right-of-way at the southeast corner of the intersection. The motorist proceeded onto the highway and was hit by a truck, which killed her husband and daughter. The bank, as personal representative of the husband's estate, brought an action against the road commission, along with others, including the landowner on whose land the corn was growing and the farmer who grew the corn. One of the issues was that the farmer who grew the corn was actually using land that was not under his exclusive control, and, in this case, inconsistent with the purpose of the right-of-way and causing a hazard to safe travel. He responded that "requiring a land user to determine exactly where the right of way lines are located places too great a burden on the land user." The court responded by stating that a landowner or occupier is under an obligation to know the boundaries of the property. The border of a right-of-way is a boundary line like any other.

Locating an Ancient Way or Proving Existence of a Road

Plot the description of the road to the scale of the map you wish to overlay on and prepare a transparent overlay. USGS quadrangle sheets often lend themselves very well to this process. Many of the descriptions will not fit exactly because of the idiosyncrasies in the measurements, which must be resolved. In addition, sometimes when a road was created and obstacles were met when it was actually located and constructed, the actual location may deviate somewhat from the description. However, if there are marked differences or wide departures, you may not have the correct description or layout.

Evidence of Road and/or Right-of-Way

Ancient Records

Willey v. Portsmouth, 35 NH 303 (1857). Reputation is competent evidence of the laying out of ancient highways. Such are recitals in ancient records and grants.

Webster v. Boscawen & a., 67 NH 111 (1891). Recitals in ancient records and grants are competent evidence of the laying out of ancient highways.

Buildings

Brooks v. Morrill, 92 Me. 172 (1898). The record of a town way properly laid out and accepted by the town establishes its limits, and is the legal boundary of lands lying adjacent thereto. Fences and buildings facing a highway can be deemed its true boundaries only in the absence of proper records and monuments.

Cellar Holes

Town of Weare v. Paquette (1981), 121 NH 653, 434 A.2d 591.

Ditches

Blake v. Hickey, 41 A.2d 707, 93 NH 318 (1945).

Foster v. Webster, 8 Misc. 2d 61, 44 NYS 2d 153 (1943).

Haby v. Hicks, 61 S.W.2d 871 (Tex. Civ. App., 1933).

Brentwood Town Records
Book 1, Page 511

We the subscribers Selectmen of Brentwood on the 23rd day of August
1793 have laid out a highway in s^d Brentwood 2 rods wide - bounded
as follows.- Beginning at the highway leading from Capt. Dudleys
dwelling house to Pick Pocket Mill so called at the South east corner
of Josiah Things homestead and the northeast corner of land owned by
the Hon. John T. Gilman and runs westerly binding on S^d Gilmans land
136½ rods, taking all of S^d way out of S^d Things land then runs
N 75° W 13 rods to Samuel Things land run N-75° W 18 rods, N 64° W
20 rods, then N 58° W 28 rods, N 60° W 50½ rods to Bartholomew Things
land, N 60° W 20 rods, N 66½° W 12 rods, N 50° W 13 rods, N 21° W
30½ rods, N 42° W 22 rods, N 50° W 51½ rods to Col John Phillips land
then the last mentioned course 14 rods then N 28° W 31 rods, then
Westerly such a point as in running 2 rods will have one rod on S^d
Trasks land, then westerly 37 rods, then N 28° W 4 rods, then N 48° W
6 rods, then N 58° W 10 rods near to the river then run westerly
binding on the bank of said river 66½ rods to S^d Robinsons Mill Grant
near S^d mill, then N 57½° W 7½ rods to the old highway leading to the
meeting house.

        Brentwood. Sept. 3, 1793

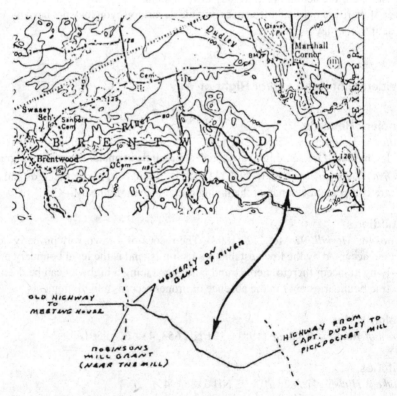

Figure 21.3 A 1793 highway layout and a "to scale" plot of the description compared
with the current USGS topographic map.

Figure 21.4 A cellar hole found alongside an ancient way.

Yturria Town & Improv. Co. v. Hildago County, 125 S.W.2d 1092 (Tex. Civ. App., 1939).

See also 76 ALR2d 548 and 76 ALR2d 525.

Fences

Hoban v. Bucklin, 184 A. 362, 88 NH 73 (1936). If fences or walls on opposite sides of a wrought road can be found to have been designed to mark the limits of use for highway purposes, a varying distance between them will not destroy their evidentiary value in properly locating the lines of the highway; and, insofar as walls are found to have been built on either side, they may be evidence tending to show the location of the side lines along that part of the highway where there are no walls.

Hull v. Richmond, 2 Woodb & M 337, F. Cas. No. 68671 (CC RI 1846). (See *usual distance*.)

Kritzberger v. Traill County, 242 N.W. 913, 62 ND 208 (1932).

Jones v. Cederquist, 1 Misc. 2d 1020, 150 NYS.2d 121 (1956).

Harmony v. Clark, 95 N.E. 47, 250 Ill 57 (1911).

Dean v. Carroll, 143 NYS 12 (1913, Sup).

Evans v. Bowman, 108 N.E. 956, 183 Ind. 264 (1915).

Commissioners of Highways of Cold Springs Twp. v. Bruner, 163 Ill. App. 657 (1911).

Fritchey v. Commonwealth, 200 A. 622, 331 Pa. 179 (1938).

Commonwealth v. Llewellyn, 14 Pa. Super. 214 (1900).

Washington v. Steiner, 25 Pa. Super. 392 (1904).

Figure 21.5   Fences on both sides of a well-traveled highway.

State v. Morse, 50 NH 9 (1870). When a road is established by use alone, it is not necessarily limited to the traveled track and the ditches on each side, but a jury would be at liberty to find that the easement extended over the whole space marked by fences that had been maintained more than 20 years, and that gave about the usual width of a highway.

Buchanan v. Wolfinger, 298 N.W. 176, 237 Wis. 652 (1941).

Newberg v. Kienle, 120 P. 3, 60 Or. 486 (1912).

People v. Quong Sing, 127 P. 1052, 20 Cal. App. 26 (1912).

Brooks v. Morrill, 92 Me. 172 (1898). The record of a town way properly laid out and accepted by the town establishes its limits, and is the legal boundary of lands lying adjacent thereto. Fences and buildings facing a highway can be deemed its true boundaries only in the absence of proper records and monuments.

Whitesides v. Green, 57 Am. St. Rep. 740, 13 Utah 341 (1896).

Riverside County v. Leslie, 292 P. 981, 109 Cal. App. 238 (1930).

Kruger v. Le Blanc, 37 N.W. 880, 70 Mich 76. (1888).

For additional decisions in which reference was made to fences or other monuments along the highway, see 76 ALR2d 535, Width and Boundaries of Public Highway Acquired by Prescription or Adverse User § 7.

Foot Path
Coffin v. Plymouth, 49 NH 173 (1870).

Hannum v. Belchertown, 36 Mass. (19 Pick) 311 (1837).

Figure 21.6   A footpath through the woods as a major byway, bordered on one side by a stone fence.

Hedges
Middletown v. Glenn, 115 N.E. 847, 278 Ill. 149 (1917).

Sidewalks
McCracken v. The City of Joliet; McCraken v. Cowing et al., 271 Ill. 270 (1915). Where none of the stakes of the original survey of a street can be located or their precise location otherwise ascertained, the conduct of lot owners in building and maintaining their improvements within a certain line and the conduct of the city in maintaining a sidewalk with reference to the same line must be given great weight in determining the location of the street line.

Stone Walls
Hoban v. Bucklin, 184 A. 362, 88 NH 73 (1936). If fences or walls on opposite sides of a wrought road can be found to have been designed to mark the limits of use for highway purposes, a varying distance between them will not destroy their evidentiary value in properly locating the lines of the highway; and, insofar as walls are found to have been built on either side, they may be evidence tending to show the location of the side lines along that part of the highway where there are no walls.

Evans v. Bowman, 108 N.E. 956, 183 Ind. 264 (1915).

Figure 21.7   Parallel stone walls between two fields, indicating the existence of a lane.

Figure 21.8   Wheel tracks indicating the traveled portion of an ancient way. This road was once a well-traveled municipal highway, today vacated and fallen into disrepair, though still used as the only access to abutting land parcels. This road has a center line and defined right-of-way lines, although they are not obvious in the photograph.

Trees

Hartford Trust Co., Adm'r v. The Town of West Hartford, 84 Conn. 646 (1911).

Jones v. Cederquist, 1 Misc. 2d 1020, 150 NYS 2d 121 (1956).

Kruger v. Le Blanc, 37 N.W. 880, 70 Mich. 76. (1888)

Wheel Tracks

Coffin v. Plymouth, 49 NH 173 (1870).

Plummer v. Ossipee, 59 NH 55 (1879). Evidence of wheel tracks is relevant to the question of where the usual travel on a highway is, and competent to show the limits of a highway by use.

Other Indications

Numerous other indications of a road, either public or private, exist. Schoolhouses, sign posts, gates, bridges, and like evidence should always be considered. Roads may be depicted on old maps and sketches, may be visible on older aerial photos, and may be described or noted by older residents, especially those who grew up in the vicinity or whose family has been there for several generations.

> *In the absence of evidence to the contrary, the earliest known center line of traveled highway is considered center line of highway right-of-way, and when original survey of such road runs in single line such line is presumed to be center line of highway.*

> —*Clark v. The State of New York*
> 246 NYS 2d 53, 41 Misc. 2d 714 (1963)

CHAPTER 22

WATER-RELATED PROBLEMS

> One's ideas must be as broad as Nature if they are to
> interpret Nature.
>
> —*Sherlock Holmes*
> A Study in Scarlet

TYPES OF PROBLEMS

Problems with water, as opposed to land, can be grouped as follows.[253]

Legal Problems of Riparian Owners of Concern to the Surveyor

Title to the bed of the body of water

Legal "riparian" or "littoral" rights, which arise by virtue of the abutting ownership

Problems arising from the washing away, emergence, or adding to the shoreline, or changes in the course of the river or stream

The use of the shore of the body of water as a boundary

Accretion versus Avulsion

Determining whether accretion or avulsion has taken place can be an extremely difficult, but not necessarily impossible, question to answer. Courts will usually try

---

[253] Curtis M. Brown, Walter G. Robillard, & Donald A. Wilson, *Boundary Control and Legal Principles*, 4th ed. (New York: John Wiley & Sons, Inc., 1995).

to reconstruct the movements of the particular body of water from the time of the original grant to the time of the suit. To do this, a wide range of evidence is relied on.

Expert and Lay Witnesses.   Since accretion cases involve the reconstruction of events that occurred several decades, sometimes more than a century, prior to trial, experts become very important.

A variety of expert testimony is used in such cases. Geologists testify as to processes by which rivers change and shape their channels and to interpret the geologic record in the particular case. Foresters state opinions on how long particular land has existed as dry land, based on the nature and age of the vegetation and tree stumps. Soil scientists may interpret the results of soil borings in the area. Several other experts may also be qualified to testify, such as surveyors or engineers.

The testimony of lay witnesses is also frequently introduced. Persons with long experience in the area may be permitted to describe their perception of how the river channel has changed over time. Lay witnesses may also testify to circumstantial evidence bearing on the question, such as the placement and movement of fences in the area, the existence of old tree stumps and other physical evidence, and the knowledge of contemporary names of land or bodies of water.

Correspondingly, the absence of eyewitness testimony or historical accounts on the part of those claiming an avulsion may create substantial doubt that an avulsion in fact occurred.

Topography, Vegetation, and Other Circumstantial Evidence.   Several topographic features may be significant in particular cases. Sandy knolls and ridges may indicate the area had been formed by accretion, and silt-filled depressions may indicate the past existence of a channel.

Soil borings and soil surveys may be indicative of how recently the land was formed. The nature and age of the vegetation may indicate how long an area has been dry land.

Evidence of signs of habitation and cultivation may indicate that the land remains identifiable and in place.

Documents and Demonstrative Exhibits.   An accurate map of the area is essential, and diagrams may be helpful. In addition, current aerial and terrestrial photographs will depict relevant features of the existing landscape. In some cases it may be worthwhile to demonstrate overall topography through the use of cross sections of one or more topographic models.

Ancient maps and surveys should be considered as helpful exhibits.

Ancient Documents.   The ancient document rule has long been recognized as an exception to the hearsay rule. Maps, surveys, plans, and plots that are twenty years old, are free on their face from suspicion, and are found in proper custody are admissible as ancient documents. Such evidence can be extremely important to show the time and manner of river channel and shoreline changes in past decades.

Official and Private Surveys.   Government and other official original surveys offer chronological documentation. References in deeds and patents make maps and plats part of the grant and are competent evidence to identify and fix the boundaries of the land. Subject to admissibility, private surveys, plans, and field notes can also be valuable demonstrative evidence.

USGS Topographic Maps.   Different editions of standard maps can show changes in the landscape as well as documenting conditions at a point in time.

Maps, Plats, and Charts.   Survey maps and other charts may be admitted to show condition or locations at particular times, depending on admissibility and testimony.

Photographs.   Photographs have the same inherent characteristics as maps, but may be useful to show topography, river or shoreline location, nature of vegetation, or other relevant matters. Old pictures, aerial or terrestrial, can be located in a number of agencies.

Newspaper Reports and Contemporaneous Accounts.   If a river has changed course by rapid movement, written reports may exist describing the event. Newspapers are also a source of weather and storm records. The weather bureau catalogs great detail concerning storms and weather-related events.

Historical Works.   Local histories, as general reputation, are usually admissible as evidence. Storm records and other dramatic natural actions and disasters are often so recorded.

Scientific Treatises.   Nearly every case concerning accretion or avulsion has one or more expert witnesses who testify as to how rivers change their shape and channels. These descriptions are based partially or entirely on the experiments and observations of other scientists. Scientific works may or may not be admissible to support the professional opinions of experts.

Burden of Proof.   Generally, the burden of proving accretion rests with the party claiming the benefit of the accretion. Not only must the process of accretion be shown; it must be shown that it in fact was gradual and imperceptible but that it attached to the claimant's land.

Presumption.   In a case where the title and rights of the litigants depend on whether a change in riparian land has occurred by reason of erosion or avulsion, it

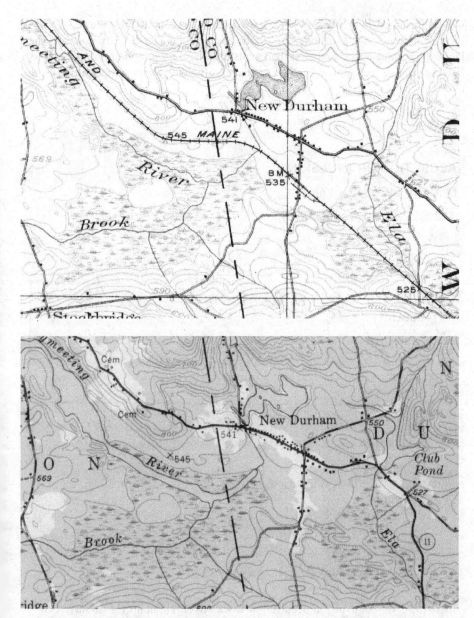

Figures 22.1 and 22.2   Dramatic changes to two bodies of water between 1919 (top) and 1957 (bottom). The river has widened significantly, and Club Pond on the later map did not exist when the earlier map was done. At that time it was a mere stream.

will be presumed, in the absence of clear evidence to the contrary, that the change was by erosion rather than avulsion. Also, in the absence of evidence to the contrary, the law will presume accretion rather than avulsion. These presumptions are rebuttable, however, and do not apply where the evidence sufficiently demonstrates an avulsive change.[254]

Case #50

ACCRETION OR AVULSION?

McClure v. Couch, et al.

188 S.W.2d 550 (Tenn., 1945)

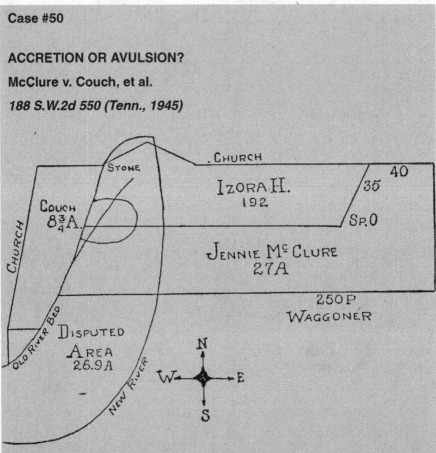

Figure 22.3   Diagram in the case of *McClure v. Couch.*

In 1913, a violent avulsion occurred in Duck River, Humphrey's County, Tennessee, that cut off the west end of a farm, creating an island of 20 or 30 acres. The owner executed a will in 1914 devising 27 acres to daughter Jennie McClure, which included 2 to 3 acres on the west end of the property and which was then west of the new channel of Duck River. In the same will he devised a long

[254]78 Am Jur2d, Waters, § 427.

rectangular tract of land to his daughter Izora, which also extended across the new channel of Duck River created by the avulsion in 1913.

Suit was filed in 1942, stating that since the 1913 avulsion, accretions had become deposited against the south bank of Jennie's 2- to 3-acre tract lying west of the new channel, and she claimed the accretions, which amounted to 26.9 acres. She stated that she and her father claimed the land and had used it openly, cultivating it and paying the taxes for more than 20 years. No one questioned her title until the Tennessee Valley Authority undertook to buy a reservoir, which would overflow the 16.9 acres as well as some of her other lands. During the negotiations, she learned from TVA that Couch claimed they were entitled to part of the accretions.

Couch claimed that the accretions attached to her farm of 8 TQF acres, which was on the west side of the old channel of Duck River previous to the avulsion. The plat filed as Exhibit No. 1 was drawn by the county surveyor and depicts the properties as they existed at the time of trial.

The court found that after the avulsion in 1913, and up to the time of trial, the channel gradually and steadily moved eastwardly and southwardly, eroding the east bank of the new channel and depositing accretions on the west and south sides thereof. Also, after the accretion, the northern part of the old channel was filled by sediment, but the southern part of the old channel was distinct, with a bank 7 to 12 feet high next to the Couch boundary on the old channel. The finding was that the change was made by erosion and accretion and not by avulsion. Consequently, no part of the accretions attached to Couch, but are attached to McClure and extend down the river therefrom without touching anyone else's property line.

The court discussed accretion and its results in adding to ownership, as opposed to an avulsion, which does not affect ownership. The court also discussed reliction and the fact that it did not apply in this case. However, the boundaries prior to the avulsion of 1913 were one thing, unchanged by the avulsion of 1913, but subsequently changed by the accretions from that point on. A complete study of both river channels over time is necessary in cases such as this to sort out the processes and their ultimate affect on boundaries.

Types of Accretion

> natural
> artificial
> natural due to artificial influence
> by owner, by adjoiner, by a removed party

Determining Earlier Shoreline.   A shoreline as a boundary is the shoreline at the time the original title was created, or the shoreline at the time the parent tract was created. Since that time, depending on the nature of the shoreline and its composition, changes likely have taken place. Many of these changes can be documented or determined, and if there is trespass, accretion or erosion, there may be an impact on the title and/or the boundaries of the ownership. Many of the techniques and resources outlined in separating accretion from avulsion are applicable to this situation as well.

When accretion is involved and spread across several owners, some apportionment method must be applied and it is necessary to compare the new shoreline with the old (original) shoreline. Determining the exact location of the old shoreline is where the difficulty lies and special techniques by other experts such as soil scientists, geologists, and botanists is required.

Case #51

ARTIFICIAL CHANGE TO SHORELINE

Seward v. Loranger

130 N.H. 570 (1988)

This dispute concerned land above the natural high water line of an artifical lake. Shortly after purchasing a shore lot, the Sewards filled a portion of a swampy area extending to the open water beyond the swamp. The predecessor in title to Loranger filled a portion of the swampy area next to their property in order to create a beach. The result, eventually, was overlapping claims to artificial fill. The court ruled that title to man-made fill vests in littoral owners, much in the way that title to natural accretion vests in riparian owners. In addition, the court allocated to the Lorangers a 40-foot frontage as their proportional share of the shoreline as diminished by the fill. The court also found that the Sewards did not have shore frontage but only a right-of-way to it, which was granted to them over land that the grantor had no title to. Therefore, they are are not littoral owners and are not entitled to frontage.

The appeals court stated: "The issue at the heart of this dispute is who has, and who has had, record title to the land lying under the fill that was deposited." The trial court concerned itself with who was entitled to the filled property as a whole, rather than with the more fundamental issue of who has, or had, title to the formerly submerged land that now lies under the fill. The court found that the predecessors in title did not convey land beyond the artificial high water line, even though they owned to the natural high water line farther out. Therefore, the artificial fill was on their submerged land, and occupied by others. Since the trial court did not address the issue of adverse possession, the case was reversed and remanded.

There are two issues that this case addresses that are significant. One is the fact that there is artificial fill not only in front of lot owners, but the fill is also on land of others. Second, the court pointed out but did not address, that if an artificial impoundment founded on flowage rights existed for a long time and was treated as a natural body of water, there may be an implied dedication in favor of the public. If that turned out to be the case, adverse possession would not be an issue.

Case #52

MEANDER LINE

Martin v. Carlin

19 Wis. 477 (1865)
Description calls for meander line and quarter line. When they
disagree, which one controls?

> Where there is a mistake in the government survey of a fractional lot, so
> that either the line of a meandered stream or a quarter section line (both of
> which are called for by the survey as constituting the boundary between two
> fractions) must be abandoned, the quarter section line should be adhered to
> as the more certain call. (*Jones v. Kimble, ante*, 452, and cases cited.)

The action was brought to recover 19 and 82/100 acres of land admitted to be in
possession of the defendant below, who is now the appellant. The land is claimed
as being a part of lot No. one in section twenty-three in township eight, range
sixteen, in Jefferson county. The section was included in the grant which was
made by the general government to the territory for the purpose of aiding in the
construction of the Rock River Canal. It is a fractional section, made so by the
Rock river, which flows through it from the northeast to the southwest. The entire
section, with the exception of three forties in the northwest quarter, is divided into
fractional lots, which are numbered from one to nine. According to the original
plat and survey of the section, lot one was bounded on the north and east by
the section line, on the south by the east and west quarter section line, and on
the west by Rock river. By the same plat and survey, lot two was bounded on
the east by the section line, on the north by the east and west quarter line and
the river, and on the west by the river likewise. The southern boundary of this
lot is immaterial to this controversy. It appears that there was a mistake in the
original survey and meandering of Rock river, and that it is further north and west
than the government survey had located it on the northeast fractional quarter, so
that there is more land in that fractional quarter than the government survey calls
for.

   The patent in plaintiff's chain describes "the southwest quarter of the northwest
quarter, and *factions one* and nine in section twenty-three of township eight, range
sixteen, containing one hundred and thirty-two and 17/100 acres according to the
official plat of the survey of the said lands returned to the general land office of
the United States by the surveyor-general." The defendant offered a patent for lots
two and three in section twenty-three, township eight, range sixteen, containing
ninety-seven and 40/100 acres according to the official plat of the survey of the
United States, with the obvious purpose of proving title to lot two, but this was

objected to and ruled out, on the ground that it was irrelevant and immaterial. It was an admitted fact in the case that the government meandered line and survey of Rock river was incorrect; that according to a correct survey, the bed of that river was further north and west than the government survey had located it, so as to leave a strip of land in the northeast fractional quarter south of that river and north of the east and west quarter section line; and the whole controversy turns upon the point whether this strip belongs to lot one or lot two. The plaintiff insists that the defendant cannot legally claim any land north of the quarter line as a part of lot two, because, by the system of government surveys, that line is the true boundary between lot one on the north and lot two on the south; and that lot one must be held to embrace all the land in that fractional quarter section, north of the quarter line and south of the river. The defendant insists that no such controlling effect should be given to the quarter section line in determining the boundary of lot two, that the meandered lines of Rock river run by the government survey must be regarded as well; and, since it appears that by the government plat and survey, lot two extended on the north to Rock river, this natural object more certainly designates the northern boundary of the lot, and other calls and lines must be subordinate to it. Besides, lot two will fall short just about the amount of land in dispute, unless it extends across the quarter section line to the river on the north, while lot one will overrun that amount. And the only question in the which we deem worthy of any particular notice or which has given us any difficulty, is, whether the quarter section line must not be considered as the boundary of lot two on the north, disregarding the delineation of Rock river on the government plat and survey, or whether this line must yield to this natural object as located on that plat? The question is one, certainly, not free from difficulty; but a majority of the court are inclined to the opinion that, under the system of government surveys, the quarter section line must control and fix the boundary of lot two on the north. In announcing this result we are not unmindful of the rule in surveys, that the law loves certainty in calls, and that a call for a natural object, such as a river, will control courses, distances and quantity. But this rule, when applied to the admitted facts of this case, supports rather than militates against the view we have taken. For, by the system adopted by the government for the surveys of the public lands, they are first surveyed into townships six miles square, the liens of which are required to correspond with the cardinal points. At the corners of the townships appropriate monuments are required to be erected. These townships are subsequently subdivided into thirty-six sections by running parallel lines each way; and at the corners where these lines intersect, monuments are erected, and also intermediate monuments equidistant between the section corners. Brightly's Dig. Laws U.S., pp. 446–7, 479, 481; Lester's Land Laws and Decisions, p. 722.

The external lines of the sections are actually run, and fixed and established. Adopting, then, the language of the court, as used in *McClintock v. Rogers*, 11 Ill., 279, 296, we say: "The original monuments, when ascertained, afford the most satisfactory and we may say conclusive evidence of the lines originally run, which are the true boundaries of the tract surveyed, whether they correspond with the plat and field notes of the survey or not. All agree that courses, distances and quantities must always yield to the monuments and marks erected and adopted by the original surveyor, as indicating the lines run by him. These monuments are facts; the field notes and plats indicating courses, distances and quantities, are by descriptions, which serve to assist in ascertaining those facts. Established monuments and marked trees not only serve to show with certainty the lines of their own tracts, but they are also resorted to, in connection with the field notes and other evidence, to fix the original location of a monument or line which has been lost, or obliterated by time, accident or design." The section and quarter posts on section twenty-three are readily ascertained. There is no difficulty whatever in running the east and west quarter line where the law requires it to be run. If the section were not a fractional one, and had been subdivided into quarter and quarter-quarter sections, this line would be recurred to as the true division line between the north and south half of the section. Why should we not resort to these established monuments to ascertain the division line between lots one and two in this case? The answer is, because on the plat Rock river is delineated as the boundary of lot two on the north. But it is admitted that the original survey or meandering of Rock river is erroneous. The meandered lines can only be run from the field notes of the original survey, and when traced according to those notes they do not mark the present location of the river. But because Rock river was designated or marked on the official plat as bounding lot two on the north; it is claimed that it must be taken as the true boundary, disregarding the east and west quarter line and the quarter posts actually established and found. Suppose the line between lots one and two were a section or township line, and there had been the same mistake in locating Rock river on the plat? Should those lines also be disregarded in determining the northern boundary of lot two? Could you override the established monuments and lines actually run, which control in all case of disputed boundaries, to reach a natural object? If such a rule were adopted, it is easy to perceive that it would result in the greatest confusion and disorder. We see no sound reason for saying that the case supposed is distinguishable from the one before us. For if the quarter section line is to be crossed to reach a natural object, why not a township or section line? What well grounded distinction exists between them? They are all lines actually run by the surveyor, or which may be run from ascertained monuments established by him. And there is as much reason for saying that the lot crosses the town line and extends to some natural object, as that it should cross the quarter section line for a like purpose. The majority therefore think that where the lines of the survey can be run from well ascertained and established monuments, they are to control and govern a description delineated on a plat. To adopt any other rule would annul the authority of the public surveys and open the door to litigation and difficulty. And as lot one embraces the land in controversy, the judgment must be affirmed.

Streambed Location

Case #53

ONCE A RIVER, NOW A POND

Several parcels were surveyed along a highway, mostly using acreage calls, since there were few measurements. However, the parcels in question were bounded by each other, the road, and the brook. One survey located the edge of the water as it existed. When one parcel came up considerably short in acreage, an investigation was in order. The titles were traced back to a 1767 probate partition:

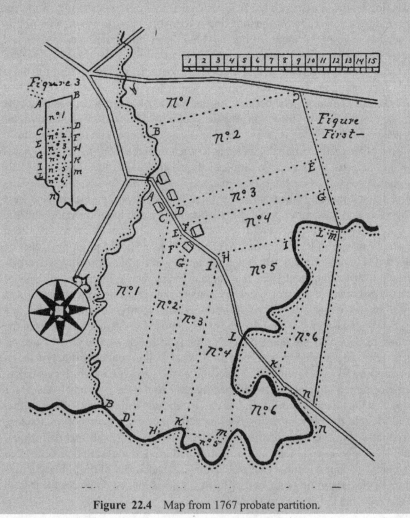

Figure 22.4 Map from 1767 probate partition.

The partition contained some measurements, but the important piece of information was the character of the river at that time. Since then, the river has been changed with a dam downstream, creating a lake. The surveyor used the edge of the lake as a boundary, which resulted in his including part of the abutting land in order to complete acreage call. When laying out the acreages according to this partition and survey, and considering "the circumstances at the time" of the creation of both the boundaries and the titles, the result is that everything fits as it should and in agreement with the deeds. The surveyor satisfied the acreage call, but not the boundary calls, to the detriment of the abutter.

Figure 22.5   Tracing made from a 1952 aerial photograph, which shows the original riverbed (black) and the extent of the flowage created by damming the river.

Case #54

THEN A BROOK, NOW A RESERVOIR

Key

300 is the parcel in question where evidence was located at the brook; it was taken in fee for reservoir purposes

    300E-1 and 300E-2 were taken as easements

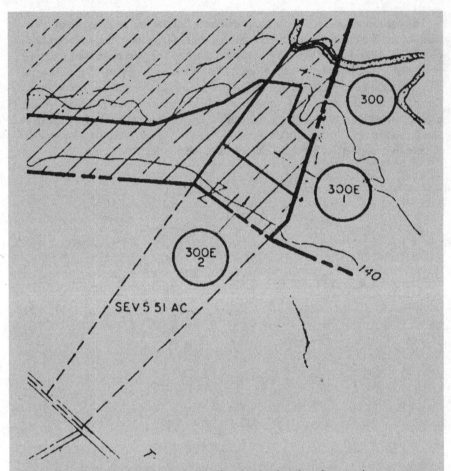

Figure 22.6   Property lines intersecting brook and reservoir.

The call was for an iron pipe at the brook; today a reservoir exists. Since the case involved a timber trespass, it was critical to locate the direction of the boundary line running back to the brook from the road, but the end point was inaccessible. Great effort was expended in an attempt to procure the field notes for the reservoir survey, without success. Since, in this case, the reservoir could not be drained, the line was retraced when the first safe ice covered the pond, before snow, at a time when the bottom could be seen. Using a metal locator and cutting a hole through the ice, the iron pipe was located and the line determined.

Other times, scuba divers have been used to find markers now covered with water. In one instance, it was important to demonstrate to the court the location of fence line. A scuba diver was sent down to swim along the fence holding onto a cord which dragged a balloon along the surface, thus demonstrating the location, direction and extent of the fence.

Case #55

ON WHICH SIDE OF THE POND IS THE BOUNDARY?

And Who Says We Can't Remove the Water?

A boundary disagreement was in the vicinity of an artificial pond wherein one surveyor believed the boundary to be on the right side, facing the pond,

Figures 22.7 and 22.8   The photograph on the left shows the pond after the water was drained from it. The photograph on the right is of the ditch that was revealed by removing the water.

while the other surveyor believed the boundary to be on the left side. This created overlapping opinions so that an important question became "Who owns the pond?" or "Who owns the land under the pond?" There was evidence in the form of fences on both sides of the pond.

The suggestion was made to drain the pond to see what, if anything, was under the water. When that was done, a 600+-foot ditch was found in the bed of the pond. This ditch, although not called for in any of the title documents, was likely the boundary between the parcels. At the least, it was strong enough evidence that the argument settled out of court.

Mill Sites

Ancient mill sites are often stand-alone titles, or have mill rights and flowage rights that are outstanding and separate from the title to the property or the water itself.

Figure 22.9 Picture postcard of an old mill, now long gone. Mill rights and flowage rights, however, may still be outstanding and in the names of others, frequently heirs and assigns of the original title holders. Occasionally later owners wish to resurrect these rights, sometimes generating their own electric power.

Figure 22.10   Site of a former mill on a small river.

Identifying the title status is for a title investigator, and the location of the site, or the extents of rights, is a survey responsibility. Either will take the investigator back in time, sometimes way back in time, to the creation of the title and/or the rights. Resurrecting the title is one thing, but finding the necessary location evidence may be quite another, due to both natural and man-made changes to the landscape through filling, shoreline protection and support, as well as flooding causing erosion, accretion, and avulsion.

Case #56

ANCIENT DAM AND MILL SITE

Back lot owners claimed a right to get to the water. Between them and the pond was a dam site described in 1826 as shown in Figure 22.11.

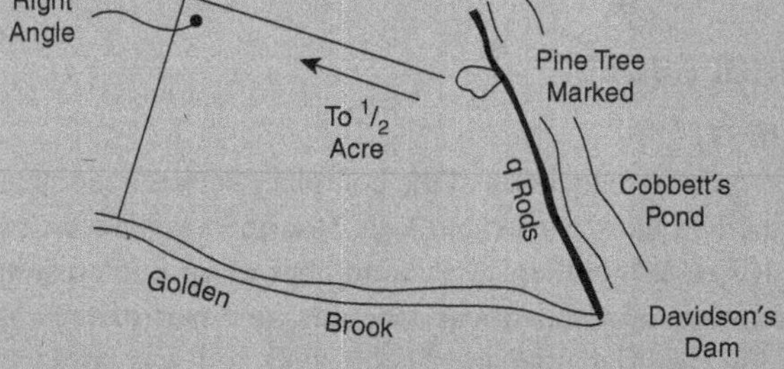

Right Angle

To ¹/₂ Acre

Pine Tree Marked

q Rods

Cobbett's Pond

Golden Brook

Davidson's Dam

Figure 22.11   Sketch from 1826 deed description.

Since rights were created in the various chains of title, it became a matter of determining the conditions at those times. The dam had been maintained, even replaced, several times, so photographic documentation was in order. With this description, along with detailed property descriptions of the surrounding parcels in 1867 and 1883, it was possible to produce a diagram of the conditions in that time frame. More than 40 vintage photographs were collected, showing the condition and location of the dam and surrounding features at various points in time.

Figure 22.12   One of the earlier dams at the site.

Case #57

HOWARD V. INGERSOLL

54 U.S. 381 (1851)

This case had to do with the location of the boundary between the states of Georgia and Alabama. The boundary, as created, was described as the western bank of the Chattahoochee River. While there were a number of very interesting exhibits in this case and many detailed historical accounts, one part of the case had to do with a mill site on the Alabama side of the river and was convincing as to its location.

Wetlands

Figure 22.13 Bogs, muskegs, and similar features are found in northern areas. Since the vegetation tends to creep inward, bogs and some smaller, boggy ponds recede over time and their perimeter changes. The bog here was a natural monument called for in a land description, "to the bog, thence along the bog...."

Natural Phenomena

Natural phenomena logically belongs in this category of water-related problems, since water- and wind-caused erosion and accretion are like processes. However, a number of other processes and events can cause subtle or dramatic changes to the landscape. In addition, some natural changes will cause movement of land and boundary evidence.

Hurricanes, tornadoes, subsidence, earth movements, earthquakes, volcanoes, and fire can each result in an event affecting the title or the boundaries to a parcel of land. If the event is a dramatic one, particularly if catastrophic, likely there will be both documentary and physical evidence of the event. However, if it is a subtle change, especially one occurring over time, its results may not be obvious. Erosion and man-made changes tend to mask the results and the evidence of the event. The word of caution is that whenever there is a disagreement with the evidence of a previous survey, do not automatically assume that someone made an error. The difference could be the result of a shift or movement in the landscape.

Natural phenomena may be classed in these ways:

- Water—rainstorms and floods
- Wind—shifting of soil and material
- Earthquakes
- Landslides and debris flows
- Subsidence
- Volcanoes
- Glaciation
- Fire

Most of these phenomena will result in one or more others soon following. For example, after a fire, there is apt to be flooding and erosion. If a major earthquake or landslide occurs under the ocean, there may be tsunami, causing devastation on the shore, sometimes far away from the underlying cause.

Use to advantage where possible.   While mostly natural disaster causes some type of destruction that affects boundaries and titles, in some instances a natural disaster can be used to advantage. For instance, in very dense brush or in impenetrable conditions, a fire can clear the area so it can be accessed. Regular seasonal fires in the Everglades, for instance, clear the way for finding government corners that likely would not be found otherwise.

Whenever a pond is drained or lowered in drought, markers often appear that could either not be found or were inaccessible prior to the decrease in water. Refer to Case #54 for an example.

CHAPTER 23

ETHICS

I can discover facts, Watson, but I cannot change them.

—*Sherlock Holmes*
"The Problem of Thor Bridge"

Ethics can be defined as (1) the study of standards of conduct and moral judgment; and (2) the system or code of morals of a particular person, religion, group, or profession. Rules of ethics, as adopted by any profession, are not intended to particularize; they are general guides of conduct and behavior.

The investigator's role is simply conducting objective investigations. It is often easy to be misled by clients who want things their way, by lawyers as advocates, by other researchers with different interpretations, by poorly written land descriptions and faulty physical evidence. It is natural to want to see the client win or get what he or she wants, but sometimes that is just not possible. *It is not a case of fitting the information to the desired result; it is a matter of deriving a result from the available information.* As the court stated in the case of *Smart v. Huckins,* "it is a matter of fitting the deed to the land, not the land to the deed."[255]

It is important to realize that investigators has no personal stake in what they find at a scene. If they do, then ethically they should not be involved with the scene. Until search is made, we cannot know what we will find. And once it is found, it cannot be ignored. Two simple principles for deciphering a land description are that an instrument must be honored in its entirely and that words in a description are presumed to have a purpose. Investigators cannot be selective by accepting evidence that they like and disregarding that which is misunderstood or contrary to a preconceived result. More important, investigators should not care what is found. If the goal is truth, then what that truth results in is of little consequence to investigators.

Investigators will never have every piece of the investigative puzzle, but that should not affect the amount of effort put forth in an investigation. Just because a

[255] 82 N.H. 342 (1926).

tree may be very old and likely gone does not mean we should not search for it, or its remains. We cannot know for certain until we investigate. To prevent a lack of knowledge from affecting their role, investigators must adhere to investigative ethics.

All facts must be treated the same and all documented according to the same procedure. Investigators do not know what the scene will tell them, and they have no personal stake in what the evidence ultimately says. Until proven otherwise, nothing defined by a scene is bad.

Investigators must be steadfast in not letting anyone—a client, an abutting property owner, a lawyer for either side of the argument, or professionals in other areas of expertise—tell them what they know. Investigators must factually and objectively report the information they have obtained. How others choose to use the information is their business, but they do not get to define what investigators know or have found.

Finally, in any book directed at professionals, it would seem unnecessary to state that altering testimony, engaging in perjury, or creating false evidence is clearly outside the scope of an investigator's duty. Seeing it done, watching it happen, ignoring when it is found may also be aiding an unauthorized or illegal practice. Sometimes professionals will not present "the whole truth" in an effort to please their client and earn a fee. There are hired guns in any field, but objective and ethical investigators must refrain from them.

This book has presented several examples where deliberate actions have attempted to mislead others. Many more examples could be presented. It does happen, and an ethical investigator must be on guard at all times to either let it happen or get caught up in it. Truth is defined by the collection of evidence, factual information that allows drawing a conclusion about an event or a series of events. Unlike testimonial evidence, physical evidence does not lie. And lies may not always be deliberate. People often tell what they think is the truth, or what they believe to be the truth with no malice intended, but in fact it is not truth at all. We may misinterpret what the physical evidence is saying, but we must realize that is our fault, or because of our shortcomings. Physical evidence will aid both the investigator and the court in corroborating or refuting much of the testimonial evidence. Used properly, physical evidence establishes a foundation of objective information that few, if any, can refute. By and large, it may well speak for itself, and therefore stand on its own.

Cases in which an expert has lied or falsified evidence only serve to erode the trust factor within the legal system and cause procedures for evaluating experts to become more stringent.

SOME ETHICAL CONSIDERATIONS

Practice in Accordance with Laws and Regulations that Govern Society and Govern Their Profession

Above all, the law must be followed. Provisions of federal, state, and local laws, as well as professional rules and regulations, must be understood and adhered to.

Depending on the particular situation involved, laws or rules from one level or source may dictate, while in another situation, different laws or rules will apply. Practicing at the federal level may be one thing, while practicing at the local level may be quite another.

Know, Understand, and Adhere to the Standards of Practice

There are two types of standards, those that are designed and accepted by some body within the profession, such as a professional association, and those that are accepted standards within the community, dictated by what the "ordinary prudent individual" would do in like circumstances.

The accepted, usually published, standards are often part of the law governing the actions of the professional. These are generally *minimum* standards, meaning that they are the absolute lowest or least level of performance that a professional should abide by. In other words, you must do at least this much, remembering that in some cases more is required.

The accepted standard in the community is that level of performance that practitioners adhere to whether part of the published standards or not. In all cases they must not be less than that required by the published or legal standards; however, in some cases a higher level of performance is required.

Standards are likely different for different practitioners as well. A consultant may be held to a higher standard than a nonconsultant, simply because he or she is a specialist and therefore should know the particular area better than the average practitioner. Standards vary, sometimes considerably, between professions or fields of endeavor. For example, the standards for title investigation are markedly different from those for land surveying, though each may be governed by the law. Practitioners should remember that fact when receiving information from professionals practicing in other areas, since the result of their work may not be of a caliber necessary for use in their own field. Working with practitioners in other fields demands an understanding of other standards of practice and an appreciation for other levels of performance.

Not Practice Outside of Their Area(s) of Expertise

Professionals in any field who practice as consultants or as experts must be careful not to offer opinions on matters outside of their expertise. They should not oversell their expertise or overstate their degree of certainty if research does not support it. Within their range, they should advance the highest level of competence possible, which means knowing the results of the latest research in their field.

When discrepancies are uncovered, it is often necessary to call in other experts, who have particular expertise or a broad understanding or the problem at hand. There are specialists in almost everything, from handwriting analysis and document examination to tree and wood identification and interpretation. Often it is a matter of understanding enough about another profession's area of practice to know when a situation is out of one's area of knowledge or responsibility and when assistance from another area of endeavor is necessary.

Never Prepare False Reports

A professional must never make or file a false report or fail to file a report where required by law. If practitioners do not have knowledge about something or do not know an answer or have a conclusion, they should not attempt to indicate that they do or state something without knowing whether it is true. It is far better either to omit that part or to state that it is unknown or unobtainable at the time.

Do Not Delegate Responsibilities to Others

A person responsible for the investigation of a problem must not delegate to assistants or others who are not qualified to perform a particular procedure. Land surveyors ultimately attaching their seal to a finished survey plan should be particularly careful about whom they assign to investigate, collect, and process evidence. Understanding what is evidence and its value or application demands considerable expertise and an understanding of the laws of evidence. Persons sent to the field to make measurements may or may not be qualified to judge other aspects of field investigation, such as boundary marks or lines of possession.

Becoming a competent title investigator requires a considerable amount of training, education, and experience. A wide range of documents that must be examined, the laws governing them have changed considerably over time, and investigators must be on guard for missing elements in instruments, missing documents, late recording, and falsified documents. Learning the process of searching a deed back in time is only the beginning of learning an area of investigation, which also includes knowledge of geography, genealogy, surveying, cartography, and law.

Conflicts of Interest

A conflict of interest is a conflict between one's obligation to the public good and one's self-interest. Many canons of ethics address this area as engaging in any activity that may give rise to the appearance of a conflict of interest. The appearance, obviously, depends on the perception of others.

It should go without saying that there must never be any bias or personal interest in the performance of the investigator's duty. Professionals must always judge matters with integrity and impartiality.

Performance Must Be Based on Accepted Scientific Methods

There are a number of accepted procedures for doing surveys, examining titles, and collecting evidence. Unorthodox methods are frowned on and may be unacceptable. Where used, they may weaken or even destroy an argument or a case.

Investigators must:

- Avoid trespass to obtain evidence
- Never use files of other investigators without their knowledge or permission
- Not mislead others for the sake of collecting information or other evidence
- Be extremely careful in the handling and processing of any evidence so that it does not become damaged or destroyed

Performance Is Bound by Ethics and Standards of the Profession

Every profession has ethics and standards, written or unwritten. Both are hallmarks of a profession; otherwise there is no profession. In addition to the foregoing, all professionals are also bound by the ethics and the standards of their own profession.

> Junk science, deceptive experts, and dubious devices occupy the ambiguous corners of the legal system.
>
> —*Katherine Ramsland*
> The C.S.I. Effect

FINAL THOUGHTS

You are limited only by your imagination. "Let's do a little dreaming."

> When you have eliminated the impossible, whatever remains, however improbable, must be the truth.
>
> —*Sherlock Holmes*
> The Sign of Four

APPENDICES

The appendices are intended to be a reference guide in support of the main text. They are not intended to be all-inclusive.

APPENDIX I

RETRACEMENT GLOSSARY

ABOUNDING. "Bounding" or "bounded by." "Bound" is synonymous with "limit" or "border." *Barney v. City of Dayton*, 4 O.C.D. 505, 8 Cir. Ct. R. 480 (1894).

ABOUT. The word "about," where the context limits and restrains its meaning, does not materially impair the certainty of a description. *Adams v. Harrington*, 14 N.E. 603, 114 Ind. 66 (1887).

Word "about" means the same as the word "at" in a phrase fixing the time within which something should be done. *In re Heine's Estate*, Ohio Prob., 100 N.E.2d, 545 (Ohio, 1950).

"About," within deed or mineral lease describing boundary as "thence about 50 varas on J.G. Survey for corner," will be disregarded in determining distance. *Humble Oil & Refining Co. v. Luther*, 40 S.W.2d 865 (Tex. Civ. App., 1931).

"About" has no effect upon the question of boundaries, nor will it control monuments, courses, and distances where the language is clear and obvious. *Wheeler v. Randall*, 47 Mass. (6 Metc.) 529 (1843); *Whitaker v. Hall*, 4 Ky. (1 Bibb) 72 (1809); *Stephens v. Heden*, 7 Ky. (4 Bibb) 107 (1815).

The term "about" cannot be used as the equivalent of exact distances. *Picharella v. Ovens Transfer Co.*, 5 A.2d 408, 135 Pa. Super. 112 (1939).

In stating the number of acres conveyed in a conveyance, it is usual to represent it as "about" so many. Yet the word "about," although it negatives the conclusion that entire precision is intended, is without any legal operation whatever. *Purinton v. Sedgley*, 4 Me. (4 Greenl.) 283 (1826).

"More or less" are words of safety and precaution and when used in deed are intended to cover some slight inaccuracy in frontage, depth or quantity in land conveyed, and ordinarily means "about" the same as terms "a little more than,, "not quite," "not more than," or "approximately," and all are often introduced into a description practically without effect. *Harries v. Harang*, 23 So.2d 786 (La. App., 1945).

ABUT. "Abutting" means to end, to border on, to touch, and "abutting" property means any property that abuts or adjoins. *Bulen v. Moody*, 63 N.E.2d 916; 77 Ohio App. 61 (1945).

ACCESSORY. See Corner Accessory.

ACRE. Ordinarily considered as containing 43,560 square feet or 160 square rods (4 roods). Other definitions may be encountered, for instance, a builder's acre or a Block Island acre, both of which contain 40,000 square feet.

ADJACENT. Lying near, close, or contiguous; neighboring; bordering on; nigh, juxtaposed, meeting, and touching. *Hall v. Gulf Ins. Co. of Dallas*, 200 S.W.2d 450 (Tex. Civ. App., 1947).

Objects are "adjacent" when they lie close to each other, but not necessarily in actual contact, but are "adjoining" when they meet at some line or point of junction. *Hauber v. Gentry*, 215 S.W.2d 754 (Mo., 1948).

ADJOINING. In the description of the premises conveyed, means "next to" or "in contact with," and excludes the idea of any intervening space. *Yard v. Ocean Beach Ass'n*, 24 A. 729, 49 N.J. Eq. 306 (1892), affirming (Ch. 1891) *Ocean Beach v. Yard*, 20 A. 763, 48 N.J. Eq. 72 (1890).

ALONG. When used in a description, means "by, on or over," according to the subject matter and content. *Church v. Meeker*, 34 Conn. 421 (1867).

APPROXIMATELY. Very nearly but not absolutely, nearly, about, close to. *In re Searl's Estate*, 186 P.2d 913, 29 Wash.2d 230 (1947), see 173 ALR 1247.

"More or less" are words of safety and precaution and when used in deed are intended to cover some slight inaccuracy in frontage, depth or quantity in land conveyed, and ordinarily means "about" the same as terms "a little more than," "not quite," "not more than," or "approximately," and all are often introduced into a description practically without effect. Civ. Code, art. 2493, *Harries v. Harang*, 23 So.2d 786 (La. App. 1945).

AREA. Usually means tract, space, region, or a broad part of land. *Maisen v. Maxey*, Tex. Civ. App., 233 S.W.2d 309 (1950).

ARPENT. A French unit of measurement, equivalent to roughly 192 feet, more or less, depending on location.

AT. In or near. *Chicago, L.S. 8 E. Ry. Co. v. McAndrews*, 124 Ill. App. 166 (1906).

"About," used to describe the location of land in an entry or grant, signifies "at" unless something can be shown to evidence a contrary intention. *Simm's Lessee v, Dickson*, 22 Fed. Cas. 158, 3 Tenn. (Cooke) 137 (1812).

BEARING, ASSUMED. A direction based on an assumed position or either true or magnetic north.

BEARING, MAGNETIC. See Magnetic Declination. A direction observed in a quadrant, or bearing system, with reference to magnetic north.

BEARING, TRUE. See Magnetic Declination. A direction observed in a quadrant, or bearing system, with reference to true north. Some compass observations are directly made with reference to true north, while others are corrected or converted.

BETWEEN. When used with reference to a period of time bounded by two other specified periods of time, such as between two days named, the days or other periods of time named as boundaries are excluded, and "between" has a like meaning when used with reference to boundaries in space. *Winans v. Thorp*, 87 Ill. App. 297 (1899).

BLOCK. A square or portion of a city or town inclosed by streets, whether partially or wholly occupied by buildings or containing only vacant lots. *Ottawa v. Barney*, 10 Kan. 270 (1872); *Fraser v. Ott*, 30 Pac. 793, 95 Cal. 661 (1892); *State v. Deffes*, 10 So. 597, 44 La. Ann, 164 (1892); *Todd v. Railroad Co., 78* Ill. 530 (1875); *Harrison v. People*, 63 N.E. 191, 195 Ill. 466 (1902); *City of Mobile v. Chapman*, 79 So. 566, 202 Ala. 194 (1918); *Commerce Trust Co. v. Blakeley*, 202 S.W. 402, 274 Mo. 52 (1917). The platted portion of a city surrounded by streets. *Cravens v. Putnam*, 165 P. 801, 101 Kan. 161 (1917). The term need not, however, be limited to blocks platted as such, but may mean an area bounded on all sides by streets or avenues. *St. Louis-San Francisco R. Co. v. City of Tulsa*, Okl. (C.C.A.) 15 F.2d 960, 963; (1926)*Commerce Trust Co. v. Keck*, 223 S.W. 1057, 283 Mo. 209, 1061 (1920) (irregular parallelograms). Yet two blocks, each bounded by a street, do not necessarily, when thrown together by the vacation of a street, constitute a single block to be included as such within an assessment district. *Missouri, K. & T. Ry. Co. v. City of Tulsa*, 145 P. 398, 401, 45 Okl. 382 (1914).

"Block" is often synonymous with "square." *Weeks v. Hetland*, 202 N.W. 807, 52 N.D. 351 (1925). As a measure of length, "block" denotes the length of one side of such a square. *Skolnick v. Orth*, 84 Misc. 71, 145 N.Y.S. 961, 962 (1914). Sometimes it means both sides of a street measured from one intersecting street to the next. *Chamberlain v. Roberts*, 253 P. 27, 81 Colo. 23 (1927). And on occasion it may be construed not to extend between two streets that completely cross the street in question, but to stop at a street running into it though not across it. *Wise v. City of Chicago*, 183 Ill. App. 215, 216 (1913).

Under a statute providing that territory sought to be excluded from a new county must be in one block, the word "block" implies the thought of solidity or compactness, and the territory sought to be excluded must be in some regular and compact form. *State v. Moulton*, 189 P. 59, 61, 57 Mont. 414 (1920).

BOUND. Synonymous with "limit" or "border." *Barvey v. Dayton*, 8 Ohio Cir. Ct. R. 480 (1894).

BY. Bounding of one piece of land "by" another piece, whether such other by long or narrow, or in any other form, locates the line at the edge, and not through the middle of the adjoining premises. *Woodman v. Spencer*, 54 N.H. 507 (1874).

"By," as indicating a terminal point of time, means "not later than; as early as." *Goldman v. Broyles*, Tex. Civ. App., 141 S.W. 283 (1911).

BY LAND OF. The words "by land of" an adjoining owner, used in the description in a deed, mean along the external boundary line of that land. *Peaslee v. Gee*, 19 N.H. 273 (1848).

CALL. A reference to, or statement of, an object, course, distance, or other matter of description, in a survey or grant, requiring or calling for a corresponding object, etc., on the land. *King v. Watkins*, C.C.Va., 98 F. 913.

In *Bouvier's Law Dictionary*, the word "call" is defined thus: "In American land law, the designation in an entry, patent, or grant of land of visible natural objects as limits to the boundary." And *Webster* defines the word as a reference to or statement of an object, course, distance, or other matter of description in a survey or grant requiring or calling for a corresponding object, etc., on the land. The meaning of "calling" for a course or boundary is well understood in the law. It is necessarily applicable to written instruments, such as entries, surveys, patents, grants, and deeds. *King v. Watkins*, 98 F. 913, C.C.Va.

CENTER. Does not necessarily mean precise geographical or mathematical center, but in common parlance means middle or central point or portion of anything. *Bass v. Harden*, 128 S.E. 397, 160 Ga. 400 (1925).

CHAIN. There are numerous types of chains, depending on the unit of measurement used and the geographical location. The most common are the Gunter's chain of 66 feet and the standard engineer's chain of 100 feet. Frequently, particularly within the Public Land Survey System, a 2-pole chain was used, consisting of 33 feet in length.

CONTIGUOUS. Two tracts of land touching at only one point are not "contiguous." *Baham v. Vernon*, 42 So.2d 141 (La. App., 1949).

"Contiguous" means in actual contact; touching; also, near, though not in contact; neighboring; adjoining; near in succession. *Ehle v. Tenney Trading Co.*, 107 P.2d 210, 56 Ariz. 241 (1940).

The word "contiguous" in its primary sense, means in actual contact or touch and when applied to tracts of land, it ordinarily conveys the idea that they border each other. *Lien v. Northwestern Engineering Co.*, 39 N.W.2d 483, 73 S.D. 84 (1949).

Two tracts of land, which touch only at common corner, are not "contiguous." *Turner v. Glass*, La. App. 195 So. 645 (1940).

Tracts which corner with one another are contiguous; "contiguous" meaning to touch. *Morris v. Gibson*, 134 S.E. 796, 35 Ga. App. 689 (1926).

CORNER. The intersection of two converging lines or surfaces; an angle, whether internal or external; as the "corner" of a building, the four "corners" of a square, the "corner" of two streets. A mere variation in a line does not constitute a "corner." *Christian v. Grant*, 64 S.W. 399 (Tenn. Ch. 1900).

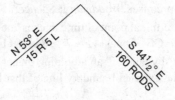

Figure AI.1 A Property Corner

A description of a tract as "lying in the southwest corner of a section" is sufficiently definite according to the rules of decision that a "corner" is a base point from which two sides of the land conveyed shall extend an equal distance so as to include by parallel lines the quantity conveyed. *Walsh's Lessee v. Ringer*, 15 Am. Dec. 555, 2 Ohio 328 (1826).

CORNER, EXISTENT. One whose position is identifiable by evidence of monument, its accessories or description in field notes, or can be located by acceptable supplemental survey record, some physical evidence, or testimony. *Reid v. Dunn,* 20 Cal. Rptr. 273 (1962).

CORNER, LOST. A point of a survey whose position cannot be determined, beyond reasonable doubt, either from traces of the original marks or from acceptable evidence or testimony that bears upon the original position, and whose location can be restored only by reference to one or more interdependent corners. *Manual of Instructions, Bureau of Land Management (1973).*See **Lost Corner.**

CORNER, NONEXISTENT. A corner which has never existed cannot be said to be lost or obliterated and established under the rules relating to the establishment of lost or obliterated corners, but should be established at the place where the original surveyor should have put it. *Lugon v. Crosier,* 240 P. 462, 78 Colo. 141.

CORNER, OBLITERATED. One at whose point there are no remaining traces of the monument or its accessories, but whose location has been perpetuated, or the point for which may be recovered beyond reasonable doubt by the acts and testimony of the interested landowners, competent surveyors, other qualified local authorities, or witnesses, or by some acceptable record evidence. *Manual of Instructions, Bureau of Land Management* (1973).

CORNER ACCESSORY. A physical object adjacent to a corner, to which the corner is referred for future identification or restoration. Accessories include bearing trees, mounds, pits, ledges, rocks, and other natural features to which distances or directions, or both, from the corner or monument are known. Accessories are part of the monument, and in the absence of the monument, carry the same weight.

COURSE. Used with reference to boundaries, is the direction of a line run with a compass or transit and with reference to a meridian. See 12 Am. Jur. 2d, Boundaries, § 10; *M'Iver v. Walker*, 9 Cranch (US) 173, 3 L Ed 694 (1815).

CURVE. In geometry, a line that changes its direction at every point; a line in which no three consecutive points are in the same direction or straight line. *Bishop & Babcock Mfg. Co. v. Fedders-Quigan Corp.,* D.C.N.Y., 159 F. Supp. 815.

DECLINATION, MAGNETIC. The difference between the location of true north and magnetic north at a given location and point in time. The difference of one degree will result in a horizontal difference in distance of 92 feet at the end of a mile.

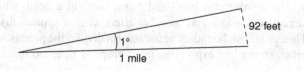

DEMARCATION/DELIMITATION. "Demarcation" is the marking of a boundary line on the ground by physical means or a cartographic representation. "Delimitation" is the defining of a boundary line in written or verbal terms. *State ex rel. Buckson v. Pennsylvania R. Co.,* 267 A.2d 455, supplemented 273 A.2d 268 (1969).

DIRECTLY OPPOSITE. Words "directly opposite" other lots described in deed mean portion that would be included in described lots if extended in straight line. *Smith v. Chappell,* 148 So. 242, 177 La. 311 (1933).

DISTANCE. A straight line along a horizontal plane from point to point and is measured from the nearest point of one place to the nearest point of another. *Evans v. United States,* C.C.A.N.Y., 261 R. 902.

DUE. In deed description, "boundary lines to continue due west from these stakes," the phrase "due" was not ambiguous and limited the course to one which traveled directly west. *Haklits v. Oldenburg,* 201 A.2d 690, 124 Vt. 199 (1964).

"Due," as used in the description in a deed, requiring the line to be run due north, means "exactly," and adds nothing to the description. The point of a compass, if due north, is exactly north, and so is simply north. *Wells v. Jackson Iron Mfg; Co.,* 47 N.H. 235, 90 Am. Dec. 575 (1866).

E. As an abbreviation, means "east." *Sibley v. Smith,* 2 Mich. 486 (1853).

EAST. Generally, the words "north," "south," "east," and "west," when used in a land description, mean, respectively, "due north," "due south," "due east," and "due west." *Plaquemines Oil & Development Co. v. State,* 23 So. 2d 171, 208 La. 425 (1945).

EASTERN ONE-HALF. In a deed conveying one-half of a tract of land, in the absence of admissible parol evidence disclosing a different intention, would mean the eastern half, formed by a line to be run due north and south through the tract; but if it appears that before the deed was executed a division into two parts, supposedly equal in area, had been made by a line, having a different bearing, actually marked on the ground by stakes and fences, according to which possession had been held for a number of years, and the parties have since held possession according to such line, the words must be taken to mean the eastern one-half as so laid off and held in severalty. *Bank v. Catzen,* 60 S.E. 499, 63 W. Va. 535 (1908).

EAST HALF. A deed conveying the east half of certain irregularly shaped lots is presumed to mean the east half in quantity. There is no presumption that the parties intend that the tract conveyed shall be ascertained by the rule of subdivision adopted in government surveys; that is, by running a line equidistant from the opposite sides of the lot. *Cogan v. Cook,* 22 Minn. 137 (1875).

Where there is nothing to suggest the contrary, the word "half," in connection with the conveyance of a part of a tract of land, is interpreted as meaning half in quantity. The words "east half" and "west half" in a deed, while naturally importing an equal division, may lose that effect when it appears that at the time some fixed line or known boundary or monument divides the premises somewhere near the center, so that the expression more properly refers to one of such parts than to a mathematical division which never has been made. *Gunn v. Brower*, 105 P. 702, 81 Kan. 242 (1909).

EAST HALF AND WEST HALF. In a description in a deed naturally import an equal division; but they may lose that effect when it appears that at the time the deed was made some fixed boundary or monument divided the premises somewhere near the center, so that the words referred more properly to one of such parts than to a mathematical division which had never been made. *People v. Hall*, 88 N.Y.S. 276, 43 Misc. 117 (1904).

EASTWARDLY. "As used in a grant, means due east, unless there be some object which can be found to control the course, in which case the course will run east, varying from that point to include the object. *Simms v. Dickson*, 3 Tenn. (1 Cooke) 137, 22 Fed. Cas. 158 (1812).

"Eastwardly," as used in a deed or grant of land to describe a call or a line to run eastwardly, is an indefinite expression, and means nothing more, necessarily, than that the land shall lie on the eastern and not on the western, side of a given line. It signifies on which side of the base of the lines marking the survey the land is to lie. *Preeble v. Vanhoozer*, 5 Ky. (2 Bibb.) 118 (1810).

ENCLOSE. The word "enclose" or "inclose" is defined as to surround, to encompass, to bound, fence, or hem in, on all sides. *White Chapel Memorial Ass'n v. Willson*, 244 N.W. 460, 260 Mich. 238 (1932).

END. Used in a tax judgement which is against an undivided third of the east end of each block, means the east half of the block. *Chiniquy v. People*, 78 Ill. 570 (1875).

The word "end" as used in a restriction in a deed providing that no building should be erected in the rear end of the lot conveyed within 10 feet from the line of the north side of the street, means the extremity, termination, limit. *Crofton v. St. Clement's Church*, 57 A. 570, 208 Pa. 209 (1904).

The word "end" is defined to be the extreme point of a line or anything that has more length than breadth. The end of a parallelogram is the line extending from one side line to the other at their extremities; and the width of the end is the length of such line. If the line connecting the extreme points of parallel side lines make an angle with one greater than that made with the other, as, for instance, one being 10 and the other 170 degrees, it might not be proper to regard this line the width of the end of the figure presented, or even as the end itself. In figures having side lines irregular and not parallel with each other, a line connecting them where they terminate may be the end, and its length the width of the end, or otherwise, according to the peculiar shape of each figure. Hence the words "end" and "width of the end," are not terms of greater precision, and the meaning of parties who may use them without any words in explanation may not always be apprehended with certainty. *Kennebec Ferry Co. v. Bradstreet*, 28 Me. 374 (1848).

END LINE. The end and side lines of a lode mining claim are not necessarily those so designated by the locator; but the "end lines" are those which are crosswise of the general course of the discovery vein on the surface, although they may have been located as the side lines. *Northport Smelting & Refining Co. v. Lone Pine-Surprise Consol. Mines Co.,* D.C.Wash., 271 F. 105.

EXTENDED LINE. Used in the description in a deed, meant a produced line. *McAndres & Forbes Co. v. Camden Nat. Bank,* 94 A. 627, 87 N.H.L. 231 (1915).

EXTEND TO. That which "extends to" does not necessarily include in. *Martin v. Hunter,* 14 U.S. (1 Wheat.) 304, 4 L. Ed. 97 (1816).

FACING. In a deed containing building restrictions applicable to lots "facing" and "having a frontage" on named street, quoted words as applied to oblong lots referred to the street which buildings to be erected on the lots were intended to face. *Aller v. Berkeley Hall School Foundation,* 103 P.2d 1052, 40 Cal. App.2d 31 (1940).

FARM. A testator, in devising his "farm at Bovingdon," meant all of the farm situated at that place, and not a part thereof. *Goodtitle v. Paul,* 2 Burrows (Eng.) 1089.

The term "farm" does not necessarily include only the land under cultivation and within a fence. It may include all the land which forms part of the tract, and may also include several connected parcels under one control; and, when one devises his "farm," he devises the whole farm, or all the land above designated. *Succession of Williams,* 61 So. 852, 132 La. 865 (1913).

FENCE. A visible or tangible obstruction, which may be a hedge, ditch, wall, or frame of wood, or any line of obstacles interposed between two portions of land so as to part off and shut in the land, and set it off as private property, and such is its meaning in defining inclosed lands. *Kimball v. Carter,* 27 S.E. 823, 95 Va. 77, 38 L.R.A. 570 (1897).

FIELD. Cleared land for cultivation or other purposes, whether inclosed or not. *Commonwealth v. Wilson,* Va., 9 Leigh. 648 (1839).

FIELD NOTES. A "map" is a picture of a survey, "field notes" constitute a description thereof, and the "survey" is the substance and consists of the actual acts of the surveyor, and, if existing established monuments are on the ground evidencing such acts, such monuments control because they are the best evidence of what surveyor actually did in making the survey and are part at least of what surveyor did. *Outlaw et al. v. Gulf Oil Corporation et al., 137 S.W.2d 787 (Texas, 1940)*

FRONT. Although it is true that a corner lot does front on both streets, only that portion of the lot which is opposite the rear and faces upon the street is properly designated as the "front" of such lot. *Staley v. Mears,* 142 N.E.2d 835, 13 Ill. App. 2d 451 (1957).

FRONTAGE. Within statute defining "business district" and "residence district," means space available for erection of buildings, and does not include cross streets or space occupied by sidewalk or any ornamental spaces in plat between sidewalks and curb. Comp. Laws 1929 & 4693 (v,w). *Wallace v. Kramer,* 296 N.W. 838, 296 Mich. 680 (1941).

FRONTING. Words "fronting" and "adjoining" are synonymous, and reference to lots "fronting" upon a street means that the lots touch boundary line of street. *Roach v. Soles,* D.C. Cal., 120 F.Supp. 400 (1954).

FRONTING AND ABUTTING. Very often, "fronting" signifies abutting, adjoining, or bordering on, depending largely on the context. *Rombauer v. Compton Heights Christian Church*, 40 S.W.2d, 545, 328 Mo. 1 (1931).

HALF. The words "the north half," used in the conveyance of a part of a platted block of land, mean the half of the block lying north of an east and west line drawn through the block, unless the surrounding facts require that these words be given a different meaning. *Lavis v. Wilcox*, 133 N.W. 563, 116 Minn. 187 (1911).

The word "half," when used in describing land, should be construed as meaning half in quantity, unless the context or surrounding facts and circumstances show a contrary intention. *Hoyne v. Schneider*, 27 P. 558, 138 Kan. 545 (1933).

"Half," as used in deeds of land according to government survey, ordinarily is used not with reference to the quantity, but with reference to a line which is equidistant from the boundary line of the parcel subdivided; but, where the government has divided a quarter of a section into what it calls "fractional 40's," the government division will govern. *Edinger v. Woodke*, 86 N.W. 397, 127 Mich. 41 (1901).

Where a deed recites that half of tract is being conveyed and there is nothing to suggest the contrary, the word "half" will be interpreted as meaning half in quantity. *McHenry v. Pence*, 212 P.2d 225, 168 Kan. 346 (1949).

"Half," when used in a conveyance which conveys the half of any particular piece of property, means an undivided half. *Baldwin v. Winslow*, 2 Minn. 213 (1858).

HALF SECTION. The general and proper acceptation of the terms "section" and "half section," as well as their construction by the general land department, denotes the land in the sections and subdivisional lines, and not the exact quantity which a perfect admeasurement of an unobstructed surface would declare. *Brown v. Harden*, 21 Ark. 324 (1860).

HEAD OF CREEK. The source of the longest branch, unless general reputation has given the appellation to another. *Davis v. Bryant*, 2 Bibb. (Ky.) 110 (1810).

HEAD OF STREAM. The highest point on the stream which furnishes a continuous stream of water, not necessarily the longest fork or prong. *Uhl v. Reynolds*, 64 S.W. 498, 23 Ky. Law Rep. 759; *State v. Coleman*, 13 N.J. Law 104 (1832).

IN RANGE WITH. A line "to range" with another, from the end of which it begins, must follow the path of the other line when extended and be a continuation of it. *Lilly v. Marcum*, 283 S.W. 1059, 214 Ky. 514 (1926).

IN SQUARES TO THE CARDINAL POINTS. Where a contract calls for a survey of land "in squares to the cardinal points," without specifying whether the true or magnetic meridian shall control in determining the cardinal points, the magnetic meridian will be held to control. *Finnie v. Clay*, 5 Ky. 351, 2 Bibb. 351 (1811).

LAID OUT. "Laid out," as used in reference to a highway, has a well-known meaning, under our statutes, and plainly includes the doing of those things by the proper local officers which are essential in creating a public highway, to authorize it to be worked and traveled, and especially the surveying, marking the course or boundaries, and ordering it to be established as a highway. The affirmative action of the public authorities is indispensible in such case. *Chicago Anderson Pressed Brick Co. v. City of Chicago*, 28 N.E. 756, 138 Ill. 628 (1891).

"Laid out," as used on the face of a map as laid out by a certain person, is equivalent to "as surveyed" by him, and embraces a reference to the monuments placed on the land by the surveyor. *Flint v. Long*, 41 P. 49, 12 Wash. 342 (1895).

LINE. In surveying and dividing grounds, means, prima facie, a mathematical line, without breadth; yet this theoretic idea of a line may be explained, by the facts referred to and connected with the division, to mean a wall, a ditch, a crooked fence, or a hedge—a line having breadth. *Baker v. Talbott*, 22 Ky. 179, 6 T.B. Mon. 179 (1827).

A testator who, intending to devise part of his farm, begins at one of the corners of it and says, "thence as the line runs," is to be understood to mean the line of the farm, whether it be straight or crooked. A crooked line is a line just as much as a straight one. We say the line of a state, a county, a coast, or of the seashore, though it have a great many bendings; and, with equal propriety, a line of a fence, a road, or a farm. *Cubberly v. Cubberly*, 12 N.J.L. 308 (1831).

That the description of a survey called for the marked "line" of another survey between designated points which was not a straight line but consisted of four lines and three marked corners was immaterial; the word "line" being often used for the plural, and vice versa, and the singular being also sufficient to describe the exterior boundary of the survey line called for between the designated points. *Bell v. Powers*, 121 S.W. 991 (Ky., 1909).

LINE OF RAILROAD. As used in a deed describing the land as lying "on the line of a certain railroad," means next to or bounded by the railroad. The line of a road must mean, as used in the contract, the boundary of the land appropriated for its use. *Burnam v. Banks*, 45 Mo. 349 (1870).

LITTLE MORE THAN. The terms "more or less," "a little more than," "not quite," "not more than," and "approximately" are terms of safety and precaution and ordinarily mean "about" when used in a deed. *Pierce v. Lefort*, 200 So. 801, 197 La. 1 (1941).

LOCATE. The words "locate" and "establish" are not synonymous. *Givens v. Woodward*, 207 S.W.2d 234 (Tex.Civ.App., 1947).

"Locate" means to designate the site or place of; to define the location or limits of, as by a survey; as, to locate a public building, a mining claim; to locate the land granted by a land warrant. *Delaware, L. & W.R.Co. v. Chiara*, CCANJ, 95 F.2d 663.

LOCATIVE CALL. Where iron pipe which marked southeast corner of surveyed section was not the northeast corner of adjoining section and call was from an iron pipe set for the southeast corner along the north boundary line of adjoining section thus creating inconsistency between calls in description, the call for pipe was a "locative call" within rule that locative call controls over descriptive call. *Outlaw et al. v. Gulf Oil Corporation et al., 137 S.W.2d 787 (Texas, 1940)*

LOST (as applied to "corner"). For corners to be lost, they must be so completely lost that they cannot be replaced by reference to any existing data or other sources of information, and before courses and distances can determine boundary, all means for ascertaining location of the lost monuments must first be exhausted. *U.S. v. Doyle*, 468 F.2d 633 (C.A. Colo., 1972).

"To be lost," when applied to section or township corners, means more than that they have been merely obliterated, tampered with, or changed, but they must be so completely lost that they cannot be replaced by reference to any existing data or other sources of information. *Mason et al v. Braught*, 146 N.W. 687 (S.D., 1914).

LOT. Defined as any portion, piece, or division of land. *Lehmann v. Revell*, 188 N.E. 531, 554 Ill. 262 (1933).

Word "lot," when used unqualifiedly, means lot in a township as duly laid out by the original proprietors. *Carney v. Dunn*, 252 S.W.2d 827, 221 Ark. 223 (1952).

"Town lots, out lots, common field lots and commons were known and recognized parts of the Spanish town or commune of St. Louis. They existed by public authority, whether by concession, custom or permission." *Vaquez v. Ewing*, 42 Mo. 247 (1868).

A "lot" is a portion of land that has been set off or allotted, whether great or small; but in common use it means simply a piece, parcel, or tract of land without regard to size and does not necessarily connect itself with buildings, but may be anywhere on the earth's surface. *Schack v. Trimble*, 157 A.2d 22, 48 N.J. Super. 45 (1957).

The term "lot" applies to any portion, piece, or division of land, and is not limited to parcels of land laid out into blocks and lots regularly numbered and platted. *Westbrook v. Rhodes*, 218 P. 873, 92 Okl. 149 (1923).

The words "lot," "piece," and "parcel" apply peculiarly to the land itself, and are never employed to describe improvements. *Canty v. Staley*, 123 P. 252, 162 Cal. 379 (1911).

Term "lot," as used in connection with urban property, is frequently defined as subdivision of block, according to plat or survey of town or city. *Lehmann v. Revell*, 188 N.E. 531, 354 Ill. 262 (1933).

MAP. A picture of a survey, "field notes" constitute a description thereof, and the "survey" is the substance and consists of the actual acts of the surveyor, and, if existing established monuments are on the ground evidencing such acts, such monuments control because they are the best evidence of what surveyor actually did in making the survey and are part at least of what surveyor did. *Outlaw et al. v. Gulf Oil Corporation et al., 137 S.W.2d 787 (Texas, 1940)*

MARKED CORNERS. (i.e., those clearly identified, and which are notorious objects) are the most satisfactory evidence of the location of a patent. *Morgan v. Renfro*, 99 S.W. 311, 124 Ky. 314 (1907).

MARKED LINE. Where bearing trees are called for at the eastern end of a northern boundary line of a survey and a stake at the western terminus, there is a presumption that the line was actually surveyed, and the corners identified by the bearing trees and the stake making the line a "marked line." *Goodson v. Fitzgerald*, 135 S.W. 696 (Tex. 1905).

MEANDER CORNERS. A so-called "meander corner" is not fixed point for measurements, as are section and quarter corners, but is a marker for courses. *Thunder Lake Lumber Co. v. Carpenter*, 200 N.W. 302, 184 Wis. 580 (1924).

"Meander corner" holds, as declared by the rules of the United States Land Office, the peculiar position of denoting a point on line between landowners without usually being the legal terminus or corner of the land owned. Where meander corners of a government survey are lost or obliterated, they must be restored in accordance with the circulars of the United States Land Office. *Kleven v. Gunderson*, 104 N.W. 4, 95 Minn. 246 (1905).

MEANDER LINE. Generally contains a call for a natural object or monument which will usually control over calls for course and distance. *State v. Arnim*, 173 S.W.2d 503 (Tex. Civ. App. 1943).

A "meander line" is described by courses and distances, and the line is thus fixed by reason of difficulty of surveying a course following the sinuosities of the shore, and the impracticability of establishing a fixed boundary along shifting sands of the ocean. *Den v. Spalding*, 104 P.2d 81, 39 Cal. App. 2d 623 (1940).

As a rule, "meander lines" relate to sinuosities and course of stream and do not constitute boundaries in absence of proof which clearly indicates a contrary intention. *Cox v. City & County of Dallas Levee Imp. Dist.*, 258 S.W.2d 851 (Tex. Civ. App. 1953).

The meander line run by the United States surveyors along a stream is not the boundary of the lands, but is run merely to determine the quantity of lands contained in the lots and the purchasers of the lots take to the center of the stream, and not merely to the meander line. *Jones v. Pettibone*, 2 Wis. 225 (1853).

MEASURE, HORIZONTAL. Measurement in a horizontal plane, that is, not including, or correcting for, slope. See **Slope Measure.**

MEASURE, SLOPE. Measurement was sometimes made along the surface of the ground, especially in some of the mountainous areas of the southeastern United States. See **Horizontal Measure.**

METES. Refers to the exact length of each line and the exact quantity of land in square feet, rods or acres. *U.S. v. 5.324 Acres of Land*, D.C. Cal., 79 F. Supp. 748. (1948)

METES AND BOUNDS. Where land was bounded in the deed by the lands of named persons, it was described by metes and bounds, which mean the boundary lines or limits of a tract. *Moore v. Walsh*, 93 A. 355, 37 R.I. 436 (1915).

"Metes and bounds" are the boundary lines of land, with their terminal points and angles. *Lefler v. City of Dallas*, Tex. Civ. App., 177 S.W. 2d 231 (1943).

MIDDLE. A call for a county boundary line as running west to the middle of a section calls for it to run to the center of the section measured from north to south as well as from east to west. *Alluvial Realty co. v. Himmelbsrger-Harrlson Lumber Co.,* 229 S.W. 757, 287 Mo. 299 (1921).

MONUMENT. "May mean anything by which the memory of a person, thing, idea, art, science or event is portrayed or perpetuated. *Odom v. Langston,* 195 S.W.2d 466, 355 Mo. 115 (1946).

"Monuments" are permanent landmarks established for the purpose of indicating boundaries. *Thompson v. Hill,* 73 S.E. 640, 137 Ga. 308 (1912).

Monuments are the visible marks or indications left on natural or other objects indicating the lines and boundaries of a survey. *Grier v. Pennsylvania Coal Co.,* 18 A. 480, 128 Pa. 79 (1889).

Where circumstances warrant, fences may be considered as "monuments." *Rodgers v. Roseville Gold Dredging Co.,* 286 P.2d 536, 135 C.A.2d 6 (1955).

Land or an adjoining proprietor is a "monument" within the rule that monuments govern measurements of land. *Di Maio v. Ranaldi,* 142 A. 145, 49 R.I. 204 (1928).

Center of section is not physical government monument, but is point capable of being mathematically ascertained, thus constituting it, in legal sense, "monument call of description." *Matthews v. Parker,* 299 P. 354, 163 Wash. 10 (1931).

MONUMENT, ARTIFICIAL. A landmark or sign erected by the hand of man. 11 C.J.S. Boundaries, § 7. Artificial monuments include stakes, stones, stone piles, pins, pipes, metal rods, concrete bounds, and building corners.

MONUMENT, NATURAL. A permanent object found on the land as it was placed by nature. 11 C.J.S. Boundaries, § 6. Natural monuments include lakes and ponds, rocks, shores and beaches, springs, streams, and rivers, and trees.

MORE OR LESS. The terms "more or less," "a little more than," "not quite," "not more than," and "approximately" are terms of safety and precaution and ordinarily mean "about" when used in a deed. *Pierce v. Lefort,* 200 So. 801, 197 La. 1 (1941).

The use of the phrase "more or less" in describing a boundary line relieves a stated distance of exactness, means the parties are to risk the quantity of land conveyed, and implies waiver of the warranty as to a specified quantity. *Salyer v. Poulos,* 122 S.W.2d 996, 276 Ky. 143 (1939).

The words "more or less," used in a contract of sale in connection with an estimated quantity, when the only measure is the estimate itself, allow only a small latitude of variation. *U.S. v. Republic Bag & Paper Co.,* C.C.A.N.Y., 250 F. 79.

The words "more or less" following a statement in a deed as to the number of acres conveyed is a matter of description only. *Maffet v. Schaar,* 131 P. 589, 89 Kan. 403 (1913).

The words "more or less," used to describe a lot of land, mean merely that the lot conveyed may be in size more or less than the dimension given, but they cannot be so extended as to include a separate and distinct lot. *McCune v. Hull*, 24 Mo. 570 (1857).

The words "more or less" usually mean "about," "substantially," or "approximately," and imply that both parties assume the risk of any ordinary discrepancy. *Alexander v. Hicks*, 5 So.2d 782, 242 Ala. 243 (1942).

The words "more or less," as used in a deed declaring that the land conveyed contained a specified number of acres, more or less, did not mean as estimated, as supposed, but should be construed to mean about the specified number of acres, and are designated to cover only such small errors or surveying as usually occur in surveys. *Crislip v. Cain*, 19 W.Va. 438 (1882).

N.E. An abbreviation of "northeast" in constant and universal use. *Sexton v. Appleyard*, 34 Wis. 235 (1874).

NEAR. Cannot be used as the equivalent of exact distances. *Picharella v. Ovens Transfer Co.*, 5 A.2d 408, 135 Pa. Super. 112 (1939).

NORTHEAST CORNER. Where description in deed was that location of square tract of land should be "in northeast corner" of M.'s league of land, description did not require that land should include extreme northeast portion of survey, where beginning point was described as northeast corner of M.'s headright league of land. *Harper v. Temple Lumber Co.*, 290 S.W. 530 (Tex. 1927).

NORTHEASTERLY. Where the terms "northerly," "northwesterly," "northeasterly," etc. are employed to designate a line in the description of a deed, such terms will be construed as equivalent to a call to run due north, due northwest, or northeast, as the case may be, in the absence of a call for visible monuments, or any other description of a line which locates it with reasonable certainty. *Irwin v. Towne*, 42 Cal. 326 (1871).

NORTHERLY. Used in describing boundary implies only a general direction. *Fosburgh v. Sando*, 166 P.2d 850, 24 Wash. 2d 586 (1946).

Where a course in a deed is described as running northerly, the word "northerly" must be construed as meaning due north. *Proctor v. Andover*, 42 N.H. 348 (1861).

NORTHERNLY. In a grant, where there is no object mentioned to direct the inclination of the course toward the east or west, means due north." *State ex rel. Chandler v. Huff*, 79 S.W. 1010, 105 Mo. App. 354 (1904), quoting and adopting definition in *Brandt ex dem. Walton v. Ogden*, N.Y., 1 Johns. 156 (1806).

NORTH HALF. Used in the conveyance of a part of a platted block of land, mean the half of the block lying north of an east and west line drawn through the block, unless the surrounding facts require that these words be given a different meaning. *Lavis v. Wilcox*, 133 N.W. 563, 116 Minn. 187 (1911).

Under a conveyance of the "north half" of a tract of land, the east line of which is of such a shape that the north line is less than one-half the length of the south line, the grantee is entitled to one-half of the area of the tract, and not merely to one-half of the north and south length of the tract. *Robinson v. Taylor*, 123 P. 444, 68 Wash. 351 (1912), Ann. Cas. 1913E, 1011.

NORTH ONE-THIRD. The description of land as the "north one-third of lots five and six" in block 77 in itself indicates a single tract. *La Selle v. Nicholls*, 76 N.W. 870, 56 Neb. 458 (1898).

NORTH PART. A deed reciting that a tract of land conveyed the "north part" of a certain lot cannot be construed to mean the north half of the lot. *Langohr v. Smith*, 81 Ind. 495 (1882).

Under deed conveying one of two adjoining lots and reserving right of way over the "south part" of the lot conveyed, the term the "south part" constituted description in contradistinction to "north part" and did not suggest the idea of a "middle part" which is no specific part but any part that is embraced in any two lines between north and south or east and west boundaries that are parallel with and equidistant from the middle lines, while the "south part" supposes a "north part," the "south part" being all that lies south, and the "north part" being all that is north of the middle line east and west. *Roberts v. Stephens*, 40 Ill. App. 138 (1891).

NORTH SIDE. A deed describing the land conveyed as the north side of the south-west quarter of a certain block should be construed to mean the north half of such quarter. *Winslow v. Cooper*, 104 Ill. 235 (1882).

NORTHWARD. In construing a patent granting five great plains, "together with the woodland around such plains, that is to say, four English miles from the said plains eastward, four English miles northward from the said plains, four English miles westward from the said plains, and four English miles southward from the said plains," the court said: "The given object to start from is the plains; the distance to run is four miles; the courses are northward, southward, eastward, and westward; and it is a settled rule of construction that, when courses are thus given, you must run due north, south, east, and west." *Jackson v. Reeves*, N.Y., 3 Caines, 293 (1805).

NORTHWARDLY. Means toward or approaching toward the north, rather than toward any of the other cardinal points. *Martt v. McBrayer*, 166 S.W.2d 823, 292 Ky. 479 (1942).

The word "north," as distinguished from the word "northwardly," conveys a definite idea that is, indicates a particular cardinal point—while the word "north-wardly" means towards or approaching towards the north, rather than towards any of the other cardinal points. *Craig v. Hawkins' Heirs*, 4 Ky. (1 Bibb) 53 (1808).

NORTHWEST. Where the base of a description in a deed was parallel of north latitude, a line called to run northwest for quantity, to adjoin a claim on the north, should not be construed to mean running north or at right angles to the base given, but the survey should be projected northwest for quantity. *Swearingen v. Smith*, 4 Ky. (1 Bibb) 92 (1809).

NORTH WITH THE HALF SECTION LINE. In a deed denoted nearness or in the same direction as the half section line and not necessarily upon or along such line. *Puntt v. Simmer*, 8 Ohiio CC (NS) 455 (1906).

NOT MORE THAN. "More or less" are words of safety and precaution and when used in deed are intended to cover some slight inaccuracy in frontage, depth or quantity in land conveyed, and ordinarily means "about" the same as terms "a little more than," "not quite," "not more than," or "approximately," and all are often introduced into a description practically without effect. *Harries v. Harang*, La. App., 23 So.2d 786 (1945).

NOT QUITE. See NOT MORE THAN; MORE OR LESS.

ON. Courts will take judicial notice that surveyors generally in making field note calls such as "with a marked line" use word "with" as having same meaning as word "on." *Carter v. Texas Co.,* 87 S.W.2d 1079, 126 Tex. 388 (1935).

ON SAID WALL. In a description reading "thence southerly on said wall ten rods," show that course of line is controlled by course of wall. *Vermont Marble Co. v. Eastman*, 101 A. 151, 91 Vt. 425 (1915).

ONE-FOURTH. The conveyance of one-fourth of a tract of land, without designating by metes and bounds or otherwise locating the part conveyed, vests in the grantee, and those claiming under him, the title to one undivided fourth of the whole tract, as tenant in common with the grantor. *McCaul v. Kilpatrick*, 46 Mo. 434 (1870).

OPPOSITE. Where a deed describes a boundary as ending at a point at one side of a street "opposite" a point on the other side, a straight line between the two points must cross the street at a right angle. *Bradley v. Wilson*, 58 Me. 357 (1870).

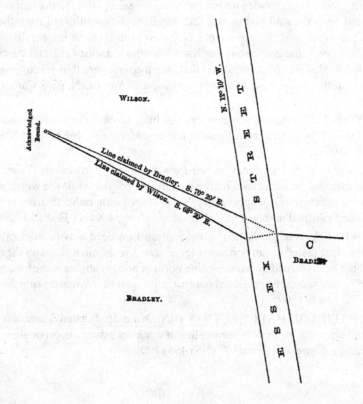

PACE. A measure of length containing two feet and a half, being the ordinary length of a step. The geometrical pace is five feet long, being the length of two steps, or the whole space passed over by the same foot from one step to another. *Black's Law Dictionary*.

"Pace," the unit of measurement employed by deed, was not precise, but did not render the deed so ambiguous as to justify resort to extraneous evidence and could be treated as approximately three feet. *Haklits v. Oldenburg*, 201 A.2d 690, 124 Vt. 199 (1964).

PARALLEL. Geometrical meaning of "parallel" is line evenly everywhere in same direction but never meeting, however far extended, and in all parts equally distant, but word also connotes with like direction or tendency, and running side by side. *Valente v. Atlantic City Elec. Co.*, 101 A.2d 106, 28 N.J. Super. 476 (1953).

By mathematical definition, parallel lines are straight lines, but, in common speech about boundaries, the words are often used to represent lines which are not straight, but photographs of each other; and courts, in passing on questions of boundaries, often use them in the latter sense. The term is used for the want of a better, and not because it in all respects affects the use to which it is applied. It is so used to avoid circumlocution, and, while such use is not exactly correct, there is no difficulty in understanding the meaning of intention. *Fratt v. Woodward*, 91 Am. Dec. 573, 32 Cal. 219 (1867).

A recital that a street runs "parallel or nearly so "with another street intimates that the streets are not absolutely parallel with one another. *Rehfuss v. Hill*, 90 N.E. 187, 243 Ill.140 (1909).

PARALLEL WITH. "Equidistant," as used in a conveyance of lots providing that the front line of all buildings thereon shall be placed equidistant from and not less than 8 feet back from the street, should be construed to mean "parallel with." *Smith v. Bradley*, 28 N.E. 14, 154 Mass. 227 (1891).

PARCEL. A part or portion of land. See *State v. Jordan*, 17 S. 742, 36 Fla. 1 (1895); *Miller v. Burke*, 6 Daly (N.Y.) 174; *Johnson v. Sirret*, 46 N.E. 1035, 153 N.Y. 51 (1897); *Chicago, M. & St. P. Ry. Co. v. Town of Churdan*, 195 N.W. 996, 196 Iowa 1057 (1923).

A part of an estate. *Martin v. Cole*, 38 Iowa 141 (1874).

It may be synonymous with lot. *Terre Haute v. Mack*, 38 N.E. 468, 139 Ind. 99 (1894).

A "parcel of land" or "parcel of real property" means a contiguous quantity of land in possession of, or owned by, or recorded as the property of, the same claimant, person, or company. *State v. Jordan*, 17 So. 742, 36 Fla. 1 (1895).

The terms "tract" and "parcel" may properly be applied to a quarter section, a half section, or a section of land. *People ex rel. Chicago General Ry. Co. v. Chase*, 70 Ill.App. 42 (1897).

The words "lot," "piece," and "parcel" apply peculiarly to the land itself, and are never employed to describe improvements. *Canty v. Staley*, 123 P. 252, 162 Cal. 379 (1911).

PARCEL OF LAND. "The terms 'tract or lot' and 'piece or parcel of real property,' or 'piece or parcel of land,' mean any contiguous quantity of land in the possession of, owned by, or recorded as the property of the same claimant, person, or company." In the connection the word "contiguous" means land which touches on the sides. Hence two quarters of the same section, which only touch at the corner. do not constitute, for the purpose of taxation, one tract or parcel of land. *Griffin v. Denison Land Co.,* 119 N.W. 1041, 18 N.D. 246 (1908).

PERCH. A perch of Paris is eighteen feet. *Sullivan v. Richardson,* 14 So. 692, 33 Fla. 1 (1884). Ordinarily considered to be $16^1/_2$ feet. See **Pole, Rod.**

POINT. In a boundary is the extremity of a line. *Tiffany v. Town of Oyster Bay,* 126 N.Y.S. 910, 141 App. Div. 720 (1910).

"Point" as used in deed descriptions held to mean tapering end of mountain or knoll. *Staley v. Richmond,* 32 S.W.2d 546, 236 Ky. 11 (1930).

POLE. Ordinarily considered to be $16^1/_2$ feet. See **Perch, Rod.**

PROLONGATION. As used in a deed, means a continued or extended line, though consisting of several angles, where such meaning would be consistent with the other words of description, rather than a direct line, which would render the next course in the deed inconsistent with the direction and monument by which it is described. *Chapman v. Hamblet,* 64 A. 215, 100 Me. 454 (1905).

PROPERTY LINE. A division between two parcels of land. *Ujka v. Sturdevant,* 65 N.W.2d 292 (N.D. 1954).

ROD. Generally considered to be $16^1/_2$ feet in length. However, in England, the rod varied from 12 to 22 feet in length, depending on what was being measured. An 18-foot rod is often encountered in the northeastern United States, particularly where settlements were made by the Scotch-Irish. In parts of the Hudson River Valley of New York, a Dutch rod of $12^1/_2$ feet is sometimes found.

ROOD. One-quarter of an acre, or 40 square rods. The unit is sometimes found in colonial surveys.

RUNNING ALONG. Description of grant as "running along Smith's line" held to have made tracts contiguous to full length of named tract, leaving no vacancy between them. *Ramsay v. Butler, Purdum & Co.,* 129 A. 650, 148 Md. 438 (1925).

RUN OUT A LINE. When used in connection with surveying, means that a person qualified to do such work shall *go* upon the ground and with proper instruments, chainmen, and necessary assistants establish and mark a line upon the surface of the earth. They clearly impose the duty of going upon the ground and actually running out the line. *Mineral County Com'rs. v. Hinsdale County Com'rs.,* 53 P. 383, 25 Colo. 95 (1898).

RUNS AND CALLS. Terms "runs and calls," "courses and distances" and "angles and distances" are synonymous; they all refer to the angles and scaled distances indicated on a plat map and must be followed in order to establish exact boundaries. *Block v. Howell,* 346 N.W.2d 441 (S.D. 1984).

S.E. Mean southeast. *Bandow v. Wolven,* 107 N.W. 204, 20 S.D. 445 (1906).

S.E. 4. The abbreviation "S.E. 4," employed in the description of the property conveyed by a tax deed, will be interpreted as meaning "southeast quarter," when it is explicitly used in another part of the same instrument as the equivalent of these words. *Kennedy v. Scott*, 83 P. 971, 72 Kan. 359 (1905).

SE QR 24. In a description of property contained in a list of delinquent real property attached to a notice of tax sale describing it as "se qr 24," the number 24 being in a column headed "sec, the letters "se" clearly meant southeast, and the description given properly described the southeast quarter of section 24. *Bandow v. Wolven*, 107 N.W. 204, 20 S.D. 445 (1906).

Sec. Mean "section." *Bandow v. Wolven*, 107 N.W. 204, 20 S.D. 445 (1906).

Section. A "section" of a township is that which is made out on ground, and a patentee takes only such land as is included within survey of plot conveyed, and he cannot later question survey as erroneous, although in fact line in question should have been placed elsewhere. *Phelps v. Pacific Gas & Elec. Co.*, 190 P.2d 209, 94 Cal. App. 2d 243 (1948).

The general and proper acceptation of the terms "section," "half section," and "quarter section" of land, as well as their construction by the general land department, denotes the land in the sectional and subdivisional lines, and not the exact quantity which a perfect admeasurement of an unobstructed surface would declare. *Brown v. Hardin*, 21 Ark. 324 (1860).

SECTION CORNER. A "quarter corner," as distinguished from a "section corner," in the government surveys means the corner or a section line midway between the section corners. *Rud v. Pope County Com'rs*, 68 N.W. 1062, 66 Minn. 358 (1896).

SECTION LINE. A "quarter line," as distinguished from a "section line," means a line running from one quarter corner to another through the center of the section. *Rud v. Pope County Com'rs*, 68 N.W. 1062, 66 Minn. 358 (1896).

SIDE. A bounding line of a geometrical figure; as, the side of a field, square, river, road, etc. *Badura v. Lyons*, 23 N.W.2d 678, 147 Neb. 442 (1946).

The word "side" has many meanings. It may not always refer to the border or edge of the water of a lake; it may refer to any part or position viewed as opposite to or contrasted with another, as the south side or north side of the pond or lake. *Webster's New International Dictionary*. Whether the one meaning or the other be intended depends entirely upon the context. *White v. Knickerbocker Ice Co.*, 172 N.E. 452, 254 N.Y. 152 (1930), see 74 A.L.R. 591, reversing 241 N.Y.S. 898, 229 App. Div. 746 (1930).

SIDE LINES. The side lines of a road or railroad are the lines which include the territory covered by the road. Public roads and highways and railroads are regarded as having three lines, the side lines and the center line equidistant between the side lines. *Maynard v. Weeks*, 41 Vt. 617 (1869).

The word "front" as applied to a house is always specific, and speaking of "side line" of house as "fronting" toward street is incorrect. "Side" may be used in a generic sense so as to include the "front," but it also has a specific meaning which distinguishes it from "front." *Howland v. Andrus*, 86 A. 391, 81 N.J. Eq. 175 (1913).

SITE. According to *Webster*, is a seat or ground plot; and a mill site is the place where a mill stands. *Miller v. Alliance Ins. Co. of Boston*, 7 F. 649.

SOUTH. Generally, the words "north," "south," "east," and "west," when used in a land description, mean, respectively, "due north," "due south," "due east," and "due west." *Plaquemines Oil & Development Co. v. State*, 23 So.2d 171, 208 La. 425 (1945).

SOUTH COURSE. In deed did not mean due south, but meant a southwardly course. *Martt v. McBrayer*, 166 S.W.2d 823, 292 Ky. 479 (1942).

SOUTHERLY. In the absence of monuments and in a deed, "southerly" means due south. *Smith v. Newell*, 86 F. 56 (1898).

The words "southerly" and "westerly" used in the identifying descriptions in deeds, are not always used to indicate a direction that is due south or west. *Brown v. McCaffrey*, 60 A.2d 792, 143 Me. 221 (1948).

There are very few words in our language more indefinite and uncertain in their meaning than the words "southerly," "easterly," and "northerly." The word "southerly," as applied to the course of a proposed highway, designating the course as "thence southerly to avoid" a certain creek, and "thence easterly and northerly through" certain lands, means nearly south, but how near, and whether east or west of south, it is impossible to tell without the use of other qualifying words; and so with regard to the words "easterly" and "northerly." It is impossible to determine with any certainty the course intended thereby. *Scraper v. Pipes*, 59 Ind. 158 (1877).

SOUTH PART. Under deed conveying one of two adjoining lots and reserving right of way over the "south part" of the lot conveyed, the term the "south part" constituted description in contradistinction to "north part" and did not suggest the idea of a "middle part" which is no specific part but any part that is embraced in any two lines between north and south or east and west boundaries that are parallel with and equidistant from the middle lines, while the "south part" supposes a "north part," the "south part" being all that is north of the middle line east and west. *Roberts v. Stephens*, 40 Ill. App. 138 (1891).

SOUTHWARD. Must be considered with reference to its subject-matter, and, as used in a conveyance describing land, may include land lying in a southwesterly direction, where that seems to be the intention. *Higgins v. Round Bottom Coal & Coke Co.,* 59 S.E. 1064, 63 W.Va. 218 (1907).

SQUARE. An open area in a city or village left between streets at their intersection. *Harvey v. Mayor and Aldermen of City of Savannah*, 199 S.E. 653, 59 Ga. App.12 (1938).

As used to designate a certain portion of land within the limits of a city or town, this term may be synonymous with "block," that is, the smallest subdivision which is bounded on all sides by principal streets, or it may denote a space (more or less rectangular) not built upon, and set apart for public passage, use, recreation, or ornamentation, in the nature of a "park" but smaller. See *Caldwell v. Rupert*, 10 Bush (Ky.) 179 (1873); *State v. Natal*, 7 So. 781, 42 La. Ann. 612 (1890); *Rowzee v. Pierce*, 23 So. 307, 75 Miss. 846, 40 LRA 402, 65 Am. St. Rep. 625 (1898); *Methodist Episcopal Church v. Hoboken*, 97 Am. Dec. 696, 33 N.J. Law 13 (1868); Rev. Laws Mass. 1902, p. 531, c. 52, §12 (Gen. Laws 1932, c. 85 §14).

A "block" or "square" is a portion of a city bounded on all sides by streets or avenues. *Missouri, K. & T. Ry. Co. v. City of Tulsa*, 45 Okl. 382, 145 P. 398, 401 (1936); *City of Mobile v. Chapman*, 79 So. 566, 571, 202 Ala. 194 (1918).

SQUARE BLOCK. In a technical sense the word "block" may and often does designate a territory, sometimes called a "square block," but it is also commonly used to designate that section of a square block, so called, fronting on a street between two intersecting streets. *City of Olean v. Conkling*, 283 N.Y.S. 66, 157 Misc. 63 (1835).

SQUARES. Generally used synonymously with "blocks" in describing urban premises, a square or block meaning a subdivision of a city or town inclosed by streets, whether occupied by buildings or inclosures or merely comprising vacant lots. *City of Mobile v. Chapman*, 79 So. 566, 202 Ala. 194 (1918).

STAKE. It is a settled rule of construction that when "stakes" are mentioned in a deed simply, or with not other added description than that of course and distance, they are intended by the parties, and so understood, to designate imaginary points. *Massey v. Belisle*, 24 N.C. 170 (1841).

SURVEY. The substance and consists of the actual acts of the surveyor. *Outlaw v. Gulf Oil Corp.*, Tex. Civ. App., 137 S.W.2d 787 (1940).

A "survey" is process by which parcel of land is measured and its contents ascertained and is also a statement of or a paper showing the result of the survey with the courses and distances and quantity of the land. *Overstreet v. Dixon*, 131 S.E. 2d 580, 107 Ga. App. 835 (1963).

The word "survey" is commonly used in old conveyances to mean the same as "tract," "boundary" or "land." *Burke v. Owens-Illinois Glass Co.*, D.C. W.Va., 86 F.Supp. 663 (1949).

To survey land means to ascertain corners, boundaries, divisions, with distances and directions, and not necessarily to compute areas included within defined boundaries; such computation being merely a matter of mathematics. *Kerr v. Fee*, 161 N.W. 545, 179 Iowa 1097 (1917).

"Survey," as used in a description in a trust deed conveying "the B. survey, lying in what is known as the I. pasture, in C. and A. counties," is synonymous with the word "land," or "grant," or "location." *Clark v. Gregory*, 26 S.W. 244 (Tex.Civ.App., 1894).

"Laid out," as used on the face of a map as laid out by a certain person, is equivalent to "as surveyed" by him, and embraces a reference to the monuments placed on the land by the surveyor. *Flint v. Long*, 41 P. 49, 12 Wash. 342 (1895).

A "map" is a picture of a survey, "field notes" constitute a description thereof, and the "survey" is the substance and consists of the actual acts of the surveyor, and, if existing established monuments are on the ground evidencing such acts, such monuments control because they are the best evidence of what surveyor actually did in making the survey and are part at least of what surveyor did. *Outlaw et al. v. Gulf Oil Corporation et al.,* 137 S.W. 2d 787 (Tex. 1940).

First survey. When a parcel or parcels are created on paper, without a survey being conducted, and the surveyor is later requested to place one of these paper-described parcels on the ground, this survey should be considered the "first" survey, in that it is the first survey to be placed on the ground after the description. *Brown, Robillard & Wilson, 1994.*

Original survey. A survey called for or presumed to have been made at the time a parcel or parcels were created. An original survey creates boundaries, it does not ascertain them *Brown, Robillard & Wilson,* 1994.

Original survey vs. first survey. The difference is whereas the original survey controls, the first survey is nothing more than an opinion by the surveyor where the description should be placed. *Brown, Robillard & Wilson,* 1994.

Resurvey means to survey again, and applies to land which has been surveyed once. *Trudeau v. Town of Sheldon,* 20 A. 161, 62 Vt. 198 (1890).

A *resurvey* is a reconstruction of land boundaries and subdivisions accomplished by rerunning and re-marking the lines represented in the field-note record or on the plat of a previous official survey. *Pointer v. Johnson,* 695 P. 2d 399, 107 Idaho 1014 (1985).

A *dependent resurvey* is a retracement and reestablishment of the lines of the original survey in their true original positions according to the best available evidence of the positions of the original corners.

An *independent resurvey* is an establishment of new section lines, and often new township lines, independent of and without reference to the corners of the original survey.

The purpose of a **resurvey** is to determine where the footsteps of the original surveyor were located, that is, to restore the original surveyor's lines in the same position as they were originally marked. The concept of "footsteps" is one of determining where the evidence of the original survey is located. *Brown, Robillard & Wilson,* 1995.

A *retracement* is a survey that is made to ascertain the direction and length of lines and to identify the monuments and other marks of an established prior survey. Retracements may be made for any of several reasons. In the simplest case it is often necessary to retrace several miles of line leading from a lost corner which is to be reestablished to an existent corner which will be used as a control. If no intervening corners are reestablished, details of the retracement are not usually shown in the record, but a direct connection between the two corners is reported as a tie. On the other hand, the retracement may be an extensive one made to afford new evidence of the character and condition of the previous survey. Recovered corners are rehabilitated, but a retracement does not include the restoration of lost corners or the reblazing of lines through the timber. The retracement may sometimes be complete in itself; but usually it is made as an early part of a resurvey.[256]

Retracement Survey. Surveyor retained to locate on the ground a boundary line which has theretofore been established performs a "retracement survey." *Rivers v. Lozeau, Fla. App. 5 Dist., 539 So. 2d 1147 (1989). Words and Phrases.*

THENCE. As used in a description of land, means "from that place." *Tracy v. Harmon,* 43 P. 500, 17 Mont. 465. (1896)

TO RANGE. Line "to range" with another held to follow its path extended, and to be a continuation of it. *Lilly v. Marcum,* 283 S.W. 1059, 214 Ky. 514 (1926).

TRACT. As ordinarily understood means contiguous bodies of land embraced in one deed. *Saulsberry v. Maddix,* 125 F.2d 430 (1942), certiorari denied 63 S.Ct. 36, two cases, 317 U.S. 643, 87 L.Ed. 518 (1942).

A lot, piece or parcel of land, of greater or less size, the term not importing, in itself, any precise dimension. See *Edwards v. Derrickson,* 28 N.J. Law 45 (1859); *Schofield v. Harrison Land & Mining Co.,* 187 S.W. 61 (Mo. Sup., 1916); *Smith v. Heyward,* 105 S.E. 275, 115 S.C. 145 (1920).

TRACT OR LOT. "The terms 'tract or lot' and 'pieces or parcel of real property,' or 'piece or parcel of land,' mean any contiguous quantity of land in the possession of, owned by, or recorded as the property of the same claimant, person or company." In this connection the word "contiguous" means land which touches on the sides. Hence two quarters of the same section which only touch at the corner do not constitute, for the purpose of taxation, one tract or parcel of land. *Griffin v. Denison Land Co.,* 119 N.W. 1041, 18 N.D. 246 (1909).

TRUE LINE. Used in a surveyor's field notes in describing the line between two sections, means a straight line. *Lillis v. Urrutia,* 99 P. 992, 9 Ca. App. 558 (1908).

VARA. The true Mexican "vara" is slightly less than 33 American inches; but by use in California it is estimated at 33 inches, and in Texas as $33\frac{1}{3}$ inches. *U.S. v. Perot,* 98 U.S. 428, 25 L. Ed. 251 (1878).

A vara, in Texas, has always been regarded as equivalent to $33\frac{1}{3}$ inches. A standard vara is somewhat less than $33\frac{1}{3}$ inches. Humboldt, in 1803, found a Mexican vara to be 839.16 millimeters, or a slight fraction over 33 inches. But it seems that a vara measure of somewhat larger dimensions obtained in Texas from an early period. The standard Mexican vara is so near to 33 inches that a standard vara measure laid on an American yard would so nearly correspond with

---

[256]In the case of *Cragin v. Powell,* 128 U.S. 691(1888), the Supreme Court of the United States cited with favor the following quotation from a letter of the Commissioner of the General Land Office to the surveyor general of Louisiana: "The making of resurveys or corrective surveys of townships once proclaimed for sale is always at the hazard of interfering with private rights, and thereby introducing new complications. A resurvey, properly considered, is but a retracing, with a view to determine and establish lines and boundaries of an original survey, but the principle of retracing has been frequently departed from, where a resurvey (so called) has been made and new lines and boundaries have often been introduced, mischievously conflicting with the old, and thereby affecting the areas of tracts which the United States had previously sold and otherwise disposed of."

33 inches that a difference could not be perceived by the naked eye. *U.S. v. Perot*, 98 U.S. 428, 25 L.Ed. 251 (1878).

WEST. Terms "east" and "west," used in description of boundary courses without modification or variation, mean due east and due west. *E.E. McCalla Co. v. Sleeper*, 288 P. 146, 105 C.A. 562 (1930).

WESTERLY. As used in an order of the county court incorporating a village, which describes the commons as "on the west side of said limits one quarter of a mile in a westerly direction," should be construed to mean due west, rendering the description definite and certain. *State ex rel. Chandler v. Huff*, 79 S.W. 1010, 105 Mo. App. 354 (1904).

WESTERLY ONE-HALF OF. The description, in a complaint in partition of the land, involved as the "westerly one-half of" a specified lot and block according to a certain recorded plat, etc., was sufficient. *Home Security Bldg. & Loan Ass'n of Alameda County v. Western Land & Title Co.*, 78 P. 626, 145 C. 217 (1904).

WEST HALF. Where there is nothing to suggest the contrary, the word "half," in connection with the conveyance of a part of a tract of land, is interpreted as meaning half in quantity. The words "east half" and "west half" in a deed, while naturally importing an equal division, may lose that effect when it appears that at the time some fixed line or known boundary or monument divides the premises somewhere near the center, so that the expression more properly refers to one of such parts than to a mathematical division which never has been made. *Gunn v. Brower*, 105 P. 702, 81 Kan. 242 (1909).

In government surveys of public lands, terms "east half" and "west half" are used, not with reference to quantity, but to a line equidistant from the boundary lines of the parcel subdivided, and those terms have the same signification in patents issued by the government. A deed of the east half of a parcel of land according to the United States survey is definite and excludes the idea of two equal quantities, and fixes the dividing line equidistant from the boundary lines of the parcel thus subdivided. *Hoyne v. Schneider*, 27 P.2d 558, 138 Kan. 545 (1933).

WESTWARDLY. Courses in a grant indicated by the term "westwardly" run due west. *Seaman v. Hogeboom*, N.Y., 21 Barb. 398 (1855).

BASIC EQUIPMENT FOR SCENE PHOTOGRAPHY

Camera(s)

Normal lens (50 mm is normal for 35 mm camera)

Wide angle lens (28 mm or similar)

Close-up lens or accessories (macro, adapter, extension tubes, etc.)

Filters (red, orange, yellow, blue, and green)

Electronic flash(es)

Remote cord for electronic flash

Extra camera and flash batteries

Locking cable release

Tripod

Film

    Color versus black and white

    Print film versus slides

Owner's manual(s)

Notebook and pen

Ruler or scale

Gray card

Index cards and felt pen

Flashlight

Other items to be considered:

    Telephoto lens (135 mm, telephoto)

    Small tools for repairs

    Clothespins and other devices for positioning evidence for close-up photography

    White handkerchief for diffusion

    Tape measure

    Color chart

BASIC EQUIPMENT FOR RETRACEMENT WORK

Aerial photos and topo maps of the area to be investigated

Any terrestrial photos or photos of evidence to be sought

Any postal card or other historical photos showing how area used to appear

Metal detectors of various types

Hand tools: shovel, trowel, screwdriver, wire brush

Rake

Ax or hatchet

Flagging

Paint

Checklist of Some Investigative Tools:

Compass

10X hand lens

Measuring tape

Mirror

Calculator

Diameter tape

Increment borer

Flashlight

GPS receiver

Cell phone

NOTEBOOK

APPENDIX IV

COURT DECISIONS DEALING WITH MAGNETIC BEARINGS, BY STATE

GA   *Riley v. Griffin,* 16 Ga. 141 (1854)

KY   *Beckley v. Bryan & Ransdale,* 2 Ky. (Ky. Dec.) 91 (1801); *Bryan v. Beckley,* 16 Ky. (Litt. Sel. Cas) 91 (1809); *Vance v. Marshall,* 6 Ky. (3 Bibb) 148 (1813)

ME   *Milliken v. Buswell,* 313 A.2d 111 (Me. 1973)

NC   *Greer v. Hayes,* 216 N.C. 396, 5 S.E.2d 169 (1939); *Cherry v. Slade,* 3 Murph 82 (1819); *Goodwin v. Greene,* 237 N.C. 244, 74 S.E. 2d 630 (1953)

NH   *Wells v. Jackson Iron Mfg. Co.* 47 N.H. 235, 90 Am. Dec. 575

OH   *McKinney v. McKinney,* 8 Ohio St. 423 (1858)

VA   *Scott v. Jessee,* 143 Va. 150, 129 S.E. 333

VT   *Brooks v. Tyler,* 2 Vt. 348 (1829)

WV   *State v. West Virginia Pulp & Paper Co.,* 108 W. Va. 553, 152 S.E. 197 (1930)

REFERENCE

70 ALR3d 1220

Chermside, Jr., Herbert B. *Boundaries: Description In Deed as Relating to Magnetic or True Meridian,* 70 ALR 3d 1220.

COURT DECISIONS DEALING WITH WOOD EVIDENCE, BY STATE

CO   *Pollard v. Shively*, et al., 5 Colo. 309 (1880)

FL   *Bridges v. Thomas*, 118 So. 2d 549 (Fla., 1960)

GA   *Riley, adminstratrix v. Griffin*, 16 Ga. 141 (1854)

ID   *Brinton v. Steele*, 25 Idaho 783 (1914)

KY   *Beckley v. Bryan & Ransdale*, 1 Ky (Ky. Dec.) 91 (1801); *Bryan v. Beckley*, 16 Ky. (Litt. Sel. Case) 91 (1809); *Vance v. Marshall*, 6 Ky (3 Bibb) 148 (1813); *Bulor's Heirs v. James M'Cawley*, 10 Ky (3 A.K. Marsh. 573 (1821); *Logan v. Evans*, 29 S.W. 636 (Ky. 1895); *May v. Wolf Valley Coal Co.*, 167 Ky 525, 180 S.W. 781 (1915); *Fletcher v. Hart et al.*, 202 Ky 485, 260 S.W. 18 (1924); Oliver v. Muncy, 262 Ky. 164, 89 S.W.2d 617 (1935); *Lewallen v. Mays*, 265 Ky 1, 95 S.W. 2d 1125 (1936)

LA   *Zeringue v. Harang*, 17 La. 349 (1841); *Boudreaux v. Shadyside Company*, 111 So. 2d 891 (1959)

MD   *Delphey v. Savage*, 177 A.2d 249, 227 Md. 37 (1961); *U.S. v. Gallas*, 269 F.Supp. 141 (D.C. Md. 1967)

ME   *Coombs v. West*, 115 Me. 489, 99 A. 445 (1916)

NC   *Rutledge v. Buchanan*, Fed. Cas. No. 12,177 (U.S. 1813); *Cherry v. Slade's Administrator*, 3 Murph (N.C.) 82 (1819); *Bowen v. John L. Roper Lumber Co.*, 153 N.C. 366, 69 S.E. 258 (1910)

NY   *Stewart v. Patrick*, 68 N.Y. 450 (1877)

OH   *Lessee of Alshire v. J.R. Hulse*, 5 Ohio (5 Ham) 534 (1832); *Hare v. Lessee of Origin Harris*, 14 Ohio 529 (1846); *Sellman v. Schaaf*, 269 N.E. 2d 60, 26 Ohio App. 2d 35 (1971)

SC   *Wash v. Holmes*, 1 Hill, Law, 12 (S.C. 1833)

TN   *James v. Brooks*; *James v. Tate*, 53 Tenn. (6 Heisk.) 150 (1871); *Morrison v. Jones*, 430 S.W.2d 668, 58 Tenn. App. 333, appeal after remand, 458 S.W. 2d 434 (1968)

TX   *Mitchell v. Burdett*, et al., 22 Tex. 633 (1858); *Stafford v. King*, 30 Tex. 257 (1867); *Sweats v. Southern Pine Lumber Company*, 361 S.W. 2d 214 (Tex., 1962); *U.S. v. Champion Papers, Inc.* 361 F.Supp. 141 (D.C. Tex. 1973)

VT   *Amey v. Hall*, 181 A.2d 69, 123 Vt. 62 (1962)

WI   *Fehrman v. Bissell Lumber Company*, 188 Wis. 82 (1925)

WV   *Bowers v. Dickinson*, 30 W. Va. 709 (1888); *Curtis v. Meadows*, 84 W.Va. 94, 99 S.E. 286 (1919)

REFERENCE

2 ALR 1428. Conveyance with reference to tree, or similar monument, as giving title to center thereof.

SOME COURT DECISIONS DEALING WITH WOOD EVIDENCE, BY TOPIC

ANNUAL RINGS

Bulor's Heirs v. James M'Cawley, 10 Ky. (3 A.K. Marsh. 573 (1821)
Bowen v. John L. Roper Lumber Company, 153 N.C. 366, 69 S.E. 258 (1910)
Curtis v. Meadows, 84 W.Va. 94, 99 S.E. 286 (1919)

BEARING TREE

Mitchell v. Burdett et al., 22 Tex. 633 (1858)

MARKED LINES

Amey v. Hall, 181 A.2d 69, 123 Vt. 62 (1962)
Bowen v. John L. Roper Lumber Company, 153 N.C. 366, 69 S.E. 258 (1910)
Curtis v. Meadows, 84 W.Va. 94, 99 S.E. 286 (1919)
James v. Brooks, 53 Tenn. (6 Heisk.) 150 (1871)
Rutledge v. Buchanan, Fed. Cas. No. 12,177 (U.S. 1813)
Sweats v. Southern Pine Lumber Company, 361 S.W.2d 214 (Tex. 1962)
Vance v. Marshall, 6 Ky (3 Bibb) 148 (1813)

POST

Fehrman v. Bissell Lumber Company, 188 Wis. 82 (1925)
Lessee of Alshire v. J.R. Hulse, 5 Ohio (5 Ham) 534 (1832)
Oliver v. Muncy, 262 Ky 164, 89 S.W. 2d 617 (1935)
Pollard v. Shively, 5 Colo. 309 (1880)
U.S. v. Gallas, 269 F.Supp. 141 (D.C. Md. 1967)
Zeringue v. Harang, 17 La. 349 (La. 1841)

STAKE

Boudreaux v. Shadyside Company, 111 So. 2d 891 (La. 1959)
Delphy v. Savage, Fed. Cas. No. 12,177 (U.S. 1813)
Sellman v. Schaaf, 269 N.E.2d 60, 26 Ohio App. 2d 35 (1971)
Stafford v. King, 30 Tex. 257 (1867)

STUMP

Pollard v. Shively, 5 Colo. 309 (1880)

TREES

Beckley v. Bryan & Ransdale, 1 Ky. (Ky. Dec.) 91 (1801)
Bowers v. Dickinson, 30 W.Va. 709 (1888)
Bridge v. Thomas, 118 So. 2d 549 (Fla., 1960)
Bryan v. Beckley, 16 Ky (Litt. Sel. Case) 91 (1809)
Cherry v. Slade, 3 Murph. (N.C.) 82 (1819)
Coombs v. West, 115 Me. 489, 99 A. 445 (1916)
Delphey v. Savage, Fed. Cas. No. 12,177 (U.S. 1813)
Fletcher v. Hunt, 202 Ky 485, 260 S.W. 18 (1924)
Jacob Hare v. The Lessee of Origin Harris, 14 Ohio 529 (1846)
James v. Brooks, 53 Tenn. (6 Heisk.) 150 (1871)
Lewallen v. Mays, 265 Ky. 1, 95 S.W.2d 1125 (1936)
Logan v. Evans, 29 S.W. 636 (Ky. 1895)
May v. Wolf Valley Coal Co., 167 Ky. 525, 180 S.W. 781 (1915)
Morrison v. Jones, 430 S.W. 2d 668, 58 Tenn. App. 333; appeal after remand, 458
S.W.2 d 434 (1968)
Riley v. Griffin, 16 Ga. 141 (1854)
Rutledge v. Buchanan, Fed. Cas. No. 12,177 (U.S. 1813)
Stafford v. King, 68 N.Y. 450 (1877)
Stewart v. Patrick, 68 N.Y. 450 (1877)
Wash v. Holmes, 1 Hill, Law, 12 (S.C. 1833)

TREE LINE

Brinton v. Steele, 25 Idaho 783 (1914)

WITNESS TREES

U.S. v. Champion Papers, 361 F.Supp. 141 (D.C. Tex. 1973)

SOME COURT DECISIONS INVOLVING FENCE VIEWERS, BY STATE

AL   *Johnson v. Frederick*, 163 Ala. 455, 50 So. 910 (1909)

CT   *Talcott v. Stillman*, 28 Conn. 193 (1859); *Edgerton v. Moore*, 28 Conn. 600 (1859); *Fox v. Beede*, 24 Conn. 271 (1855)

IA   *Scott v. Nesper*, 194 Iowa 538, 188 N.W. 889 (1922); *McKeever v. Jenks*, 59 Iowa 300, 13 N.W. 295 (1882)

IL   *Hill v. Tohill*, 225 Ill. 384, 80 N.E. 253 (1907)

IN   *Tomlinson v. Bainaka*, 163 Ind. 112, 70 N.E. 155 (1904); *Bruner v. Palmer,* 108 Ind. 397, 9 N.E. 354 (1886)

KS   *Robertson v. Bell*, 36 Kan. 748, 14 P. 160 (1887)

MA   *Day v. Dolan*, 174 Mass. 524; 55 N.E. 384 (1899); *Butman v. Fence Viewers of Chelsea,* 327 Mass. 386, 99 N.E..2d 44 (1951)

ME   *Megquier v. Bachelder*, 112 Me. 340, 92 A. 187 (1914); *Lamb v. Hicks*, 11 Met. (Me.) 496 (1846); *Scott v. Dickinson*, 14 Pick. (Me.) 276 (1833); *Harris v. Sturdivant*, 29 Me. 366 (1849); *Emery v. Maguire*, 87 Me. 116, 32 A. 781 (1895)

MI   *Vincent v. Ackerman*, 155 Mich. 614, 119 N.W. 1085 (1909); *Gilson v. Munson*, 114 Mich. 671, 72 N.W. 994 (1897)

MN   *McClay v. Clark*, 42 Minn. 363, 44 N.W. 255 (1890)

MO   *McNaughton v. Schaffer* (Mo. App.), 314 S.W..2d 245 (1958)

NE   *Meyer v. Perkins*, 89 Neb. 59, 130 N.W. 986 (1911); *Schnakenbert v. Schroeder*, 219 Neb. 813, 367 N.W..2d 692 (1985)

NH   *Gallup v. Mulvah*, 24 N.H. 204 (1852); *Hartshorn v. Schoff,* 51 N.H. 316 (1871)

NJ   *State, Titman, Prosecutrix v. Smith*, 61 NJL 191, 38 A. 810 (1897)

NY   *Bromley v. Mollnar*, 179 Misc. 713, 39 N.Y.S.2d 424 (1942)

PA   *Shriver v. Stephens*, 20 Pa. 138 (1852)

OH   *Clark v. Chambers*, 81 Ohio L. Abs. 57; 160 N.E. 2d 870 (1957); *Robb v. Brachmann*, 24 Ohio St. 3 (1873)

VT   *Barber v. Vinton*, 82 Vt. 327, 73 A. 881 (1909).

COLLOQUIAL TREE NAMES

EASTERN UNITED STATES

Proper Name	Colloquial Names
Baldcypress, *Taxodium distichum*	baldcypress, black cypress, buck cypress, cow cypress, gulf cypress, knee cypress, Louisiana black cypress, Montezuma baldcypress, pecky cypress, pond baldcypress red cypress, river cypress, southern cypress, swamp cypress, tidewater red cypress, white cypress, yellow cypress
Eastern White Pine, *Pinus strobus*	apple pine, balsam pine, black pine, Canadian white pine, conk pine, cork pine, New England pine, northern pine, pumpkin (punkin) pine; sapling pine, silver pine, soft pine, Weymouth pine, white pine
Jack Pine, *Pinus banksiana*	black pine, blackjack pine, bull pine, check pine, grey pine, juniper, princess pine, scrub pine, spruce pine
Loblolly Pine, *Pinus taeda*	Arkansas pine, bastard pine, black pine, bog pine, buckskin pine, bull pine, cornstalk pine, foxtail pine, heart pine, Indian pine, longshucks pine, longstalk pine, maiden pine, meadow pine, North Carolina pine, old pine, oldfield pine, prop pine, sap pine, shortleaf pine, southern pine, torch pine
Longleaf Pine, *Pinus palustris*	broom pine, brown pine, fat pine, Florida pine, Georgia pine, Gulf coast pitch pine, hard pine, heart pine, hill pine, longstraw pine, pitchpin, Rosemary pine, sump-tall, tea pine, Texas yellow pine, turpentine pine, yellow pine
Pitch Pine, *Pinus rigida*	black pine, hard pine, longshot pine, mountain pine, ridge pine, sap pine, southern pine, southern yellow pine, torch pine
Red Pine, *Pinus resionsa*	hard pine, northern pine, Norway pine, pig iron pine, red deal, shellbark Norway pine, Quebec pine, yellow deal
Sand Pine, *Pinus clausa*	Alabama pine, Florida spruce pine, northern sand pine, oldfield pine, scrub pine, spruce pine.

Proper Name	Colloquial Names
Shortleaf Pine, *Pinus echinata*	Arkansas pine, bull pine, Carolina pine, forest pine, North Carolina pine, poor pine, Rosemary pine, southern pine, yellow pine
Slash Pine, *Pinus elliottii*	bastard pine, Cuba pine, Gulf Coast pitch pine, meadow pine, saltwater pine, she pine, spruce pine, swamp pine
Spruce Pine, *Pinus glabra*	black pine, bottom white pine, cedar pine, kings-tree, lowland spruce pine, poor pine, southern white pine, spruce lowland pine, spruce pine, Walter pine, white pine.
Table-Mountain Pine, *Pinus pungens*	black pine, bur pine, hickory pine, mountain pine, poverty pine, prickly pine, ridge pine, yellow pine
Virginia Pine, *Pinus virginiana*	alligator pine, bastard pine, black pine, cedar pine, hickory pine, nigger pine, Jersey pine, North Carolina pine, poor pine, poverty pine, river pine, scrub pine, spruce pine, river pine
EasternHemlock, *Tsuga canadensis*	black hemlock, Canada hemlock, hemlock pine, hemlock spruce, Huron pine, red hemlock, spruce hemlock, spruce pine, suga, water spruce, white hemlock
Eastern Larch, *Larix laricina*	black larch, hack, hackamatack, juniper, red larch, tamarack
Balsam Fir, *Abies balsamea*	balm-of-gilead fir, blister fir, blister pine, Eastern fir, fir pine, Gilead fir, sapin, silver fir, silver pine, single spruce
Fraser Fir, *Abies fraseri*	balsam fir, double spruce, eastern fir, Fraser balsam fir, healing balsam, mountain balsam, she-balsam, southern balsam fir, southern fir
Black Spruce, *Picea mariana*	bog spruce, Canadian spruce, double spruce, he-balsam, spruce pine, water spruce, yew pine
Red Spruce, *Picea rubens*	Adirondack spruce, Canadian spruce, double spruce, he balsam, spruce pine. West Virginia spruce, yellow spruce
White Spruce, *Picea glauca*	Adirondack spruce, bog spruce, Canadian spruce, cat spruce, double spruce, eastern spruce, he-balsam, juniper, Maritime spruce, skunk spruce, spruce pine, water spruce, wit-spar
Eastern redcedar, *Juniperus virginiana*	coast juniper, coast red cedar, eastern red juniper, juniper, pencil cedar, pencil juniper, post cedar, red juniper, redcedar, red juniper, sand cedar, savin, southern juniper, southern red cedar, Tennessee red cedar, Virginia juniper
Atlantic White Cedar, *Chamaecyparis thyoides*	Coast white cedar, post cedar, southern white cedar, swamp cedar, white cedar, white cypress
Northern White Cedar, *Thuja occidentalis*	arborvitae, Atlantic red cedar, cedar, Eastern cedar, Michigan white cedar, New Brunswick cedar, swamp cedar, white cedar
Flowering Dogwood, *Cornus florida*	arrow-wood, boxwood, dogwood, Virginia dogwood
Black Maple, *Acer saccharum v. nigrum*	black sugar maple, hard maple, rock maple, sugar maple, white maple

Proper Name	Colloquial Names
Boxelder, *Acer platanoides*	ash maple, California boxelder, ashleaf maple, Manitoba maple, Red River maple, stinking ash, sugar ash, three-leaf maple
Red Maple, *Acer rubrum*	branch maple, Carolina red maple, Drummond maple, scarlet maple, soft maple, swamp maple, water maple, white maple,
Silver Maple, *Acer saccharinum*	creek maple, river maple, silver-lonn, silverleaf maple, soft maple, swamp maple, water maple, white maple, creek maple
Sugar Maple, *Acer saccharum*	hard maple, river maple, rock maple, rough maple, sapwood, socker-lonn, sugar tree, sweet maple, white maple
Black Ash, *Fraxinus nigra*	basket ash, brown ash, hoop ash, splinter ash, swamp ash, water ash
Blue Ash, Fraxinus quadrangulata	Virginia ash
Carolina Ash, Fraxinus caroliniana	Florida Ash, Pop Ash, Swamp Ash, Water Ash
Green Ash, *Fraxinus pennsylvanicum*	bastard ash, black ash, blue ash, brown ash, Darlington ash, gray ash, piss ash, pumpkin ash, red ash, river ash, swamp ash, water ash, white ash
White Ash, *Fraxinus americana*	Biltmore ash, Biltmore white ash, Canadian ash, cane ash, green ash, smallseed white ash, southern white ash
Ohio Buckeye, *Aesculus glabra*	fetid buckeye, smooth buckeye, stinking buckeye, Texas buckeye, white buckeye; American horsechestnut
Red Buckeye, Aesculus pavia	firecracker-plant, scarlet buckeye, woolly buckeye
Sweet Buckeye, *Aesculus octandra*	big buckeye, large buckeye, yellow buckeye
Black Walnut, *Juglans nigra*	American walnut, Burbank walnut, gunwood, round-nut, Virginia walnut
Butternut, *Juglans cinerea*	lemon walnut, oilnut, white walnut
Black Hickory, *Carya texana*	buckley hickory, pignut hickory, Texas hickory
Bitternut Hickory, *Carya cordiformis*	bitter hickory, bitter pecan, bitternut, butternut hickory, pecan, pignut, pig hickory, redheart hickory, swamp hickory, white hickory, yellow-bud hickory
Mockernut Hickory, *Carya tomentosa*	big-bud, big hickory, black hickory, bullnut, common hickory, hardbark hickory, hickory-nut, hognut, mockernut, red hickory, true hickory, white hickory, whiteheart hickory

Proper Name	Colloquial Name
Nutmeg Hickory, *Carya myristicaeformis*	bitter water hickory, bitter waternut, blasted pecan, scalybark hickory, shagbark, shagbark hickory, shellbark hickory, swamp hickory, upland hickory.
Pignut Hickory, *Carya glabra*	bitternut, black hickory, broom hickory, brown hickory, coast pignut hickory, false shagbark, hard shell, little pignut, little shagbark, nutmeg hickory, oval pignut hickory, pignut, red hickory, redheart hickory, small fruited hickory, small pignut, smoothbark hickory, swamp hickory, sweet hickory, sweet pignut, sweet pignut hickory, switch-bud hickory, true hickory, white hickory.
Sand Hickory, *Carya pallida*	pale hickory, paleleaf hickory, pallid hickory, pignut hickory
Shagbark Hickory, *Carya ovata*	bird's eye hickory, Carolina hickory, curly hickory, littlenut shagbark hickory, little pignut, little shagbark, mockernut hickory, red hickory, redheart hickory, scalybark hickory, shagbark walnut, shellbark, shellbark hickory, shellbark tree, skid hickory, small pignut, small pignut hickory, southern hickory, southern shagbark hickory, southern shellbark, sweet walnut, upland hickory, white hickory, whiteheart hickory, white walnut.
Shellbark Hickory, *Carya lacinosa*	big shagbark, big shellbark, bigleaf shagbark hickory, bottom shellbark, king nut, ridge hickory, thickbark hickory, thick shellbark hickory, true hickory, western shellbark
Water Hickory, *Carya aquatica*	bitter pecan, bitter water hickory, faux hickory, lowground hickory, lowland hickory, noot hickory, not hickory, pecan, pecan hickory, pignut hickory, swamp hickory, water bitternut, wild pecan.
Honeylocust, *Gleditsia triacanthos*	common honeylocust, Confederate Pintree, honey, honeyshucks, shucks honeylocust, squeak-bean, sweet-bean, sweetlocust, thornlocust, thorn-tree, thorny Acacia, thornylocust, three-thorned Locust
Waterlocust, *Gleditsia aquatica*	black locust, swamp honey locust, waterlocust
Black Locust, *Robinia pseudoacacia*	Acacia, bastard locust, black laurel, locust, false acacia, green, red, white or post locust, peaflower locust, shipmast locust, white locust, white honey-flower, yellow locust
Hawthorn, *Crataegus spp.*	Arabic denim, blackthorn, blue haw, cockspur, haw tree, hog apple, May haw, may tree, Newcastle thorn, parsley haw, red thorn, river haw, scarlet haw, thorn apple, thorn pear, whitethorn
Osage Orange, *Maclura pomifera*	bodare, bodark, bodeck, bodock, bois d'arc, bowwood, hedge, hedge apple, horse apple, mock orange, osage, rootwood, wildorange, yellow-wood

Proper Name	Colloquial Name
Sassafras, *Sassafras albidum*	aguetree, cinnamonwood, common sassafras, red sassafras, saxifrax, smelling-stick, white sassafras
Paper-Mulberry, *Broussonetia papyrifera*	mulberry, paper mulberry
Red Mulberry, *Morus rubra*	black mulberry, moral, mulberry, Virginia mulberry
White Mulberry, *Morus alba*	mulberry
Sweetgum, *Liquidamber styraciflua*	alligator-tree, alligatorwood, ambarwood, American mahogany, blisted, delta redgum, figured gum, gum, gumtree, gumwood, hazel, hazel pine, hazelwood, incense-tree mulberry, opossum-tree, plain redgum, quartered redgum, redgum, sapgum, sapwood hazel pine, satin walnut, satinwood, splint sapgum, splinted sapgum, starleaf gum, sycamore gum, whitegum.
American Sycamore, *Platanus occidentalis*	button-ball tree, buttonwood, cotonier, lacewood, planetree, sycamore, Virginia maple, water beech, whitewood
Serviceberry, *Amelanchier spp.*	apple shadbush, Indian cherry, Indian pear, juiceplum, juneberry, lancewood, pigeonberry, service, shadbush, sugar pear, sugarplum, wild pear
Sourwood, *Oxydendrum arboretum*	arrow-wood, elk-tree, lily-of-the-valley tree, sorrel-tree, sour gum
American mountain-ash, *Sorbus americana*	dogberry, elder-leaved sumach, life-of-man, missey-moosey, mountain sumac, roundwood, rowan tree, sarvice-berry, winetree
Yellow-Poplar, *Liriodendron tulipifera*	basswood, blue poplar, canarywood, canoewood, cucumbertree, hickory-poplar, old wives shirt, poplar, saddle-tree, sap poplar, secoya, tulip-poplar, tuliptree, tulipwood, white-poplar, whitewood
Cucumbertree, *Magnolia acuminata*	black linn, blue magnolia, cowcumber, cucumber, cucumber magnolia, cucumberwood, elkwood, Indian-bitter, magnolia, mountain magnolia, pointed-leaved magnolia, yellow cucumbertree, yellow-flower magnolia, yellow linn, yellow poplar, wahoo
Common Persimmon, *Diospyrus virginiana*	boawood, butterwood, date plum, persimmon, possum wood, seeded plum, simmon, Virginia date palm, date plum
Black Tupelo, *Nyssa sylvatica*	bay poplar, blackgum, bowl gum, cotton gum, gum, pepperidge, sour-gum, stinkwood, swamp tupelo, tupelo gum, yellow gum, yellow gumtree, wild pear-tree

Proper Name	Colloquial Name
Rusty Blackhaw, *Viburnum rufidulum*	blackhaw, bluehaw, nannyberry
Sourwood, *Oxydendrum aboreum*	arrowwood, elk-tree, lily-of-the-valley tree, sorreltree, sour gum
Pawpaw, *Asimina triloba*	banana, black mangrove, blackwood, custard apple, false banana, fetid-shrub, jasmine, wild banana
Redbay, *Persea borbonia*	false mahogany, galls bay, laurel tree, magnolia wild banana, shore bay, swamp bay, sweet bay, tisswood, white bay
American Holly, *Ilex opaca*	boxwood, evergreen holly, holly, prickly holly, white holly
American Basswood, *Tilia Americana*	basswood, beetree, limetree, linden, linn, spoonwood, whitewood, yellow basswood.
White Basswood, *Tilia heterophylla*	basswood, beetree linden
American Elm, *Ulmus americana*	ellum, Florida elm, gray elm, river elm, rock elm, soft elm, springwood, swamp elm, water elm, weeping elm, white elm
Rock Elm, *Ulmus thomasii*	cork elm, cliff elm, corkbark elm, hickory elm, rock elm, swamp elm, Thomas elm, wahoo, white elm
Slippery Elm, *Ulmus rubra*	gray elm, Indian elm, moose elm, red elm, soft elm
Winged Elm, *Ulmus alata*	cork elm, wahoo, mountain elm, red elm, southern elm, wahoo, water elm, witch elm
Hackberry, *Celtis occidentalis*	bastard elm, beaverwood, false elm, hacktree, hardhack, hoop ash, huck, nettletree, oneberry, sugarberry
Eastern Cottonwood, *Populus deltoides*	aspen cottonwood, big cottonwood, Carolina poplar, cotton tree, eastern poplar, great plains cottonwood, Missourian poplar, necklace poplar, palmer cottonwood, plains cottonwood, Rio Grande cottonwood, river cottonwood, river poplar, southern cottonwood, Tennessee poplar, Texas cottonwood, valley cottonwood, Vermont poplar, Virginia poplar, water poplar, western cottonwood, whitewood, yellow cottonwood
Swamp Cottonwood, *Populus heterophylla*	bigleaf cottonwood, black cottonwood, cotton gum, cotton tree, cottonwood, downy cottonwood, downy poplar, river cottonwood, swamp poplar
Bigtooth Aspen, *Populus grandidentata*	Canadian poplar, poplar, popple, whitewood
Quaking Aspen, *Populus tremuloides*	aspen, golden aspen, leaf aspen, mountain aspen, popple, quiver-leaf, white poplar
Balsam Poplar, *Populus balsamifera*	balm, balm of Gilead, balm cottonwood, balsam, balsam cottonwood, bam, black balsam poplar, black cottonwood, black poplar, California poplar, Canadian balsam poplar, Canadian poplar, cottonwax, hackmatack, hairy balm of Gilead, heartleaf balsam poplar, northern black cottonwood, Ontario poplar, tacamahac, tacamahac poplar, toughbark poplar, western balsam poplar

Proper Name	Colloquial Name
Black Birch, *Betula lenta*	black birch, black cherry birch, cherry birch, mahogany, mahogany birch, mountain birch, mountain mahogany, red birch, river birch, spice birch, yellow birch, sweet birch
Gray Birch, *Betula populifolia*	blue birch, blueleaf birch, broom birch, fire birch, gray birch, oldfield birch, pin birch, poplar-leaved birch, poverty birch, small white birch, white birch, wire birch
Paper Birch, *Betula papyrifera*	canoe birch, gray birch, large white birch, northwestern paper birch, paper birch, red birch, silver birch, white birch
River Birch, *Betula nigra*	black birch, red birch, water birch
Yellow Birch, *Betula alleghaniensis*	bitter birch, Canadian silky wood, gray birch, hard birch, Quebec birch, silver birch, swamp birch white birch, witch hazel
Hophornbeam, *Ostrya virginiana*	deerwood, hardhack, ironwood, leverwood, roughbark ironwood
Ironwood, *Carpinus caroliniana*	blue beech, broomwood, musclewood, water-beech
Black Willow, *Salix nigra*	Dudley willow, Goodding willow, Southeastern black willow, swamp walnut, swamp willow, tall black willow, willow
Black Cherry, *Prunus serotina*	black wild cherry, cabinet cherry, cherry, mountain black cherry, rum cherry, wild cherry, whiskey cherry
American Beech, *Fagus grandifolia*	beech, beechnut, Carolina beech, gray beech, red beech, ridge beech, stone beech, white beech, winter beech
American Chestnut, **Castanea dentata**	chestnut, prickly o-heh-yah-bur, sweet chestnut, white chestnut, wormy chestnut
Northern Red Oak, *Quercus rubra*	black oak, buck oak, Canadian red oak, gray oak, leopard oak, Maine red oak, mountain red oak, red oak mountain red oak, Spanish oak, spotted oak, southern red oak, swamp red oak, water oak, West Virginia soft red oak
Black Oak, *Quercus velutina*	blackjack, Dyer's oak, jack oak, quercitron oak, redbush, red oak, smooth-bark oak, spotted oak, tanbark oak, yellow oak, yellowbark oak
Blackjack Oak, *Quercus marilandica*	blackjack, barren oak, black oak, jack oak, scrub oak
Pin Oak, *Quercus palustris*	red oak, Spanish oak, swamp oak, swamp Spanish oak, water oak
Scarlet Oak, *Quercus coccinea*	bastard oak, black oak, buck oak, red oak, Spanish oak, spotted oak
Scrub Oak, *Quercus ilicifolia*	barren oak, bear oak, black dwarf oak, black scrub oak
Shumard Oak, *Quercus shumardii*	red oak, Schneck oak, southern red oak, spotted oak, swamp red oak, Texas oak
Southern Red Oak, *Quercus falcata*	bottomland red oak, cherrybark oak, Elliott oak, red oak, Spanish oak, swamp red oak, swamp Spanish oak, turkeyfoot oak, water oak

Proper Name	Colloquial Name
Turkey Oak, *Quercus laevis*	barren scrub oak, blackjack, Carolina red oak, forked-leaf, sand jack
Willow Oak, *Quercus phellos*	black oak, laurel oak, peach oak, pin oak, red oak, swamp willow oak water oak, willow swamp oak
White Oak, *Quercus alba*	blue oak, fork-leaf white oak, Louisiana white oak, mantua oak, ridge white oak, stave oak, West Virginia soft white oak
Basket Oak, Quercus michauxii	cow oak, swamp oak
Bluejack Oak, Quercus incana	bluejack, cinnamon oak, high-ground willow oak, sandjack, shin oak, turkey oak
Bur Oak, *Quercus macrocarpa*	blue oak, burr oak, mossycup oak, scrub oak, white oak
Chestnut Oak, *Quercus prinus*	cow oak, mountain oak, rock chestnut oak, rock oak, swamp oak, tanbark oak white oak, white chestnut oak
Chinquapin Oak, *Quercus muehlenbergii*	chestnut oak, chinkapin oak, dwarf chestnut oak, dwarf chinkapin, pin oak, rock oak, rock chestnut oak, running white oak, scrub oak, shrub oak, white oak yellow oak, yellow chestnut oak
Laurel Oak, *Quercus laurifolia*	Darlington oak, diamond-leaf oak, roble laurel, swamp laurel, water oak, willow oak
Overcup Oak, *Quercus lyrata*	swamp post oak, swamp white oak, water white oak
Post Oak, *Quercus stellata*	barren white oak, bastard oak, bastard white oak, box oak, box white oak, brash oak, Delta post oak, Durand oak, iron oak, pin oak, ridge oak, rough oak, rough white oak, southern oak, turkey oak, white box oak, white oak
Shingle Oak, *Quercus imbricaria*	glossy oak, jack oak, laurel oak, pin oak, swamp oak, turkey oak, water oak, white oak
Swamp White Oak, *Quercus bicolor*	blue oak, cherry oak, curly swamp oak, swamp oak, white oak
Swamp Chestnut Oak, *Quercus michauxii*	basket oak, cow oak, mountain oak, swamp oak, white oak
Water Oak, *Quercus nigra*	barren oak, blackjack, possum oak, punk oak, spotted oak
Willow Oak, *Quercus phellos*	laurel oak, peach oak, pin oak, red oak, water oak

WESTERN UNITED STATES

Proper Name	Colloquial Name
Bristlecone Pine, *Pinus aristata*	Balfour pine, foxtail pine, hickory pine, jack pine, wind pine
Coulter Pine, *Pinus coulteri*	bigcone pine, bull pine, large-cone pine, nigger pine (see Digger Pine), nut pine, pitch pine
Digger Pine, *Pinus sabiniana*	blue pine, bull pine, grey pine (see Coulter Pine), nut pine, round-top, Sabine pine, silver pine, Wythe pine
Jeffrey Pine, *Pinus jeffreyi*	blackbark pine, blackwood pine, bull pine, peninsula black pine, peninsula pine, redbark pine, redbark sierra pine, sapwood pine, truckee pine, western black pine, western yellow pine
Knobcone Pine, *Pinus attenuate*	Mount Shasta pine, narrow-cone pine, prickly-cone pine, sandy-slope pine, scrub pine, snow-line pine, sunny-slope pine
Limber Pine, *Pinus flexilis*	bull pine, jack pine, Rocky Mountain white pine, pitch pine; scrub pine, white pine
Lodgepole Pine, *Pinus contorta*	beach pine, black pine, Bolander's pine, coast pine, cypress, Henderson pine, jack pine, knotty pine, Murray pine, prickly pine, sand pine, scrub pine, shore pine, spruce pine, tamarack pine, twisted pine, western jack pine, white pine
Monterey Pine, *Pinus radiata*	insular pine, remarkable pine, small-cone pine, smooth-cone pine, spreading-cone pine
Pinyon, *Pinus cembroides*	Arizona nut pine, Colorado pinyon pine, Mexican nut pine, Mexican stone pine, nut pine, pinyon pine
Ponderosa Pine, *Pinus ponderosa*	Arizona pine, big pine, bird's-eye pine, blackjack pine, bull pine, foothills yellow pine, heavy pine, knotty pine, red pine, rock pine, western yellow pine, pino real; pitch pine, yellow pine, western pitch pine, western soft pine, western yellow pine
Singleleaf Pinyon, *Pinus monophylla*	grey pine, Nevada nut pine, pinyon
Sugar Pine, *Pinus lambertiana*	big pine, California Sugar Pine purple-coned sugar, shade pine, sockertall, sugar pine
Western White Pine, *Pinus monticola*	finger-cone pine, Idaho White Pine, mountain pine, silver pine, white pine (may have been confused with larch), yellow pine
Whitebark pine, *Pinus albicaulis*	alpine pine, creeping pine, pitch pine, scrub pine, white-stem pine, yellow pine
Pacific Yew, *Taxus brevifolia*	Canadian yew, mountain mahogany, Oregon yew, Western Yew, Yew (sometimes confused with hemlock)

Proper Name	Colloquial Name
Mountain Hemlock, *Tsuga mertensiana*	alpine hemlock, alpine spruce, black hemlock, mountain hemlock, Olympic fir, Pacific Coast hemlock, Patton's hemlock, Patton's spruce, Prince Albert's fir, weeping spruce, western hemlock, western hemlock spruce, Williamson's spruce
Western Hemlock, *Tsuga heterophylla*	laska pine, alpine hemlock, alpine spruce (may have been confused with spruce or fir), berg-hemlock, black hemlock, gray fir, hemlock, mountain hemlock, Olympic fir, Pacific hemlock, Patton's hemlock, Patton's spruce, Prince Albert's fir, silver fir, weeping spruce, West Coast hemlock, Williamson's spruce
Subalpine Larch, *Larix lyallii*	Lyall's larch, mountain larch, tamarack, timberline larch, woolly larch
Western Larch, *Larix occidentalis*	British Columbia tamarack, hackmatack, juniper, Montana larch, mountain larch, Oregon larch, red American larch, roughbarked larch, tamarack, Western tamarack
Blue Spruce, *Picea pungens*	balsam, Colorado spruce, Colorado blue spruce, Parry's spruce, prickly spruce, silver spruce, water spruce, white spruce
Engelmann Spruce, *Picea engelmanni*	Arizona spruce, Columbian spruce, Engelmann elm, Engelmann spar, mountain spruce, Sequoia silver spruce, Sitka spar, silver spruce, white spruce
Sitka Spruce, *Picea sitchensis*	coast spruce, menzies spar, silver spruce, tideland spruce, West Coast spruce, Western spruce, yellow spruce
Douglas-fir, *Pseudotsuga menziesii*	alpine hemlock, black fir, British Columbia pine, Columbian pine, Doug Fir, Coast Douglas-fir, cork-barked Douglas spruce, Douglas Spruce, fir, golden rod fir, gray Douglas, green Douglas, Montana fir, Oregon Douglas-fir, Oregon fir, Oregon Pine (occasionally misnamed "pine"), Oregon spruce, Patton's hemlock, Puget Sound pine, red fir, red pine, red spruce, yellow fir, yellow national fir
Grand Fir, *Abies grandis*	balsam fir, California great fir, giant fir, great silver fir, lowland fir, Oregon fir, Puget Sound fir, rough-barked fir, silver fir, Vancouver den, white fir, yellow fir
Noble Fir, *Abies procera*	bracted fir, California red fir, feather-coned fir, red fir, tuck-tuck, white fir
Red Fir, *Abies magnifica*	alpine fir, Arizona cork fir, Arizona fir, balsam, black balsam, caribou fir, cork fir, corkbark fir, downey-cone fir, golden fir, mountain balsam, mountain fir, Oregon balsam fir, pumpkin-tree, red fir, Rocky Mountain fir, Shasta fir, western balsam, white fir (sometimes confused with Douglas-fir

Proper Name	Colloquial Name
Silver Fir, *Abies amabalis*	alpine fir, balsam fir, Cascade fir, lovely silver fir, lovely fir, red fir, red silver fir, silver fir, western fir, western balsam fir, white balsam, white fir
Subalpine Fir, *Abies lasiocarpa*	alpine fir, balsam, Arizona fir, balsam fir, black balsam, caribou fir, cork fir, corkbark, downey-cone fir, mountain balsam, mountain fir, Oregon balsam fir, pumpkin-tree, Rocky Mountain fir, western balsam, white balsam, white fir
Redwood, *Sequoia sempervirens*	California cedar, California redwood, coast redwood, giant-of-the-forest, Humboldt redwood, ledwood, Mexican cherry
Giant Sequoia, *Sequoiadendron gigantea*	Bigtree, Sierra Redwood
Alaska-Cedar, *Chamaecyparis nootkatensis*	Alaska cypress, Alaska ground cypress, Alaska yellow-cedar, Nootka false-cypress, Pacific Coast yellow cedar, Sitka cypress, yellow cypress, yellow-cedar
Arizona Cypress, *Cupressus arizonica*	cedro, Cuyamaca cypress, Piute cypress, red-bark cypress, rough-bark Arizona cypress, smooth cypress, yew-wood
Incense-cedar, *Calocedrus decurrens*	bastard cedar, California calocedar, California incense-cedar, California post cedar, juniper, pencil cedar, post cedar, red cedar, roughbark cedar, white cedar
Port Orford-Cedar, *Chamaecyparis lawnoniana*	ginger pine, Lawson cypress, Lawson false-cypress, matchwood, Oregon-cedar, Oregon cypress, pencil cedar, Port Orford white-cedar spruce gum, white cedar, white cypress
Western Redcedar, *Thuja plicata*	British Columbia redcedar, California cedar, canoe-cedar, giant arborvitae, giant-cedar, Idaho cedar, Oregon cedar, Pacific redcedar, red cedar pine, shinglewood
Alligator Juniper, *Juniperus deppeana*	Cedro Chino, checkered-bark juniper, mountain cedar, oakbark juniper, thick-barked juniper, Western juniper
Rocky Mountain Juniper, *Juniperus scopulorum*	Cedro Rojo, Colorado juniper, redcedar, river juniper, Rocky Mountain redcedar, Western juniper, Western redcedar
Utah Juniper, *Juniperus osteosperma*	bigberry juniper, cedar, desert juniper, sabina; Utah cedar, Western juniper, Western redcedar, white cedar
Western Juniper, *Juniperus occidentalis*	California juniper, Canada juniper, cedar, pencilwood, San Bernadino juniper, Sierra Juniper, Western cedar, yellow cedar
Pacific Dogwood, *Cornus nuttalli*	andubon, California dogwood, flowering dogwood, kornel, mountain dogwood, western flowering dogwood
California Buckeye, *Aesculus californica*	horsechestnut
Bigleaf Maple, *Acer macrophyllum*	big-leaf, broadleaf maple, Californian maple, Oregon maple, pacific maple, Rocky Mountain maple, shrubby maple, Sierra maple, soft maple, white maple

Proper Name	Colloquial Name
Rocky Mountain Maple, *Acer glabrum*	bark maple, Douglas maple, dwarf maple, mountain maple, New Mexico maple, Sierra maple
Western Mountain Maple, *Acer glabrum*	bark maple, Douglas maple, dwarf, New Mexico maple, Rocky Mountain maple, shrubby maple, Sierra maple, soft maple
Green Ash, *Fraxinus pennsylvanica*	bastard ash, black ash, blue ash, brown ash, Canadian ash, Darlington ash, gray ash, piss ash, pumpkin ash, red ash, rim ash, river ash, soft ash, swamp ash, water ash, white ash
Oregon Ash, *Fraxinus latifolia*	(may have been confused with cherry), basket ash, water ash, white ash
California-laurel	Bayberry, Bay, California bayberry, California myrtle, Myrtle, Oregon-myrtle, Pacific bayberry, Pacific-myrtle, pepperwood, spice-tree, waxmyrtle
Quaking Aspen, *Populus tremuloides*	aspen, golden aspen, leaf aspen, mountain aspen, popple, quiver-leaf, trembling aspen, Vancouver aspen, white poplar
Black Cottonwood, *Populus trichocarpa*	balsam cottonwood, California poplar, cottonwood, Western balsam poplar
Plains Cottonwood, *Populus deltoides*	aspen cottonwood, big cottonwood, Carolina poplar, cotton tree, cottonwood, Fremont cottonwood, Missourian poplar, necklace poplar, Palmer cottonwood, Rio Grande cottonwood, plains poplar, river cottonwood, river poplar, southern cottonwood, Texas cottonwood, valley cottonwood, water poplar, yellow cottonwood
Willow, *Salix spp.*	Golden-osier
Red Alder, *Alnus rubra*	Oregon alder, Pacific Coast alder, Western alder (may have been confused with mulberry)
Cascara Buckthorn, *Rhamnus purshiana*	bearberry, bearwood, Cascara, Cascara Sagrada, chittam, coffee-tree shittum, shittumwood
Pacific Madrone, *Arbutus menziesii*	Coast Madrone, Laurel, Laurelwood, Madrone Tree, Manzanita, Strawberry-tree
Netleaf Hackberry, *Celtis reticulata*	Sugarberry, Thick Leaved Hackberry, Western Hackberry
Western Juneberry, *Amelanchier alnifolia*	juneberry, Pacific serviceberry, pigeon berry, service, Saskatoon serviceberry, western shadbush, western service
Cascara Buckthorn, *Frangula purshiana*	barberry, bearberry, bearwood, bitterbark, bitterboom, bittertrad, buckthorn, California coffee, cascara, chittern, chittum, coffeeberry, coffeebush, Oregon bearwood, pigeonberry, shittumwood, wahoo, wild coffee, yellow-wood
Golden Chinquapin, *Castanopsis chrysophylla*	chestnut, chink, chinkapin, chinquapin, evergreen chestnut, giant evergreen-chinkapin, giant chinkapin, goldenleaf chestnut

Proper Name	Colloquial Name
Tanoak, *Lithocarpus densiflorus*	chestnut oak, tanbark oak
Blue Oak, *Quercus douglasii*	California blue oak, California rock oak, California white oak, Douglas oak, encina, Hill oak, iron oak, mountain oak, mountain white oak, post oak, rock oak, white mountain oak, white oak
California Black Oak, *Quercus kelloggii*	black oak, Kellogg oak, mountain black oak
California Live Oak, *Quercus agrifolia*	coast live oak, encina, evergreen oak, live oak
California White Oak, *Quercus lobata*	California oak, hollyleaf oak, Roble, valley oak, valley white oak, water oak, weeping oak, white oak
Canyon Live Oak, *Quercus chrysolepis*	black live oak, canyon oak, goldcup oak, hickory oak, iron oak, live oak, maul oak, white live oak
Emory Oak, *Quercus emoryi*	Bellota, black oak, blackjack oak, Roble Negro
Gambel Oak, *Quercus gambelii*	encino, mountain oak, Rocky Mountain white oak, Utah white oak, white oak
Interior Live Oak, *Quercus wislizenii*	black live oak, highland live oak, live oak, Sierra live oak
Oregon White Oak, *Quercus garryana*	Brewer oak, British Columbia oak, Garry oak, Oregon oak, Pacific post oak, prairie oak, shin oak, western oak, white oak
Valley Oak, *Quercus lobata*	California oak, California white oak, California white valley oak, roble, valley oak, water oak, weeping oak, white oak

Compiled from Little, *Check List of Native and Naturalized Trees of the United States (including Alaska)*; C. Albert White, comp., *Durability of Bearing Trees*; USDA, Trees, *Yearbook of Agriculture* 1949; miscellaneous sources (colloquial names are apt to show up anyplace).

TREE IDENTIFICATION

Get one or more good manuals.

Baldcypress	Grows in wet areas or in water; leaves featherlike
Eastern White Pine	Needles 5 in a cluster; bark gray, deeply fissured into broad ridges; cones long
Pitch Pine	Needles 3 in a cluster; cones short and broad
Shortleaf Pine	Needles 2 or 3 in a cluster; bark reddish-brown; cones small
Loblolly Pine	Needles 3 in a cluster; bark reddish brown and deeply fissured in broad scaly plates; cones 3–5 inches long, with stiff prickles
Table-Mountain Pine	Needles 2 in a cluster; bark has loose dark brown scales tinged with red; cones very knobby
Virginia Pine	Small (scrubby) tree; needles 2 in a cluster; cones curved, very prickly; bark dark brown, thin, with scaly plates
Eastern Hemlock	Needles short, flat, blunt pointed; cones small, brownish; bark brown or purplish (bright *inside*), furrowed into scaly ridges
Fraser Fir	Bark gray or brown with resin blisters; needles flat
Eastern redcedar	Bark reddish brown, thin and shreddy; leaves scalelike; but on new growth needlelike and very sharp-pointed
Flowering Dogwood	Bark dark reddish brown broken into small square or rounded blocks; leaves paired, elliptical or oval, slightly hairy beneath
Sugar Maple	Bark furrowed into irregular ridges or scales; leaves paired 3 or 5 long-pointed lobes, smooth beneath
Black Maple	Bark gray, becoming deeply furrowed; leaves paired, 3 or 5 lobed, short-pointed, hairy beneath
Red Maple	Bark gray, thin, smooth; twigs reddish; leaves paired, lobes sharp-pointed, irregularly and sharply toothed, hairy beneath
Silver Maple	Bark gray, thin, smooth; leaves paired, deeply 5-lobed, lobes long-pointed, irregularly toothed, green above, silvery beneath
Boxelder	Bark gray or brown, thin, with narrow ridges or fissures; twigs green; leaves paired, *compound,* 3 or 5, sometimes 7 or 9, leaflets
White Ash	Bark gray with diamond-shaped fissures; leaves paired, compound, short-pointed, smooth or lightly toothed, smooth or hairy beneath

Green Ash	Bark gray, fissured; leaves paired, compound, 7 or 9 slightly toothed leaflets, long-pointed, smooth or hairy beneath
Yellow Buckeye	Bark gray, separating into thin scale; leaves paired, compound, leaflets 5
Ohio Buckeye	Bark gray, much furrowed, broken into scaly plates; leaves paired, compound; leaflets 5
Black Walnut	Bark dark brown to black, thick, with deep furrows and narrow, forking ridges; compound leaves, 12–24 inches long, 15–23 leaflets, finely toothed, hairy beneath
Butternut	Bark light gray, furrowed into broad, flat ridges; compound leaves, 15–30 inches long, finely toothed slightly hairy above, soft hairy beneath
Bitternut Hickory	Bark light brown, shallowly furrowed, with narrow, forking ridges or thin scales; compound leaves, 5–9 leaflets, finely toothed, more or less hairy beneath; winter buds bright yellow
Mockernut Hickory	Bark gray, irregularly furrowed into flat ridges; compound leaves, 7–9 leaflets, finely toothed, dark yellow green and shiny above, pale and densely hairy beneath
Shellbark Hickory	Bark gray, shaggy with long, thin, straight plates; compound leaves, leaflets usually 7, finely toothed, dark green and shiny above, pale and soft-hairy beneath
Shagbark Hickory	Bark gray, shaggy with long, thin, curved plates; leaves compound, leaflets usually 5, finely toothed
Pignut Hickory	Bark dark gray, with furrows and forking ridges; leaves compound, leaflets usually 5, or 5 and 7, finely toothed
Red Hickory	Bark gray, furrowed, often scaly or shaggy; leaves compound, leaflets 7 or 5, finely toothed, hairy at first but becoming smooth
Honeylocust	Bark grayish brown or black, fissured into long, narrow, scaly ridges; trunk and branches with large, stout, usually branched spines, rarely absent; leaves compound, once or twice divided leaflets blunt or rounded at apex.
Waterlocust	Branches armed with stout spines 3–5 inches long; compound leaves with oval leaflets; prefers deep swamps and rich bottomlands
Black Locust	Bark brown, thick, deeply furrowed, with rough, forked ridges. Twigs with a pair of spines about 1/2 inch long at the base of each leaf; compound leaves, leaflets rounded at apex
Sassafras	Bark reddish brown, deeply furrowed; leaves oval or elliptical, often two or three-lobed (mitten-shaped); bright green above, paler and smooth or hairy beneath; distinct odor to scraped twig
Red Mulberry	Bark dark brown, fissured and scaly; leaves broadly oval or heart-shaped, abruptly long-pointed, coarsely toothed, sometimes 2- or 3-lobed; rough above, soft-hairy beneath
Sweetgum	Bark gray, deeply furrowed; twigs reddish brown, developing *corky ridges*; leaves maplelike but *star-shaped* with 5 long-pointed, finely toothed lobes. Fruit distinct as a brownish spiny ball

American Sycamore Frequents wet soils; bark of branches whitish, thin, smooth; bark of *trunk peeling off in large flakes*, smoothish with patches of brown, green and gray. Leaves like maple except veins meet above stem rather than where stem joins the leaf

Serviceberry *Inside of two-toned buds with whitish hair; gray bark often streaked with darker lines*

American mountain-ash Often an ornamental; 13–17 sharply serrate leaflets and clusters of bright red-orange fruits; bark like cherry—nearly black with lenticels

Yellow-Poplar Bark brown, becoming think and deeply furrowed; leaves of unusual squarish shape with broad, slightly notched or nearly straight apex and 2 or 3 lobes on each side (*saddle-shaped*); shiny dark green above, pale green beneath

Cucumbertree Bark dark brown, furrowed, with narrow, scaly, forking ridges; leaves short pointed, yellow green and smooth above, light green and soft-hairy or nearly smooth beneath

Common Persimmon Bark dark brown, thick, deeply divided into small, square, scaly blocks; leaves oval or elliptical, long-pointed, rounded at base, shiny dark green above, pale green and smooth or hairy beneath

Black Tupelo Bark reddish brown, deeply fissured an block-shaped ridges; leaves elliptical, short or blunt-pointed, wedge-shaped or rounded at base, pale and often hairy beneath

Sourwood Leaves sour tasting, elliptical

American Holly Bark light gray, thin, smoothish, with wartlike projections; leaves *evergreen*, elliptical, spine-pointed and coarsely spiny-toothed, stiff and leathery, shiny green above and yellow-green beneath

American Basswood Bark gray, deeply furrowed into narrow, scaly ridges; leaves in two rows, heart-shaped, long-pointed, coarsely toothed with long-pointed teeth, dark green above, light green beneath with *tufts of hair in angles of main veins. Fruit characteristic, borne on bract*

White Basswood Similar to above, except leaves smaller, unequal at base, and white or brownish beneath with dense hairy coat

American Elm Bark gray, deeply furrowed, with broad, forking, scaly ridges; twigs soft-hairy becoming smooth; leaves elliptical, long-pointed, unequal sided, doubly toothed with unequal teeth, slightly rough (sandpapery) above and soft-hairy beneath

Slippery Elm Similar to above except leaves tend to be larger

Winged Elm Bark light brown, thin, irregularly fissured; twigs becoming corky winged; leaves as in other elms

Hackberry Bark light brown to gray with corky warts or ridges becoming scaly; leaves long-pointed, the two sides unequal, sharply toothed except in lower part, 3 main veins from base, bright green

Sugarberry Similar to above except *edges of leaves smooth or with very few teeth*

Eastern Cottonwood Bark at first yellowish green and smooth, becoming gray and deeply furrowed; *leaves triangular,* long-pointed, coarsely toothed with curved teeth, smooth, light green and shiny; *leafstalks flat*

Yellow Birch
Bark (aromatic on young branches, smelling like wintergreen) yellowish or silver gray, shiny, separating into papery, curly strips, on old trunks reddish brown. Leaves oval and doubly toothed, dull dark green above, yellow green below

River Birch
Bark reddish brown or silvery gray, shiny, becoming fissured and separating into papery scales; leaves oval, wedge-shaped at base, doubly toothed, shiny dark green above, whitish and usually hairy beneath

Black Willow
Bark dark brown or blackish, deeply furrowed, with scaly, forking ridges. *Leaves lance-shaped,* long pointed, finely toothed, green on both sides, shiny above and pale beneath

Black Cherry
Bark dark reddish brown, smooth at first, becoming irregularly fissured and scaly; leaves oblong, long-pointed and finely toothed, shiny dark green above, light green beneath. *Scraped twigs with bitter almond odor; glands on stem where it joins leaf*

American Beech
Bark blue gray, thin, smooth. Leaves long-pointed, with coarse, curved teeth, dark blue green above and light green beneath, usually smooth, *turning copper-brown in fall and hanging on through winter like oaks*

Chestnut
Bark dark brown, irregularly fissured into broad, flat ridges, eaves narrowly oblong, long-pointed, coarsely toothed with slightly curved teeth, many parallel lateral veins, yellow green, smooth. Species in nearly extinct but small-diameter trees remain from sprouts. Very durable wood, so remnants of trees, stumps, posts and fences may still be found

Oaks are often difficult to distinguish, so calls in descriptions may be erroneous.

RED OAK GROUP
Leaves typically lobed, and bristle-tipped; broken brown acorn shells (not cups) have hairy inner surfaces; meat of acorn is yellow, bitter and usually inedible

Northern Red Oak
Bark dark brown, fissured into broad, flat ridges; leaves 7- to 11-lobed with a few irregular bristle-pointed teeth, dark green above, pale yellow green beneath

Scarlet Oak
Bark dark brown or gray, fissured into irregular, scaly ridges; leaves deeply 7-lobed, lobes broader toward tip, with a few bristle-pointed teeth, *edges rounded between the lobes turn scarlet in fall*

Shumard Oak
Bark gray or reddish brown, fissured into scaly plates; leaves oval or elliptical, 5- to 9-lobed with a few bristle-pointed teeth, edges rounded or pointed between the lobes, beneath light green with *tufts of hairs along midrib*

Pin Oak
Bark grayish brown, smooth, becoming fissured with low, scaly ridges; leaves 5- to 7-lobed with a few bristle-pointed teeth, *dark green and very shiny above,* light green and nearly smooth beneath.

Black Oak	Bark blackish, thick, deeply furrowed, with blocklike ridges; *inner bark yellow (stick a knife blade into the bark and the part making contact will turn black)*; leaves 7- or 9-lobed, the lobes broad and with a few bristle-pointed teeth, shiny dark green above, *usually brown-hairy beneath, turning dull red or brown in fall*
Southern Red Oak	Bark dark brown, thick, fissured into narrow ridges; leaves deeply 3- to 7-lobed the lobes with 1 to 3 bristle-pointed teeth, or slightly 3-lobed near broad apex, dark green, smooth, and shiny above, *rusty or grayish hairy beneath, turning brown or orange in fall*
Blackjack Oak	Bark blackish, thick and rough, divided into small squarish blocks; *leaves broadest and 3-lobed at apex,* lobes shallow and broad with 1 or few bristle-pointed teeth, dark green, smooth, and shiny above, *brownish or rusty-hairy beneath, turning brown or yellow in fall*
Water Oak	Bark gray, fissured into irregular, scaly ridges; leaves broadest at the 3-lobed or smooth apex or sometimes with several lobes, *dull blue green,* paler beneath, becoming smooth except for *tufts of hairs along axis, turning yellow in fall and shedding in winter*
Willow Oak	bark gray or brown, smoothish, on large trunks becoming fissured into scaly ridges; leaves very narrowly oblong or lance-shaped, short-pointed with smooth or slightly wavy edges, light green and shiny above, beneath dull and slightly hairy or nearly smooth, turning pale yellow in fall *(leaves look like willow, except not toothed, hence the name)*
WHITE OAK GROUP	Many species have leaves with short, rounded lobes. <u>None</u> is bristle-tipped; inner acorn shells are hairless; acorn meat is white, relatively sweet and often edible
White Oak	Bark light gray, fissured into scaly ridges; leaves deeply or shallowly 5- to 9-lobed, smooth, bright green above, *pale or whitish beneath, turning deep red in the fall*
Chestnut Oak	Bark brown or blackish, on large trunks becoming deeply furrowed into broad ridges. Leaves short- or long-pointed, narrowed and pointed or rounded at base, *edges wavy with rounded teeth,* shiny yellow green above, *paler and hairy or nearly smooth beneath, turning dull orange in fall*
Swamp Chestnut Oak	Bark light gray, fissured and scaly; leaves short- or long-pointed, wedge-shaped or rounded at base, *edges wavy with rounded teeth,* shiny dark green above, *grayish hairy beneath, turning crimson in the fall*
Chinkapin Oak	Bark light gray, thin, fissured, and flaky; leaves oblong or *broadly lance-shaped,* short- or long-pointed, *usually rounded at base, edges way with coarse, slightly curved teeth,* dark or yellowish green above, *whitish hairy beneath, turning orange and scarlet in the fall*

Post Oak	Bark reddish-brown, fissured into broad, scaly ridges; *leaves usually wedge-shaped at base,* deeply 5- to 7-lobed, *lobes broad and middle lobes largest,* dark green and rough above, *grayish hairy beneath, turning brown in the fall*
Bur Oak	Bark light brown, deeply furrowed into scaly ridges; leaves oblong, *wedge-shaped at base, broadest above middle, the lower part deeply lobed nearly to middle and the upper half with shallow lobes,* dark green and usually shiny above, *grayish or whitish hairy beneath, turning yellow or brown in the fall*
Overcup Oak	Bark brownish-gray, fissured into large irregular, scaly ridges; leaves oblong, wedge-shaped at base, *deeply lobed nearly to middle with 7 to 9 rounded or pointed lobes, the 2 lowest lobes on each side much smaller,* dark green and smooth above, *white hairy beneath, turning yellowish orange, or scarlet in the fall*

Descriptions adapted and compiled from: Harlow & Harrar, *Textbook of Dendrology; Trees, Yearbook of Agriculture,* 1949; Neelands, *Important Trees of Eastern Forests;* Petrides, *A Field Guide to Trees and Shrubs;* Peattie, *A Natural History of Trees of Eastern and Central North America,* and Wilson, personal notes and observations.

TREE IDENTIFICATION ON AERIAL PHOTOGRAPHS

White Pine Largest crown of any conifer; clearly outlined star shape. Hemlock has a denser body, which splits the crown to a contrasting black and white side. The crown of aspen often has a jagged outline; however, the commonly pure occurrence and even stand pattern and the pure grouping habit help to differentiate from the crown of white pine. Red oak commonly appears in a scattered fashion on dry hilltops; trees are usually open-grown, have a low height and a wide crown, which are always adequate characteristics for correct identification. Dead deciduous trees with naked limbs and branches can form a star-shaped image. These crowns, however, are very open because of the absence of foliage.

Fraser Fir Very symmetrical tree; top is sharply pointed in a very narrow convergence with a characteristically dense, rigid and spire-shaped tip. A highlight, caused by reflections from the dense foliage, near the top is common. Can be confused with spruce or white cedar.

Eastern Hemlock Open-grown hemlock has large prominent branches, but in dense stands it has narrow, pointed, and ovally cone-shaped crowns. Can be difficult to distinguish from white pine in the case of overmature trees, where the irregular crowns and the bright prominent branches produce a star-shaped outline similar to that of white pine. However, even in extreme cases, hemlock is more regular, outlining a dense body, while white pine has a relatively smaller center for the long prominent branches. When both species occur together, white pine shows a distinctly darker photographic tone. There is also a similarity between hemlock and basswood. Both have a hard contrast between their highlight and shadow, but basswood is always smoothly rounded on the top, outlining a large and dense body, while hemlock is sharply pointed and narrower.

Sugar Maple	A young sugar maple stand has a very solid and even canopy, formed by very dense, coarse, and irregular crowns. The crown of mature or overmature sugar maple is usually large and very compact. The crown texture is a conspicuous feature, because the *billowy* surface of the bright sugar maple crown casts a hard, but distinct, shadow pattern, resulting in a characteristic crown relief. The sharp outline is also helpful in distinguishing sugar maple from most of the other tolerant hardwood species. Young sugar maple stands may be confused with immature stands of aspen or white birch, because of the very dense canopy and the even height. Red oak also has billowy crowns, similar to sugar maple, but the tone of red oak is darker and the crown texture is smoother than that of sugar maple.
Red Maple	Red maple has a moderately wide and usually irregular crown. The crown texture is tufted, because of the outstanding and *upright-growing* small branches, which can be detected.
Silver Maple	Generally grows in moist and swampy locations. Its trunk divides near the ground into several ascending, big branches, which form a wide-spreading crown. Smaller branches are slender and twig structure is dense. On hot days and in a light breeze, the leaves have a tendency to turn to expose the whitish green undersides, resulting in the crowns having a lighter tone and color than under other conditions.
Yellow Birch	Mature yellow birch is a large tree with a finely *tufted*, dense, and distinctly outlined crown, which usually appears very *dark* on the photographs because of the dark green foliage. White elm and red oak often resemble yellow birch.
American Elm	White elm usually produces a large, flat, umbrella-shaped crown top. The spreading main limbs form a fairly (sometimes very) open crown, whose texture is finely *tufted*. Often a fine, regular pattern of *light dots* is evident in the crown, whose overall photographic tone is usually dark.
Red Oak	The rounded crown of red oak is compact and dense with a smooth velvety texture that becomes *billowy* at maturity. It usually appears *dark* on aerial photographs due to its lustrous, bright-green leaves and its compact crown.
White Oak	Develops a massive, broad crown when growing in the open, while in the forest it has a straight trunk and a smaller crown. Differences from red oak are slight.
Ash	Has an *irregular* narrow, very *open* crown. The light green compound foliage produces a light and fuzzy image, with is almost transparent on poor, very wet sites. Except when growing in the open, ashes have straight trunks with little taper. Found in the same locations as elm and silver maple, ash can be distinguished by its crown size or by the fact it has no drooping branches.
Basswood	The crown is large, usually symmetrical, closed and oval. Crown texture is smooth with a contrasting photographic tone. Because of its oval shape and contrasting photographic tone, basswood may be compared to hemlock.

Black Cherry Has a very *smooth* crown texture, a round crown top, and a very *dark* photographic tone. It is easily mistaken for beech, because of the same fuzzy crown texture, although some fine differences do exist. Black cherry usually has a regular rounded crown top with a smaller diameter; the outline of the crown is more pronounced and the tone is definitely darker than that of beech. Because of this very dark tone, the odd sporadic black cherry stands out clearly from the lighter-toned associates.

Willow When the species occurs in swamps that do not produce merchantable timber, the stand pattern is irregular and the canopy is usually very dense. The individual crowns usually do not appear clearly, and the stand profile therefore shows a fuzzy texture, outlining a fairly coarse contour. The height of the stand changes gradually, decreasing towards wetter areas of the swamp. Some scattered trees usually stand out from the thick shrubbery.

SILHOUETTES AND SHADOWS

Figure AX.1 Shadows cast on the ground are strong indications of certain crown features. Where it is possible to see the entire shadow of a tree, more detail of crown shape and overall tree shape than from an overhead view of the crown. Images away from the center of a photograph and closer to the edge, as well as oblique photography, will allow the viewer to see overall tree shapes.

In the identification of tree species on aerial photos, these mistakes are commonly made:

- White pine is occasionally mistaken for hemlock.
- Red pine is sometimes confused with balsam poplar.
- Jack pine is sometimes mistaken for black spruce or white birch.
- White spruce is often interpreted as balsam fir, black spruce, or white cedar.
- Black spruce is frequently mistaken for balsam fir, white spruce or white cedar, and sometimes for jack pine.
- Balsam fir is often identified as black spruce, white spruce, or white cedar.
- Hemlock is occasionally mistaken for white pine.
- White cedar is frequently confused with balsam fir, black spruce, white spruce, or larch.
- Larch is occasionally mistaken for black spruce, aspen, or black spruce.
- Sugar maple is sometimes interpreted as red maple or red oak.
- Red maple is frequently mistaken for sugar maple, sometimes for white elm, and in young age classes is often confused with white birch.
- White birch is commonly mistaken for aspen and balsam poplar.
- Yellow birch is frequently confused with white elm, sometimes with red oak or beech.
- Beech is frequently mistaken for black cherry or red oak.
- White elm is often interpreted as yellow birch.
- Red oak sometimes is confused with yellow birch, sugar maple, beech, or aspen.
- Aspen is commonly confused with white birch.
- Balsam poplar is often mistaken for red pine, sometimes for white birch.
- Black ash is frequently confused with aspen.
- Black cherry is commonly mistaken for beech.

From Victor G. Zsilinszky, *Photographic Interpretation of Tree Species in Ontario* (1966). *Source*: Ontario Ministry of Natural Resources. Copyright: 2007 Queens Printer Ontario.

IDENTIFICATION OF STUMP HOLES

Species	Characteristics
Basswood	Leaves a large stump hole, and sprouts around the stump. Fallen tree leaves shallow stump hole and 3 to 7 main root holes around perimeter. New growth sprouts from stump. Color of wood varies, but inner portions are often brown, distinguishing it from sugar maple, yellow birch, and elm.
American Beech	Decayed tree leaves a fairly deep stump hole. Advanced decay causes an unconnected mass of thin, brittle, black ribbons.
White Birch	Fair-size stump hole. Sprouts easily, so may remain in parent location for many generations. Wood decays into a white, stringy mass held together by a sheath of bark.
Yellow Birch	Decayed stump is shallow in areas with a high water table, and 1 to 2 feet deep in well-drained soils. Badly decayed stump on wet ground has a greasy mushy feel. Outer shell bark can remain when interior stump is completely decayed. A blue-green stain is often associated with yellow birch in the black decay stage. Reddish-brown to orange decay occurs in moist sites. Color, substance, and feel of decayed wood is similar to that of elm.
Butternut	Does not leave a prominent stump hole. Decays rapidly into a dark-colored mass that disappears rapidly.
Black Cherry	Decays rapidly. Color of wood changes from dark red to reddish black to yellowish orange.

Species	Characteristics
American Elm	Does not leave a distinct stump hole. Fallen trees often leave a large, trench-shaped hole and earth mound. Wood is decay resistant, but in advanced stages of decay, black flakes are reduced to fragile ribbons. Inner wood is light brown to buff-colored and occasionally gray to off-white, while wood under bark is dark.
Red Maple	Decays rapidly and sprouts from stump. Leaves a distinct stump hole if the tree was of any size before decay began. Light gray to grayish tan. Decay is fibrous and stringy. On wetter sites, decay is orange to reddish orange. Decayed wood is blackish and becomes powdery when exposed to sunlight. On rare occasions, decay may be black to charcoal gray.
Sugar Maple	Stump hole is distinct. Hole is about 50 percent larger than the tree's diameter and relatively deep. In a decayed stump the bark is gone, the wood turns black, and flakes off in scales.
Northern Red Oak	Leaves a deep, distinct stump hole slightly larger than the diameter of the stump. Tree is decay resistant and a prolific sprouter. Dark gray to mottle brownish black and often stained blue-green deep into firm portions of the wood. Reddish brown to bright orange. Drier portions of stump are dark gray to brownish black and blue-green stained.
Black Oak	Little chance of recovery of original trees due to a very fast decay rate. Leaves a large stump hole. Second growth oak usually found in vicinity.
White Oak	Leaves a deep stump hole. Very resistant to decay.

Descriptions adapted and compiled from: Drahn and Stefan, *Decayed Wood Identification of Bearing Trees of Public Land Surveys*; Lapalla, *Retracement and Evidence of Public Land Surveys*; and Wilson, personal notes and observations.

WOOD IDENTIFICATION

A key to wood identification is the distinguishable cellular features found in each tree species. The presence of vessels, ducts, and rays of various sizes and locations, along with cellular variations, provides a means for species identification. As wood decays due to moisture changes, soil conditions, and the spread of fungi, cellular structure features are altered. It is important to expose or extract the least decayed part of the tree stump to examine key identifiers.

To obtain a sample from a tree stump, make a cross-sectional cut using a sharp razor blade or knife or, if possible, break off a piece of stump by hand. A sharp cut or a broken sample will result in the least disturbed and best end view of the wood cells and vessels. The sample should then be examined under a 10X or better hand lens.

SIMPLE, GENERAL KEY TO WOOD IDENTIFICATION

Softwoods

Resin Canals

Large: Pines
Small:
    Abrupt transition: Douglas-fir, Larch
    Gradual transition: Spruces

No Resin Canals

No odor
    Abrupt transition: Hemlocks
    Less than abrupt transition: Firs
Odor: Cedars

Other

Redwood
Baldycypress

Hardwoods

Ring Porous

Late wood pores in small multiples or solitary:
    Single row: Hickory
    Multiple row: Ash
Late wood pores in tangential bands: Elm
Late wood pores in clusters: Sassafras, Locust
Other: Magnolia

Semi-Ring Porous

Large pores: Walnut, Butternut
Rays indistinct under lens: Willow, Cottonwood, Aspen

Diffuse Porous

Pores very numerous and crowded: Cherry
Pores crowded: Sycamore, Beech
Pores not crowded: Dogwood, Maple
Pores not in radial lines: Birch
Rays distinct (naked eye): Basswood
Rays distinct (under lens): Gum, Apple, Alder

Eastern Species

Softwoods

Baldcypress	Red-brown, almost black; solid wood has a greasy feel and *rancid odor*; latewood amber to dark brown
Eastern White Pine	Soft in texture; *large resin canals*, mostly solitary, evenly distributed; sapwood yellowish, heartwood reddish; early to latewood transition gradual
Pitch Pine	A hard pine; transition abrupt

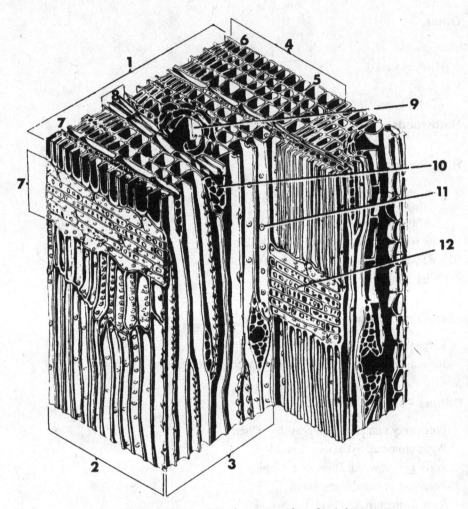

Figure AXII.1 Wood structure of a softwood.

1. cross-sectional face	7. wood ray
2. radial face	8. fusiform ray
3. tangential face	9. vertical resin duct
4. annual ring	10. horizontal resin duct
5. earlywood	11. bordered pit
6. latewood	12. simple pit

From A.N. Foulger, *Classroom Demonstrations of Wood Properties,* US Department of Agriculture, Forest Service, Forest Products Laboratory, PA-900.

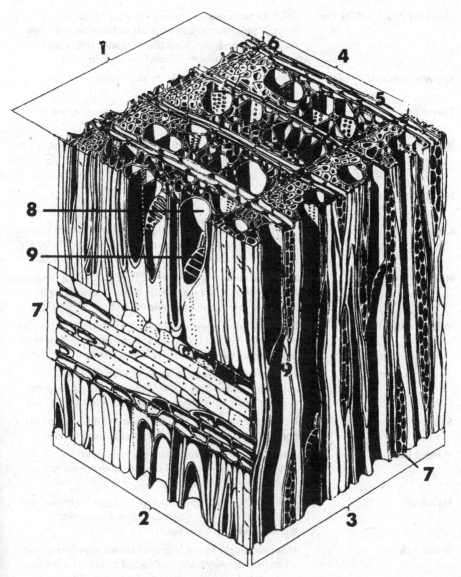

Figure AXII.2   Wood structure of a hardwood.

1. cross-sectional face	6. latewood
2. radial face	7. wood ray
3. tangential face	8. vessel
4. annual ring	9. perforation plate
5. earlywood	

From A.N. Foulger, *Classroom Demonstrations of Wood Properties,* US Department of Agriculture, Forest Service, Forest Products Laboratory, PA-900.

Shortleaf Pine/Loblolly Pine	Southern pines usually identified as a group; resinous; "pitchy" pine odor; yellowish; early to latewood transition abrupt; rings and resin ducts highly visible, resin canals large, numerous, mostly solitary, evenly distributed
Eastern Hemlock	Latewood reddish brown; fairly abrupt to gradual transition in rings; no resin canals
Fraser Fir	Gradual transition; no resin ducts; rays very fine, not distinct to naked eye
Eastern redcedar	Fine texture; *rose red or purplish*; *sweet "cedar-chest" odor*

Ring-Porous Woods

Chestnut	Grayish brown; *mild odor*; bitter taste due to tannin; tyloses in pores; latewood pores very numerous; rays barely visible with lens; *will turn blue when wet and rubbed with steel*
RED OAK GROUP	Latewood pores few and distinct; large pores open (tyloses absent or sparse; wood generally darker in color; largest rays conspicuous; tallest less than 1 inch
WHITE OAK GROUP	Latewood pores small, numerous, and indistinct; large pores abundantly clogged with tyloses; wood generally lighter in color; largest rays conspicuous, tallest greater than $1\frac{1}{4}$ inch
American Elm	Heartwood light brown to brown or reddish brown; earlywood pores large, in continuous row, latewood pores in wavy bands; rays not distinct without lens
Slippery Elm	Heartwood red to dark brown or reddish brown; earlywood pores 2–6 pores wide; latewood pores in wavy bands
Winged Elm	Heartwood light brown to brown or reddish brown; earlywood pores small and indistinct in an intermittent row, latewood pores in wavy bands
Hackberry	Heartwood cream, light brown, or light grayish brown, with a yellowish cast; earlywood more than 1 pore wide, latewood pores in wavy bands
White Ash	Heartwood light brown or grayish brown; sapwood creamy white (may be very wide); earlywood 2–4 pores wide, pores moderately large, surrounded by lighter tissue; latewood pores solitary and in radial multiples of 2–3
Sassafras	Heartwood color gray or grayish brown; spicy odor and taste; earlywood 3–8 pores wide; latewood pores solitary and in multiples of 2–3; tyloses abundant, with sparkle; rays conspicuous
Red Mulberry	Heartwood orange-yellow to golden or russet brown, turning deep russet on exposure; earlywood 2–8 pores wide, pores may contain white deposits; latewood pores in nestlike groups merging laterally to form wavy or interrupted bands; tyloses visible with sparkle

Honeylocust	Heartwood light red to reddish or orange-brown; earlywood 3–5+ pores wide, pores large; latewood pores solitary, in radial multiples, and in nestlike groups; no tyloses; rays conspicuous
Black Locust	Heartwood olive or yellow-brown to dark yellow-brown, dark russet brown with exposure; sapwood never more than 3 growth rings wide; earlywood 2–3 pores wide, pores large; latewood pores in nestlike groups, merging into interrupted or somewhat continuous bands; tyloses extremely abundant with yellow cast and sparkle; *very sweet smell when moistened*
Shagbark Hickory; Bitternut Hickory; Mockernut Hickory; Shellbark Hickory; Pignut Hickory and Red Hickory not readily distinguishable	Heartwood light to medium brown or reddish brown; early mostly an intermittent single row of thick-walled pores; latewood pores not numerous, solitary and in radial multiples of 2–5; tyloses moderately abundant; spiderweb pattern in latewood

Semi-Ring Porous or Semi-Diffuse Porous Woods

Live Oak	Heartwood brown to gray brown; growth rings not always distinct; largest pores distinct without lens, smaller pores in latewood, generally in radial arrangement; tyloses sparse; largest rays conspicuous to eye
Common Persimmon	Heartwood core dark, nearly black; sapwood creamy white, darkening to yellow or light gray with age (very wide); growth rings not distinct; pores medium to large, thick-walled, appear to be very few, solitary or in radial multiples; occasional tyloses; rays fine, visible with lens; *ripple marks on tangential surface*
Black Walnut	Heartwood medium brown to deep chocolate brown; earlywood pores fairly large, decreasing gradually to quite small in outer latewood; pores solitary or in radial multiples of 2 to several; tyloses moderately abundant; short tangential lines of parenchyma; rays fine, visible but not conspicuous with lens; *pores in a diagonal line*
Butternut	Heartwood medium or cinnamon brown, often with uneven streaks of color; growth rings usually fluted; earlywood pores fairly large, decreasing gradually in size to quite small in outer latewood, pores solitary or in radial multiples of 2 to several; tyloses moderately abundant; short tangential lines of banded or diffuse-in-aggregates parenchyma visible with lens; rays fine, visible but not conspicuous with lens; *can groove with fingernail*

Diffuse Porous Woods

American Beech	Heartwood creamy white with reddish tinge to medium reddish brown; growth rings distinct; pores small and solitary and in irregular multiples and clusters, numerous and evenly distributed throughout most of the ring, narrow but distinct latewood in each ring due to fewer, smaller pores; largest rays conspicuous on all surfaces, rays are of two sizes
American Sycamore	Heartwood light to dark brown, usually with a reddish cast; growth rings distinct due to unusual lighter color of latewood; pores small and in irregular multiples and clusters, numerous and evenly distributed throughout most of the growth ring; latewood zone evident by fewer, smaller pores and lighter color; rays easily visible without lens on all surfaces, *all are large*
HARD MAPLES	rays 5 to 7 cells wide
Sugar Maple/Black Maple	Heartwood creamy white to light reddish brown; growth rings distinct due to darker brown narrow latewood line; pores small, with *largest approximately equal to maximum ray width in cross section*; solitary or in radial multiples, very evenly distributed; rays visible to eye on tangential surface as very fine, even-sized, evenly distributed lines; *pith flecks infrequent*
Boxelder	Pores smaller and more numerous; pink streaks in wood due to a fungus growth
SOFT MAPLES	rays 3–5 cells wide
Red Maple/Silver Maple	Heartwood creamy white to light reddish brown, commonly with grayish cast or streaks; pores small, solitary, and in radial multiples, very evenly distributed; *largest as large or slightly larger than widest rays on cross section*; rays may be visible on tangential surface as very fine, even-sized, and evenly spaced lines; *pith flecks frequent*
Flowering Dogwood	Heartwood dark brown; *sapwood creamy with flesh or pinkish cast (fairly wide sapwood)*; pores very small, mostly solitary with some radial multiples; rays approximately as wide or wider than largest pores
Sourwood	Sapwood yellowish brown to light pinkish brown, wide; heartwood brown tinged with red when first exposed, becoming duller with age; pores numerous, very small (indistinct to the naked eye); rays not distinct without a hand lens
American Holly	Heartwood near white with gray or blue-gray cast or streaks; rings barely distinct; pores extremely small, *mostly arranged in chains that commonly extend across growth-ring boundaries*

Black Cherry   Heartwood light to dark cinnamon or reddish brown; growth rings sometimes distinct because of narrow zone or row of numerous slightly larger pores along initial earlywood; pores through growth ring solitary and in radial or irregular multiples and small clusters; rays not visible on tangential surface; distinct bright lines across transverse surface, conspicuous with lens; *has gum ducts*

Yellow-Poplar   *Heartwood green or yellow to tan with greenish cast,* sapwood creamy white (often wide); growth rings delineated by distinct light cream or yellowish line of marginal parenchyma; pores small, solitary, but mostly radial or irregular multiples and small clusters; rays distinct with lens on cross section

Cucumbertree   Sapwood whitish, narrow or wide; heartwood yellow, greenish yellow to brown, or greenish black; growth rings distinct, delineated by a whitish line of terminal parenchyma; pores small, indistinct without a hand lens; rays distinct to the naked eye

American and White Basswood (indistinguishable)   Heartwood creamy white to pale brown; *faint but characteristic musty odor*; growth rings indistinct or faintly delineated by marginal parenchyma, sometimes with blurry whitish spots along the growth-ring boundary; pores small, mostly in irregular multiples and clusters; rays distinct by not conspicuous with lens on transverse surface; *occasional pith flecks*; ripple marks on tangential surface

Yellow Birch/River Birch   Heartwood light to dark brown or reddish brown; sapwood white or pale yellow; pores small to medium, solitary or in radial multiples of 2 to several, with lens, pore diameters clearly greater than ray width; *pores appear to be filled with whitish substance; pith flecks sparse or absent in Yellow Birch, abundant in River Birch, otherwise same appearance*

Black Willow   Heartwood light brown to pale reddish or greyish brown; pores medium to small, usually with apparent size gradation from early wood to latewood, solitary and in multiples of 2 to several; *rays very fine, barely visible with hand lens, rays "weave" through the pores*

Eastern Cottonwood   Heartwood grayish to light grayish brown, sometimes with olive cast; *moist wood with foul odor*; pores medium to small, usually with apparent size gradation from earlywood to latewood, solitary and in radial multiples of 2 to several; *rays very fine, not easily seen with lens*

Yellow Buckeye   Heartwood creamy white to pale yellowish white; pores very small, solitary in multiples and small clusters; *ripple marks on tangential surface*

Ohio Buckeye	Sapwood white to grayish white, gradually merging into heartwood; heartwood creamy white to pale yellowish white, frequently with darker (grayish) streaks of oxidative sap stain; pores numerous, minute, not visible without a hand lens; *rays very fine, scarcely visible with a hand lens, close and seemingly forming half of the area on the transverse surface*; ripple marks wanting or sporadic
Sweetgum	Heartwood gray or reddish brown, *with variegated pigment*; pores very small, numerous, solitary and in multiples and small clusters, often in intermittent radial chains; *rays very fine, not distinct even with hand lens*
Black Tupelo	Heartwood medium gray *or gray with green or brown cast*; pores very small, numerous, solitary and in multiples and small clusters; *rays barely visible even with hand lens*

Western Species

WesternWhite Pine	Pleasant, "piney" odor; resin canals large, numerous, mostly solitary and evenly distributed
Sugar Pine	Faint, but distinct, resinous odor; resin canals *very* large, numerous, mostly solitary and evenly distributed
Ponderosa Pine	Mild "piney" odor; resin canals numerous, medium, mostly solitary and evenly distributed
Sitka Spruce	No odor; gradual earlywood/latewood transition; medium-size resin canals, sparse to numerous, variable in distribution, solitary or up to several in tangential groups
Douglas-fir	Unique, resinous odor; distinct reddish-brown heartwood; abrupt earlywood/latewood transition; resin canals medium to small, relatively few and variable in distribution, solitary or up to several in tangential groups
Western Larch	Distinct heartwood, russet to reddish brown; abrupt earlywood to latewood transition; resin canals small, relatively few and variable in distribution, solitary or up to several in tangential groups
Western Hemlock	No odor, heartwood indistinct, light in color; gradual earlywood/latewood transition; no resin canals
White Fir; Grand Fir	No odor, may have a bitter taste; indistinct heartwood, may be light in color; gradual early wood/latewood transition, no resin canals
Incense-cedar	"Pencil-cedar" aroma; distinct, medium-reddish to purplish-brown heartwood; gradual earlywood/latewood transition
Western Redcedar	Pungent cedar odor; distinct, medium to dark brown or reddish-brown heartwood; abrupt earlywood/latewood transition (gradual in wide rings)

Port Orford-Cedar	Gingerlike odor; distinct, yellowish or pinkish brown heartwood; gradual earlywood/latewood transition
Alaska-Cedar	Distinct odor, like raw potatoes; distinct, yellow to dark-yellow heartwood; gradual earlywood/latewood transition, rings usually very narrow
Redwood	Distinct heartwood, medium to deep reddish brown; earlywood/ latewood transition abrupt, coarse to very coarse texture
Pacific Yew	No odor; distinct, orange to russet heartwood; gradual earlywood/ latewood transition, very fine texture

Ring-Porous Hardwoods

Golden Chinquapin	Heartwood light brown, often pinkish; earlywood pores large, round, in single row, intermittent with dark fiber; latewood pores in distinct fanlike patches, appearing to erupt from each earlywood pore; tyloses present; rays fine, barely visible with lens
Tanoak	Sapwood light reddish brown when first exposed, turning darker with age and then difficult to distinguish from heartwood; growth rings scarcely distinct or wanting, when visible, delineated by a faint narrow line of darker (denser) fibrous tissue at the outer margin; pores barely visible to the naked eye, unevenly distributed, inserted in light-colored tissue in streamerlike clusters that extend radially for some distance across several to many rings; two types of rays
RED/BLACK OAKS	Latewood pores few and distinct; large pores open; tyloses absent or sparse; wood generally darker in color; largest rays conspicuous; tallest less than 1 inch
WHITE OAKS	Latewood pores small, numerous and indistinct; large pores abundantly clogged with tyloses; wood generally lighter in color; largest rays conspicuous, tallest greater than $1\frac{1}{4}$ inch
Oregon Ash	Sapwood nearly white, wide; heartwood grayish brown, light brown or pale yellow streaked with brown; pores large, distinctly visible to the naked eye, forming a band 2–4 pores in width; transition from earlywood to latewood abrupt; parenchyma visible with hand lens in summerwood, frequently uniting laterally; rays barely visible with the naked eye

Diffuse-Porous Hardwoods

Pacific Dogwood	Sapwood light pinkish brown, very wide; pores small, not visible without a hand lens, solitary and in multiples of 2 to several; parenchyma not visible, or barely visible with lens; rays of two widths

Bigleaf Maple	Sapwood reddish white, sometimes with a grayish cast; heartwood pinkish brown; growth rings not very distinct, delineated by a narrow light line of fibrous tissue; pores moderately small to medium size, indistinct without a hand lens; rays visible to the naked eye, broadest about as wide as the largest pores
California-laurel	Characteristically spicy odor; heartwood light brown to grayish brown with darker streaks; pores small to medium, evenly distributed, solitary or in multiples of 2 to several, each pore or multiple surrounded by a lighter sheath of vasicentric parenchyma; rays fine, clearly visible with hand lens
Quaking Aspen	Creamy-white to light grayish-brown heartwood; pores small to very small; rays very fine, not easily seen with hand lens
Black Cottonwood	Sapwood whitish, frequently merging into the heartwood; heartwood grayish white to light grayish brown; wood odorless or with a characteristic disagreeable odor when moist; growth rings distinct but inconspicuous; pores small, the largest barely visible to the naked eye in the springwood; terminal parenchyma, the narrow, light-colored line more or less distinct; rays very fine, scarcely visible with a hand lens
Red Alder	Heartwood pale tan when freshly cut (not distinct from sapwood) darkening with age to light reddish brown or flesh color; pores small, solitary, and in mostly radial multiples; large aggregate rays widely scattered but easily seen without lens
Pacific Madrone	Sapwood white or cream colored, frequently with a pinkish tinge; heartwood light reddish brown; growth rings barely visible with a hand lens, delineated by a continuous, uniseriate row of pores in the early springwood; pores minute, barely visible with a hand lens; rays barely to readily visible with a hand lens

Descriptions compiled from: Hoadley, *Identifying Wood* Brown, Panshin and Forsaith, *Textbook of Wood Technology*, and Wilson, personal notes and observations.

FOR FURTHER INFORMATION

International Wood Collectors Society is a nonprofit organization, founded in 1947 to promote wood collecting, the exchange of specimens, and wood craftsmanship. It also promotes a standardized sample size and assists in the identification of wood.

Secretary Treasurer, IWCS
2913 Third Street
Trenton, MI 48183

U.S. Forest Products Laboratory will identify up to five wood samples free of charge for those with "a clear need for wood identification." Samples at least 1 inch × 3 inches × 6 inches are recommended, although any size large enough to be hand-held during sectioning is acceptable. Samples should be labeled individually and accompanied by any information on origin or popular name.

Center for Wood Anatomy Research
U.S. Forest Products Laboratory
1 Gifford Pinchot Drive
Madison, WI 53705-2398

Agencies Offering Wood Identification on a Fee Basis

Write or call these agencies before sending samples:

Ethnobotanical Research Services
Department of Anthropology
Oregon State University
Corvallis, OR 97331
(503) 737-0123.
Program Manager, Tropical Timber Identification Center
College of Environmental Science and Forestry
Syracuse, NY 13210
(315) 473-8788.

Wood Sample Kits

Bruce T. Forness
International Wood Collectors Society
Drawer B, Main Street
Chaumont, NY 13622
Norman Jones
Colonial Hardwoods, Inc.
212 North West Street
Falls Church, VA 22046

SELECTED SOURCES AND RESOURCES

I am an omnivorous reader with a strangely retentive
memory for trifles.

—*Sherlock Holmes*
"The Lion's Mane"

Education never ends. It is a series of lessons, with the
greatest for the last.
   All knowledge comes useful to the detective.

—*Sherlock Holmes*
"The Red Circle"

Aldisert, Ruggero J. *Logic for Lawyers. A Guide to Clear Legal Thinking.* New York: Clark Boardman Company, Ltd., 1989.

Allport, Susan. *Sermons in Stone. The Stone Walls of New England and New York.* New York: W.W. Norton & Company, 1990.

Ang, Tom. *Digital Photographer's Handbook*, 3rd ed. London: DK Publishing, Inc., 2006.

Barker, Noah. *Land Surveyor. An Essay on the Cardinal Points: being a collection of authorities in explanation of the terms, "due north," "due south," "due east,"and "due west," as applied to land surveying.* Bangor, ME: Printed by Samuel S. Smith, 1864.

Bedini, Silvio A. *Thinkers and Tinkers.* Rancho Cordova, CA: Landmark Enterprises, 1975.

Bedini, Silvio A. *Early American Scientific Instruments and Their Makers.* Rancho Cordova, CA: Landmark Enterprises, 1986.

Bedini, Silvio A. *The Jefferson Stone.* Frederick, MD: Professional Surveyors Publishing Co., Inc., 1999.

Bouchard, Harry, and Francis H. Moffitt. *Surveying*, 5th ed. Scranton, PA: International Textbook Company, 1965.

Brinker, Russell C. *Elementary Surveying*, 5th ed. Scranton, PA: International Textbook Company, 1969.

Brinker, Russell C., and Paul R. Wolf. *Elementary Surveying*, 7th ed. New York: Harper & Row, Publishers, 1984.

Brown, Curtis M., Walter G. Robillard, and Donald A. Wilson. *Evidence & Procedures for Boundary Location*, 5th ed. Hoboken, NJ: John Wiley & Sons, Inc., 2006.

Brown, H. P., A. J. Panshin, and C. C. Forsaith. *Textbook of Wood Technology*, vol. 1. New York: McGraw-Hill Book Company, Inc., 1949.

Buckner, R. B. *Fundamentals of Measurement Theory and Analysis*. Jefferson, NC: Land Surveyor's Publications, Buckner Surveying Publications, 2004.

Bureau of Land Management. *Manual of Instructions for the Survey of the Public Lands of the United States*. Technical Bulletin 6. Washington, DC: U.S. Government Printing Office, 1973.

Bureau of Land Management. *Restoration of Lost and Obliterated Corners & Subdivision of Section. A Guide for Surveyors*. Washington, DC: U.S. Government Printing Office, 1975.

Cadastral Survey Training Staff. *Corner Point Identification*. Washington, DC: Bureau of Land Management, 1976.

Cameron, Layne. *The Geocaching Handbook*. Guilford, CT: The Globe Pequot Press, 2004.

Chisum, W. Jerry, and Brent E. Turvey. "Evidence Dynamics: Locard's Exchange Principle & Crime Reconstruction," *Journal. of Behavioral Profiling* 1, no. 1 (January 2000).

Clifton, Robert T. *Barbs, Prongs, Points, Prickers & Stickers. A Complete and Illustrated Catalogue of Antique Barbed Wire*. Norman: University of Oklahoma Press, 1970.

Copi, Irving M., and Carl Cohen. *Introduction to Logic*, 11th ed. Upper Saddle River, NJ: Pearson Education, Inc., 2002.

Coyle, Heather Miller, ed. *Forensic Botany*. Boca Raton, FL: CRC Press, 2005.

Crowe, Elizabeth Powell. *Genealogy Online*, 6th ed. New York: Osborne/McGraw-Hill, 2002.

Cuomo, Paul, and Roy Minnick. *Advanced Land Descriptions*. Rancho Cordova, CA: Landmark Enterprises.

Davis, Harold. *Digital Photography. Digital Field Guide*. Hoboken, NJ: John Wiley & Sons, Inc., 2005.

Davis, Raymond E., and Francis S. Foote. *Surveying Theory and Practice*, 4th ed. New York: McGraw-Hill Book Company, Inc., 1953.

Davis, Raymond E., and Joe W. Kelly. *Elementary Plane Surveying*, 4th ed.. New York: McGraw-Hill Book Company, 1969.

Dines, Jess E. *Document Examiner Textbook.* Grand Cayman Island: Pantex International Ltd., 1998.

Drahn, Richard, and Milo Stefan. *Decayed Wood Identification of Bearing Trees of Public Land Surveys.* USDA—Forest Service, 1981.

Dyer, Mike. *The Essential Guide to Geocaching.* Golden, CO: Fulcrum Publishing, 2004.

Eastman Kodak Company. New York: Rochester.

    Publication M-2. *Using Photographs to Preserve Evidence.* 1976.

    Publication M-27. *Ultraviolet and Fluorescence Photography.* 1972.

    Publication M-28. *Applied Infrared Photography.* 1977.

    Publication N-9. *Basic Scientific Photography.* 1970.

    Publication N-12A. *Close-Up Photography.* 1969.

Edlin, Herbert L. *What Wood Is That? A Manual of Wood Identification.* New York: Penguin Putnam Inc., 1969.

Ellen, David. *The Scientific Examination of Documents. Methods and Techniques.* Boca Raton, FL: CRC Press, 1997.

Erzinçlioglu, Zakaria. *Forensics. True Crime Scene Investigations.* New York: Barnes & Noble Books, 2000.

Forest Products Laboratory. *Wood: Colors and Kinds.* Agricultural Handbook No. 101. Washington, DC: U.S. Government Printing Office, 1956.

Foulger, A. N. *Classroom Demonstrations of Wood Properties.* Washington, DC: U.S. Government Printing Office, 1969.

Gardner, Ross M. *Practical Crime Scene Processing and Investigation.* Boca Raton, FL: CRC Press, 2005.

Garrison, D. H., Jr. Protecting the Crime Scene. FBI Law Enforcement Bulletin, September 1994.

Genge, N. E. *The Forensic Casebook.* New York: Ballantine Books, 2002.

Glover, Jack. *The Bobbed Wire Bible.* Cow Puddle Press, 1996.

Greene, James, and David Lewis. *Handwriting Analysis.* London: Treasure Press, 1990.

Greenhood, David. *Mapping.* Chicago: The University of Chicago Press, 1964.

Griffin, Robert J. "Retracement and Apportionment as Surveying Methods for Reestablishing Property Corners," *Marquette Law Review* **43** (1960): 484–510.

Gurney, Alan. *Compass.* New York: W.W. Norton & Company, 2004.

Hagemeier, Harold. *Barbed Wire Identification Encyclopedia,* 3rd ed. Amarill, TX. Devil's Rope Museum Publishing Co., 2001.

Harlow, William M., and Ellwood S. Harrar. *Textbook of Dendrology.* New York: McGraw-Hill Book Company, Inc., 1958.

Hermanson, Knud. "When Is a Rod Not 16.5 feet (More times than Not)," *Probate and Property.* 6, No. 5 (September-October 1992).

Hoadley, R. Bruce. *Identifying Wood.* Newtown, CT: The Taunton Press, 1990.

Hughes, Sarah. *Surveyors and Statesmen. Land Measuring in Colonial Virginia.* Richmond, VA: The Virginia Surveyors Foundation, Ltd. and The Virginia Association of Surveyors, Inc., 1979.

James, Stuart H., and Jon J. Nordby, eds. *Forensic Science. An Introduction to Scientific and Investigative Techniques.* Boca Raton, FL: CRC Press, 2003.

Jones, Otis A., executive coordinator. *Following in Their Footsteps. Land Surveying in North Carolina.* Chapel Hill, NC: Chapel Hill Press, Inc., 2006.

Josephson, John R., and Susan G. Josephson. *Abductive Inference. Computation, philosophy, technology.* Cambridge: Press Syndicate of the University of Cambridge, 1996.

Kelly, Jan Seaman, and Brian S. Lindblom. *Scientific Examination of Questioned Documents,* 2nd ed. Boca Raton, FL: CRC Press, 2006.

Kiely, Terrence F. *Forensic Evidence: Science and the Criminal Law,* 2nd ed. Boca Raton, FL: CRC Press, 2006.

Kirkham, E. Kay. *The Handwriting of American Records for a Period of 300 Years.* Logan, UT: The Everton Publishers, Inc., 1973.

Kissam, Philip. *Surveying Practice.* New York: McGraw-Hill Book Company, 1966.

Lappala, Donald D. *Retracement and Evidence of Public Land Surveys.* USDA: U.S. Forest Service, Eastern Region, 1974

Lee, Henry C. "Advances in Forensic Science: Legal and Ethical Responsibilities." Paper presented at the Montreal Conference, July 1999.

Lee, Henry C. with Thomas W. O'Neil. *Cracking Cases. The Science of Solving Crimes.* Amherst, MA: Prometheus Books, 2002.

Leybourn, William. *The Compleat Surveyor,* 2nd ed. Printed by R. and W. Leybourn, for G. Sawbridge, at the signe of the Bible upon Ludg-gate-hill, 1657.

Little, Elbert L., Jr. *Check List of Native and Naturalized Trees of the United States (including Alaska).* USDA: Forest Service, 1953. (for colloquial names).

Love, John. *Geodaesia: or, The Art of Surveying and Measuring of Land, Made Easie.* London: Printed for John Taylor at the Ship in St. Paul's Church-Yard, 1688.

Makower, Joel, ed. *The Map Catalog,* 2nd ed. New York: Tilden Press Inc., 1990.

Martin, George A. *Fences, Gates & Bridges. A Practical Manual.* Brattleboro, VT: The Stephen Greene Press, 1974.

McCallum Henry D., and Frances T. *The Wire That Fenced the West.* Norman: University of Oklahoma Press, 1965.

McNichol, Andrea. *Handwriting Analysis. Putting It to Work for You.* Chicago: Contemporary Books, 1991.

Neelands, R.W. *Important Trees of Eastern Forests.* USDA: Forest Service, 1968.

Nordby, Jon J. *Dead Reckoning. The Art of Forensic Detection.* Boca Raton, FL: CRC Press, 2000.

Orn, Clayton L. "Vanishing Footsteps of the Original Surveyor," *Baylor Law Review* **4,** no. 3 (Spring 1952).

Petrides, George A. *A Field Guide to Trees and Shrubs.* The Peterson Field Guide Series. Boston: Houghton Mifflin Company, 1958.

Petrides, George A., and Olivia. *Western Trees.* The Peterson Field Guide Series. Boston: Houghton Mifflin Company, 1998.

Plisga, Stanley J., Jr., David A. Tyler, Donald A. Wilson, and Carroll F. Merriam. *Field Surveying Techniques and Plane Coordinate Computations.* Department of Civil Engineering, University of Maine. 1974.

Pye, Kenneth. *Geological and Soil Evidence. Forensic Applications.* Boca Raton, FL: CRC Press, 2007.

Quimby, E. T. *A Paper on Terrestrial Magnetism designed for The Use of Surveyors.* Concord, MA: Printed by the Republican Press Association, 1874.

Ramsland, Katherine. *The Science of Cold Case Files.* New York: The Berkeley Publishing Group, 2004.

Ramsland, Katherine. *The C.S.I. Effect.* New York: The Berkeley Publishing Group, 2006.

Rayner, William Horace, and Milton O. Schmidt. *Fundamentals of Surveying.* New York: Van Nostrand-Reinhold Company, 1969.

Reader's Digest. *Crime Scene Investigation.* London: Elwin Street Limited, 2004.

Robillard, Walter G., and Lane J. Bouman. *Clark on Surveying and Boundaries.* Fifth Edition. Charlottesville, VA: The Michie Company, 1987.

Robillard, Walter G., and Donald A. Wilson. *Evidence and Procedures for Boundary Location,* 5th ed. Hoboken, NJ: John Wiley & Sons, Inc., 2006.

Robinson, William F. *Abandoned New England. Its Hidden Ruins and Where to Find Them.* Boston: New York Graphic Society, 1976.

Redsicker, David R. *The Practical Methodology of Forensic Photography,* 2nd ed. Boca Raton, FL: CRC Press, 2001.

Schwid, Bonnie L. *Forensic Document Examination.* Milwaukee, WI: Anagraphics, Inc., 1986.

Scott, Charles C. *Photographic Evidence.* St. Paul, MN: West Publishing Company, 1969.

Shigo, Alex L. *A Tree Hurts, Too.* Northeast Forest Experiment Station, Forest Service, USDA. Washington, DC: U.S. Government Printing Office, nd.

Sloane, Eric. *Eric Sloane's America.* New York: Promontory Press, 1956.

Sloane, Eric. *Our Vanishing Landscape.* New York: Ballantine Books, 1955.

Sloane, Eric. *A Reverence for Wood.* New York: Wilfred Funk, Inc., 1965.

Staggs, Steven. *Crime Scene and Evidence Photographer's Guide.* Temecula, CA: Staggs Publishing, 1997.

Stokes, Marvin A., and Terah L. Smiley. *An Introduction to Tree-Ring Dating.* Chicago: The University of Chicago Press, 1968.

Stryka-Rodda, Harriet. *Understanding Colonial Handwriting.* Baltimore, MD: GenealogicalPublishing Co., Inc., 1987.

Sullivan, Daniel F. "Change in Shoreline by Accretion or Avulsion," *American Jurisprudence Proof of Facts*, 2nd, vol. 21. Rochester, NY: Lawyers Cooperative Publishing Company. Pp. 147–250.

Symons, Harry. *Fences.* Toronto: McGraw-Hill Ryerson Limited, 1958.

Taylor, Troy. "Rules for Investigations." Paper presented to the American Ghost Society, 2004.

USDA. Bureau of Land Management Cadastral Training Staff. *Durability of Bearing Trees.* n.d.

Van Sickle, Jan. *GPS for Land Surveyors*, 2nd ed. Chelsea, MI: Ann Arbor Press, 2001.

Wattles, William C. *Land Survey Descriptions.* Los Angeles: Title Insurance and Trust Company, 1968.

Wattles, William C. *Writing Legal Descriptions.* Self-published. 1976.

Wecht, Cyril H., and John T. Rago, eds. *Forensic Science and Law*. Boca Raton, FL: CRC Press, 2006.

Welch, William Lewis. *Francis Lyford of Boston, and Exeter, and some of his descendants.* Salem, MA: Essex Institute, 1902.

Wessels, Tom. *Reading the Forested Landscape. A Natural History of New England.* Woodstock, VT: The Countryman Press, 1997.

Wilson, Donald A. "Accretion or Avulsion?" *The Benchmark* (official publication of the New Hampshire Land Surveyors Association), 7, no. 3 (1976): 79–83.

Wilson, Donald A. "Rules of the Game: Rules for Investigations," *Professional Surveyor* 25, no. 11 (November 2005).

Wilson, Donald A. *Interpreting Land Records.* Hoboken, NJ: John Wiley & Sons, Inc., 2006.

Zsilinszky, Victor G. *Photographic Interpretation of Tree Species in Ontario*, 2nd edition. Ontario. Department of Lands and Forests, 1966.

INDEX